Nanofillers

Analyzing the modifying effects of various inorganic nanofillers on the mechanical properties of polymer nanocomposites, this book covers processing, characterization, properties, and applications to analyze how these materials allow for innovative multifunction.

This volume looks at various synthesis methods available within inorganic nanofillers and what characterizes them, covering design, manufacturing processes, and end-user results. The chapters focus on metal oxides, energy storage, and devices, alongside polymers used for packaging. This book details the role inorganic nanofillers have to play in cost-effective manufacturing processes due to their strength and efficiency, and their subsequent relevance to low-cost yet high-performance materials. Covering topics such as corrosion resistance, wear resistance, strength, and characterization, this book is an essential companion for any engineer working with inorganic nanofillers.

This book will be of interest to engineers involved with inorganic nanofillers in a variety of industries, including automotive, aerospace, and biomedical engineering.

Nanofillers

Fabrication, Characterization and Applications of Inorganic Nanofillers

Edited by
Bhasha Sharma, Vijay Chaudhary,
Shashank Shekhar, and Partha Pratim Das

CRC Press
Taylor & Francis Group
Boca Raton London New York

CRC Press is an imprint of the
Taylor & Francis Group, an informa business

First edition published 2024
by CRC Press
6000 Broken Sound Parkway NW, Suite 300, Boca Raton, FL 33487-2742

and by CRC Press
4 Park Square, Milton Park, Abingdon, Oxon, OX14 4RN

CRC Press is an imprint of Taylor & Francis Group, LLC

ISBN: 9781032245898 (hbk)
ISBN: 9781032245904 (pbk)
ISBN: 9781003279389 (ebk)

DOI: 10.1201/9781003279389

Typeset in Times
by codeMantra

Contents

Preface

Polymer nanocomposites have been investigated for about three decades. To get deep insights into the modifying effects of various nanofillers on the mechanical and physical properties of polymer nanocomposites, the four basic aspects of processing, characterization, properties, and applications will be critically discussed. Nanofillers do not represent only a creative alternative to design new materials and compounds for academic research, but their improved or unusual features allow the development of innovative multifunctional industrial applications. The development of these high-performance nanofillers requires thorough design, material selection, and synthesis. It involves the selection of fillers out of the vast number of fillers available today in the search for the required property. It also involves the determination of the best production process, considering the end-use requirements, and finally the fabrication of the material. This underscores the importance of the design and synthesis phases of polymer nanocomposite production. This book will provide researchers and industries to work on selecting the type of nanofiller (organic/inorganic) based on the targeted applications. The detailed discussion on synthesis, characterization, and application of nanofillers. The limitations and future perspectives of using nanofillers have been highlighted. The readers will get help understanding the selection criteria of nanofiller before utilizing them for various applications. This book will also address the advantages of inorganic nanofillers to promote applications of resultant nanocomposites. This book will also address the limitations of using nanofiller to promote sustainability and an eco-friendly environment, the economic aspects of using nanofillers, and their future perspectives.

The contents of this book will be beneficial for students of mechanical engineering, civil engineering, material science, chemistry, physics, and researchers both working in industry and academia. In nutshell, this book will focus on the synthesis, characterization, and applications of inorganic nanofillers.

We, the editors, would like to pay gratitude to contributors who have participated in the volume of this book. We are also thankful to the entire team at CRC Press (Taylor & Francis Group) for publishing this book in the fastest possible time and in the most efficient manner.

Preface

Polymer nanocomposites have been investigated [illegible] the effects of various nanofillers [illegible] properties of polymer nanocomposites [illegible] aspects of processing, characterization, properties, and applications [illegible] critically discussed [illegible] [illegible]

[illegible] this book [illegible]

[illegible]

Editors

Dr. Bhasha Sharma is currently working as an Assistant Professor in the Department of Chemistry at Shivaji College, University of Delhi, India. She received her BSc in Polymer Sciences in 2011 from the University of Delhi. Dr. Sharma completed her Ph.D. in Chemistry in 2019 under the guidance of Prof. Purnima Jain from the University of Delhi. She has more than 7 years of teaching experience. She has published more than 40 research publications in reputed international journals. Her recently edited book titled *Graphene-based biopolymer nanocomposites* has been published in Springer Nature. Her authored book *3D Printing Technology for Sustainable Polymers* and edited books *Biodegradability of Conventional Plastics: Opportunities, Challenges, and Misconceptions* and *Sustainable Packaging: Gaps, Challenges, and Opportunities* have been accepted by Wiley, Elsevier, and Taylor & Francis, respectively. Her research interests revolve around sustainable polymers for packaging applications, environmentally benign approaches for biodegradation of plastic wastes, fabrication of bionanocomposites, and finding strategies to ameliorate the electrochemical activity of biopolymers.

Dr. Vijay Chaudhary is currently working as an Assistant Professor (Grade-I) in the Department of Mechanical Engineering, Amity School of Engineering and Technology (A.S.E.T.), Amity University Uttar Pradesh, Noida (India). He received his B. Tech. in 2011 from the Department of Mechanical Engineering, Uttar Pradesh Technical University, Lucknow, India and then completed M. Tech (Hons) in 2013 from the Department of Mechanical Engineering, Madan Mohan Malaviya Engineering College, Gorakhpur, India. He received his Ph.D. in 2019 from the Department of Mechanical Engineering, Netaji Subhas University of Technology, University of Delhi, India. His research areas of interest include processing and characterization of polymer composites, tribological analysis of biofiber-based polymer composites, water absorption of biofiber-based polymer composites, and surface modification techniques related to polymer composite materials. Dr. Chaudhary has over 8 years of teaching and research experience. He has published more than 40 research papers in peer-reviewed international journals as well as in reputed international and national conferences. He has published 16 book chapters with reputed publishers. More than 25 students have completed their Summer Internships, B.Tech. Projects and M.Tech. Dissertations under his guidance. Currently, he is working in the field of bio-composites, nanocomposites, and smart materials.

Dr. Shashank Shekhar is currently an Assistant Professor at Netaji Subhas University of Technology and is also associated with the Quantum Research Centre of Excellence as Associate Director in the Department of Renewable Energy. He completed his PhD in Chemistry at the University of Delhi. Dr. Shekhar has been working on biopolymers and Schiff base metal complexes for the last 5 years and has published more than 20 articles in reputed international journals. He has 6 years of research and teaching experience. Presently, he is working on several projects, including the circular economy approach to plastic waste, synthesis of nanomaterials and nanocomposites for energy harnessing, biodegradation of plastic wastes, electrochemical analysis of resultant biodegradable nanocomposites for employment in supercapacitor applications, and polymer technology for packaging applications.

Mr. Partha Pratim Das is pursuing Master of Technology (Materials Science and Metallurgical Engineering) in the Department of Materials Science and Metallurgical Engineering at Indian Institute of Technology Hyderabad (IITH), India. Currently, he is working in the field of active food packaging to extend the shelf-life of fresh produce with Cellulose and Composites Research Group at IIT Hyderabad. He completed his B.Tech in Mechanical Engineering with first-class distinction in the year 2021 from Amity University Uttar Pradesh, Noida, India. During his B.Tech, he worked on various projects with the Indian Institute of Technology Guwahati and Indian Oil Corporation Limited, Guwahati, Assam. He has presented several research papers at national and international conferences and published a good number of research papers in SCI journals and book chapters to his credit with reputed publishers including Springer, CRC press (T&F), and Elsevier. He also served as a reviewer in Materials Today: Proceedings and Applied Composite Materials, Springer. In the year 2020, he received the Innovative researcher of the year award. He is also a Certified Executive of Lean Management and Data Practitioner (Minitab and MS-Excel) from the Institute for Industrial Performance and Engagement (IIPE), Faridabad, India. He is a community associate member at American Chemical Society (ACS). His area of research includes natural fibre-based composites, processing and characterization of polymer matrix composites, nanofiller-based composites for various applications and biodegradable food packaging.

Contributors

Qazi Adfar
Vishwa Bharti Degree College
Srinagar, India

Mohammad Aslam
Materials Chemistry & Engineering Research Laboratory, Department of Chemistry
National Institute of Technology
Srinagar, India

Yashdi Saif Autul
Department of Mechanical and Materials Engineering
Worcester Polytechnic Institute (WPI)
Worcester, Massachusetts

Puspendu Barik
Technical Research Centre
S. N. Bose National Centre for Basic Sciences
Salt Lake City, India

B. C. Bhadrapriya
School of Pure and Applied Physics
Mahatma Gandhi University
Kottayam, India

Ashis Bhattacharjee
Department of Physics
Institute of Science
Visva-Bharati University
Santiniketan, India

Litty Theresa Biju
St. Berchmans College
Changanassery
Ernakulam, India

Bosely Anne Bose
School of Pure and Applied Physics
Mahatma Gandhi University
Kottayam, India

G. L. Devnani
Department of Chemical Engineering
Harcourt Butler Technical University
Kanpur, India

Mohammed Enamul Hoque
Department of Biomedical Engineering
Military Institute of Science and Technology (MIST)
Dhaka, Bangladesh

M. Iqbal Ishra
Postgraduate Institute of Science
University of Peradeniya
Peradeniya, Sri Lanka
and
MRSC
Institute of Materials Engineering and Technopreneurships
Kandy, Sri Lanka

M. G. R. Tamasha Jayawickrama
Postgraduate Institute of Science
University of Peradeniya
Peradeniya, Sri Lanka
and
MRSC
Institute of Materials Engineering and Technopreneurships
Kandy, Sri Lanka

Feba Anna John
St. Berchmans College
Changanassery
Ernakulam, India

Ajith James Jose
St. Berchmans College
Changanassery
Ernakulam, India

Nandakumar Kalarikkal
School of Nanoscience and Nanotechnology
Mahatma Gandhi University
Kottayam, India

Neeraj Kumar
Centre for Nano Science and Engineering (CeNSE)
Indian Institute of Science
Bangalore, India

Shrikant S. Maktedar
Biofuels Research Laboratory
Department of Chemistry
National Institute of Technology
Srinagar, India

M. M. M. G. P. G. Manthilaka
Postgraduate Institute of Science
University of Peradeniya
Peradeniya, Sri Lanka
and
MRSC
Institute of Materials Engineering and Technopreneurships
Kandy, Sri Lanka

Syed Talha Muhtasim
Department of Mechanical Engineering
Military Institute of Science and Technology (MIST)
Dhaka, Bangladesh

B. S. S. Nishadani
Postgraduate Institute of Science
University of Peradeniya
Peradeniya, Sri Lanka
and
MRSC
Institute of Materials Engineering and Technopreneurships
Kandy, Sri Lanka

P. A. Nizam
School of Chemical Science
Mahatma Gandhi University
Kottayam, India

Suresh Kumar Patel
Department of Chemical Engineering
Government Polytechnic, Lakhimpur Kheri
Lakhimpur, India

Tousif Reza
Department of Mechanical Engineering
Military Institute of Science and Technology (MIST)
Dhaka, Bangladesh

Deepak Singh
Department of Chemical Engineering
Institute of Engineering & Technology
Lucknow, India

Dhananjay Singh
Department of Chemical Engineering
Government Polytechnic, Lakhimpur Kheri
Lakhimpur, India

Mohammad Fahim Tazwar
Department of Mechanical Engineering
Military Institute of Science and Technology (MIST)
Dhaka, Bangladesh

Abhishek Tevatia
Department of Mechanical Engineering
Netaji Subhas University of Technology
New Delhi, India

Prakash Chander Thapliyal
Advanced Structural Composites and Durability Group
CSIR-Central Building Research Institute
Roorkee, India

Sabu Thomas
School of Energy Materials
Mahatma Gandhi University
Kottayam, India

1 Outline and Classification of Inorganic Nanofillers in Polymers

M. G. R. Tamasha Jayawickrama,
M. Iqbal Ishra, B. S. S. Nishadani, and
M. M. M. G. P. G. Manthilaka
University of Peradeniya
Institute of Materials Engineering and Technopreneurships

CONTENTS

DOI: 10.1201/9781003279389-1

1.1 INTRODUCTION

Twenty years ago, nanocomposite was not popular, and 'hybrid' or 'molecular composite' was used instead (Okada & Usuki, 2006). In those days, inorganic fillers were used as additives for polymers to enhance thermal, mechanical, and chemical stability. Those traditional fillers did not possess the superior properties of nanoparticles because most of the time they were in micron size. In 1990, the word 'nanocomposite' appeared for the first time on paper, which is a polymer-clay hybrid in the polymer field for the first time. That polymer-clay nanocomposite was clay/nylon-6 nanocomposite, and it is to produce timing belt covers for a Toyota car.

Based on their origin, fillers may be classified into two types:

- organic fillers
- inorganic fillers. (Kango et al., 2013)

Caseri claimed (Caseri, 2000, 2009) that the first reported polymer-inorganic nanocomposite (PINC) in the literature was probably a polymer-Au nanocomposite by Lüdersdorff. A purple solid nanocomposite was obtained with the coprecipitation of gum arabic and gold in ethanol. And that resulting nanocomposite, known as the gold ruby glasses due to their red color, is still produced commercially. In particular, gold nanoparticles have been used as colorizing agents for glass for a long time (Caseri, 2009).

Over the last two decades, scientific and technological interest has focused on polymer-inorganic nanocomposites (PINCs) due to their unique properties and numerous applications in modern technology (Hong & Chen, 2014; Li et al., 2010). The PINCs are mostly a simple combination of inorganic nanoparticles (nanometer-sized inorganic nanoparticles, typically in the range of 1–100 nm), which are uniformly dispersed in and fixed to a polymer matrix. It allows to combine/enhance the properties of the inorganic phase and the properties of polymers, and advanced new functions can also be generated for the PINCs (Ajayan et al., 2003; Li et al., 2010). As an example, the selectivity and thermal stability of the inorganic fillers and the flexibility and processability of polymers (Cong et al., 2007). Also, it leads to the new functionality of polymer-based materials or new processing methods for inorganic materials and it can provide high-performance novel materials that find applications in many industrial fields. Those nanoparticles are also termed 'nanofillers' or 'nano-inclusions' and in this kind of combination inorganic nanoparticles are acting like 'additives' to increase polymer performance (Ramanathan et al., 2007; Vaisman et al., 2007).

Also, the nanoparticles have a strong tendency to disperse insufficiently in the polymer matrix and undergo agglomeration. This reduces the mechanical and optical properties of the nanocomposites (Kruenate et al., 2004; Yang et al., 2006). To improve the dispersion and reduce the agglomeration, surface modifications to the nanoparticles can be done to generate a strong repulsion between nanoparticles (Figures 1.1 and 1.12).

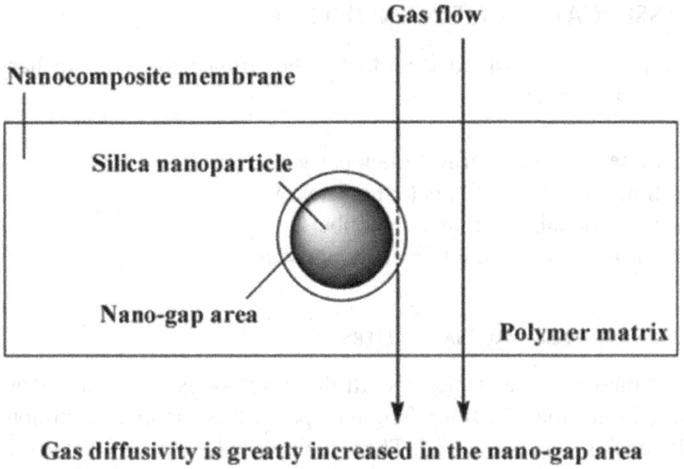

FIGURE 1.1 Illustration of nanogap formation in the BPPOdp/silica nanocomposite membranes. (Adopted from Cong, Hu, et al., 2007.)

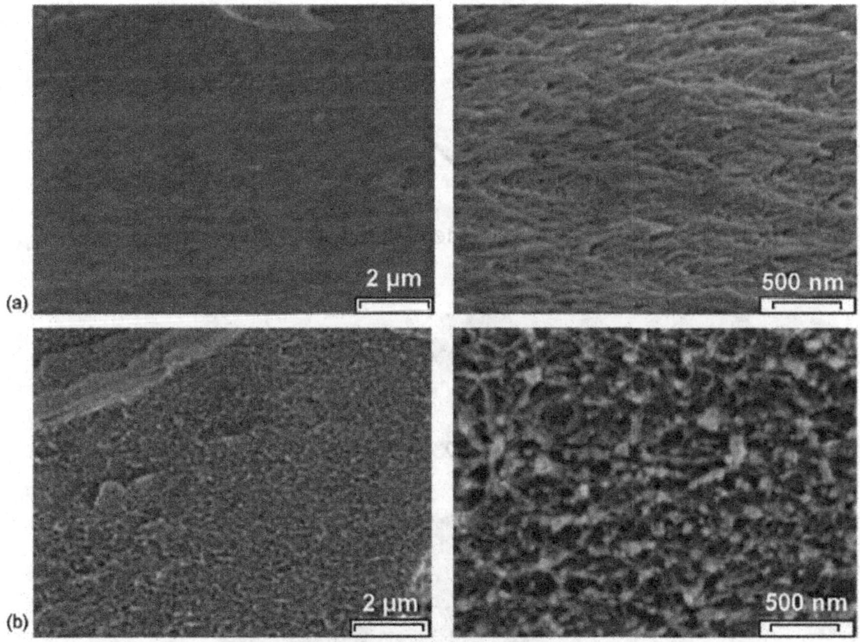

FIGURE 1.2 SEM photomicrographs of cross-sections of: (a) poly(1-trimethylsilyl-1-propyne) (PTMSP) membrane and (b) PTMSP/silica nanocomposite membrane prepared by sol–gel process. (Adopted from Gomes et al., 2005.)

1.2 CLASSIFICATION OF NANOFILLERS

Nanofillers are mainly classified according to their dimensions (Bhattacharya, 2016). These include (Figure 1.3):

1. Zero-dimensional nanofillers (nanoparticle)
2. One-dimensional nanofillers (nanoplatelet)
3. Two-dimensional nanofillers (nanofiber)
4. Three-dimensional nanofillers (nanoparticulate)

1.2.1 Zero-Dimensional Nanofillers

Zero-dimensional nanomaterials have all the dimensions within the nanoscale (no dimension is larger than 100 nm). The nanoparticle is the most common example for zero-dimensional nanomaterial. These nanoparticles can be crystalline or amorphous, ceramic, metallic, or polymeric (Khan et al., 016).

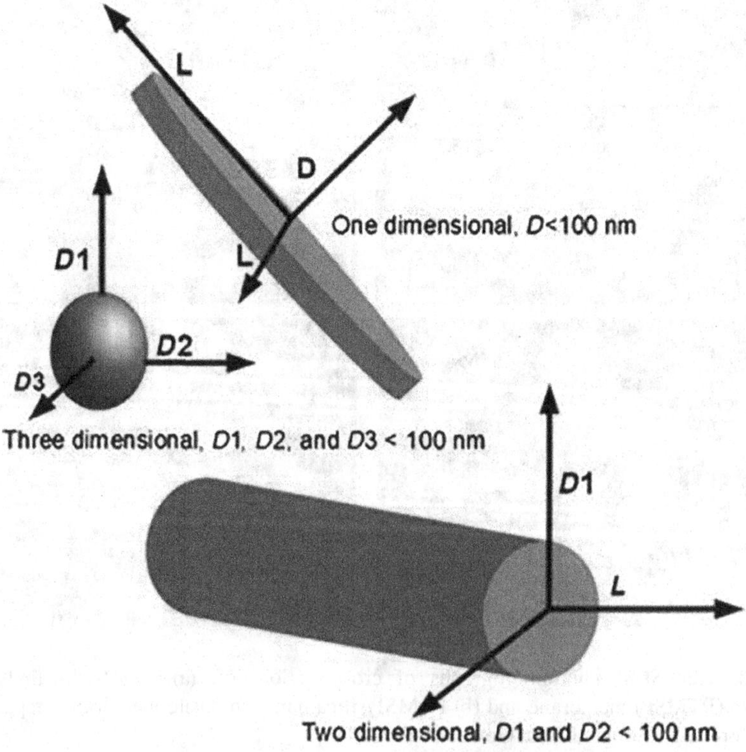

FIGURE 1.3 Classification of nanofillers. (Adopted from Akpan et al., 2018.)

1.2.2 One-Dimensional Nanofillers

Nanofillers that have one of their dimensions less than 100 nm are called one-dimensional (1D) nanofillers (Khan et al., 2016). They are typically in the form of sheets of one to a few nanometers thick to hundreds and thousands of nanometers long and are also found in the form of nanodisks, nanoprism, nanosheets, branched structures, nanoplates, and nanowalls (Dong et al., 2010; Jung et al., 2008; Mann & Skrabalak, 2011; Siril et al., 2009; Tiwari et al., 2012; Vizireanu et al., 2010). Those 1D nanofillers useful in the formation of nanodisks because they exhibit unique shape-dependent characteristics (Jung et al., 2008; Kim et al., 2009). And also, 1D nanofillers are useful in electrical and thermal applications such as biosensors, microelectronics, biomedical, sensors, and coatings because of their dimensionality and because of their excellent electrical, magnetic, and optic properties (Li et al., 2017).

For example, nanographene platelets and montmorillonite clay layered silicates, MMT, graphite nanoplatelets (GNP), ZnO nanoplatelets, ZnO nanodisks, carbon nanowall, amphiphilic graphene platelets, ZnO nanosheets, and Fe_3O_4 nanodisks.

1.2.3 Two-Dimensional Nanofillers

Nanofillers that have two dimensions less than 100 nm are called two-dimensional (2D) nanofillers (Verdejo et al., 2011). They are mostly in the shape of fibers, tubes, or filaments. 2D nanofillers have better flame-retardant property compared with 3D and 1D nanofillers (Isitman et al., 2012), and they are very useful in sensors (Shahjamali et al., 2014), catalysis (Sanjeeva Rao et al., 2014), photocatalysts, energy, nanocontainers, nanoreactors (Pradhan et al., 2013), electronics, and optoelectronics. Compared with 3D fillers, 2D nanofillers found to provide a higher degree of reinforcement (Bhattacharya, 2016; Brune & Bicerano, n.d.).

For example, Carbon nanotubes (CNTs), cellulose whiskers, boron nitride (BN) tubes, boron carbon nitride tubes, gold or silver nanotubes, 2D graphene, black phosphorus (Pradhan et al., 2013), and clay nanotubes.

1.2.4 Three-Dimensional Nanofillers

Three-dimensional (3D) nanofillers relatively have particles in all three dimensions in the nanoscopic (nanometer) scale. They are generally referred to as nanoparticles or zero-dimensional or isodimensional nanoparticles and usually found in cubical and spherical shapes (Bhattacharya, 2016). Also, 3D nanofillers are commonly called as nanospheres, nanogranules, and nanocrystals. They are used in coatings (Olad & Nosrati, 2013), separation and purification (Maximous et al., 2009), and biomedicine (Matos et al., 2015) because they possess good stability, high refractive index, hydrophilicity, ultraviolet (UV) resistance and excellent transparency to visible light, nontoxicity, high photocatalytic activity, and low cost (Chen & Liu, 2011).

For example, nanosilica, nanotitanium oxide, nanoalumina, semiconductor nanoclusters, carbon black, polyhedral oligomeric silsesquioxane (POSS), nanomagnesium hydroxide, silicon carbide, silica, and quantum dots (QDs) (Kumar et al., 2009).

1.3 PROPERTIES OF POLYMERS, INORGANIC NANOFILLERS, AND NANOCOMPOSITES

The addition of a certain amount of inorganic nanofillers to an organic polymer can increase the electrical conductivity, dielectric and mechanical properties, flame retardancy, thermal stability, chemical resistance, UV resistance, water repellency, environmental stability, radar absorption, magnetic field resistance, etc. (Sanchez et al., 2005).

Those properties are strongly dependent on some characteristics of nanoparticles.

- Size/aspect ratio (length/diameter) of the filler
- Shape (e.g., spheres, rods, wires, polyhedrons, and core-shell structures)
- Composition (e.g., metals, metal oxide, and semiconductors) (Burda et al., 2005; Daniel & Astruc, 2004).
- Degree of dispersion of the filler and orientation in the matrix, and
- The adhesion at the filler-matrix interface (Cho & Paul, n.d.)

Inorganic nanoparticles (NPs) probably is one of the premier and broad applications in storage (Kramer & Sargent, 2014; Stratakis & Kymakis, 2013) catalysis (Masala & Seshadri, 2004), sensor (Cho et al., 2012), semiconductor, and medicine (Chan & Nie, 1998; Hu et al., 2012; Zhang et al., 2012). Because of the ability to enhance optical, mechanical, magnetic, and optoelectronic properties, those nanocomposites have been extensively used in various fields, such as safety, military equipment, automotive, protective garments, aerospace, optical devices, and electronics (Figure 1.4).

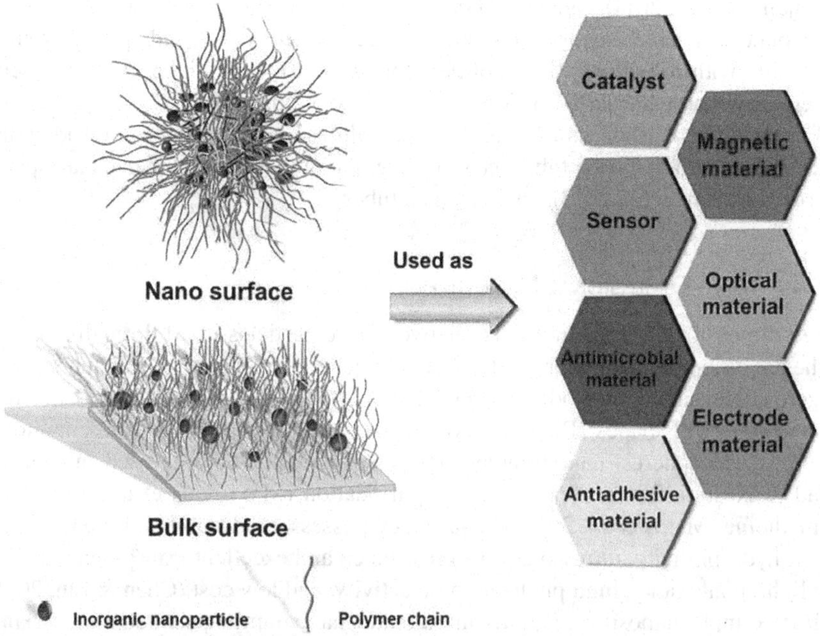

FIGURE 1.4 Illustration of uses of inorganic nanoparticles. (Adopted from Nie et al., 2016.)

TABLE 1.1
Synthesis, Properties of Selected Inorganic Nanoparticles

Sr. No.	Nanoparticle	Synthesis	Properties
1	TiO_2	Hydrothermal, sonochemical, solvothermal, reverse micelles, sol–gel, flame spray pyrolysis, and nonhydrolytic approach	Optical, electronic, spectral, structural, mechanical and anti-corrosion properties
2	ZnO	Sol–gel, homogeneous precipitation, mechanical milling, organometallic synthesis, microwave method, spray pyrolysis, thermal evaporation, and mechanochemical synthesis	Optical properties, thermal conductivity, electrical, sensing, transport, magnetic, and electronic properties
3	Al_2O_3	Flame spray pyrolysis, reverse microemulsion, sol–gel, precipitation, and freeze-drying	Optical, transport, mechanical, and fracture properties
4	SiO_2	Sol–gel, flame synthesis, water-in-oil microemulsion processes	Physicochemical, luminescent, optical, thermal, and mechanical properties
5	Magnetic	Coprecipitation, microemulsions, sol–gel techniques, solvothermal, electrochemical, pulsed laser ablation, and sonochemical method	Magnetic, caloric, physical, and hydrodynamic properties
6	Ag	Microwave processing, laser ablation, gamma irradiation, ultrasonic spray pyrolysis, photochemical method, chemical reduction by inorganic and organic reducing agents, thermal decomposition of silver oxalate in water and in ethylene glycol and electrochemical synthesis	Optical properties due to surface plasmon resonance (SPR), antiangiogenic, thermal, structural, electrical, and catalytic properties
7	Au	Chemical reduction, physical	Optical and photothermal properties

Source: Adopted From Kango et al. (2013).

When choosing a polymer, mechanical, thermal, electrical, optical, and magnetic behaviors, hydrophobic/hydrophilic balance, chemical stability, bio-compatibility, optoelectronic properties, and chemical functionalities (i.e., solvation, wettability, templating effect, etc.) are being considered. Fillers are not only used to reduce the cost of the composites. Generally, the reason for adding inorganic particles into polymers is to improve their mechanical properties such as the tensile strength, modulus, or stiffness via reinforcement mechanisms described by theories for nanocomposites (Fornes & Paul, 2003; Lee & Lin, 2006; Min et al., 2006; Pattanayak & Jana, 2005; Wang et al., 2005) (Table 1.1).

1.4 APPLICATIONS OF INORGANIC NANOPARTICLES AND NANOCOMPOSITES

See Tables 1.2 and 1.3.

TABLE 1.2
Applications of Some Selected Inorganic Nanoparticles

Sr. No.	Nanoparticle	Applications
1	TiO_2	Photocatalysis, dye-sensitized solar cells, gas sensor, nanomedicine, skincare products, wastewater treatment by removal of organic and inorganic pollutants, and antimicrobial applications
2	ZnO	Electronic and optoelectronic device applications, gas sensor, photocatalytic degradation of organic and inorganic pollutants for wastewater treatment, cosmetics, medical filling materials, antimicrobial, and anticancerous applications
3	Al_2O_3	Wastewater and soil treatment by removal of heavy metal ions and antimicrobial applications, ceramic ultrafilters and membranes to remove pathogenic microorganisms, for gas separation, in catalysis and absorption processes, drug delivery, etc.
4	SiO_2	Drug delivery, tissue engineering, carrier for antimicrobial applications, biosensing
5	Magnetic	Biomedicine, cancer treatment, MRI, drug delivery, removal of toxic metal ions, and antimicrobial applications
6	Ag	Antibacterial and antifungal applications in water purification systems, paints and household products, antiviral applications against HIV-I and monkeypox virus, biosensing
7	Au	Antibacterial and antiviral applications, biosensing, MRI, and cancer diagnosis

Source: Adopted From Kango et al. (2013).

TABLE 1.3
Potential Applications of Polymer-Based Inorganic Nanocomposites

Nanocomposites	Applications
Polycaprolactone/SiO_2	Bone-bioerodible for skeletal tissue repair
Polyimide/SiO_2	Microelectronics
PMMA/SiO_2	Dental application, optical devices, Bioactive bone substitute
Poly(ethyl acrylate)/SiO_2	Catalysis support, stationary phase for chromatography
Poly(p-phenylene vinylene)/SiO_2	Nonlinear optical material for optical waveguides
Poly(amide-imide)/TiO_2	Composite membranes for gas separation
Poly(3,4-ethylene-dioxythiphene)/V_2O_5	Cathode materials for rechargeable lithium batteries
Polycarbonate/SiO_2	Abrasion resistant coating
Shape memory polymers/SiC	Medical devices for gripping or releasing therapeutics within blood vessels
Nylon-6/LS	Automotive timing-belt-TOYOTA
Nylon-6/clay	Barrier films – Bayer AG
Nylon-6/clay	Films and bottles – Honeywell

(Continued)

TABLE 1.3 (*Continued*)
Potential Applications of Polymer-Based Inorganic Nanocomposites

Nanocomposites	Applications
Nylon-6, 12, 66/clay	Auto fuel systems – ube
Nylon-6/PP/clay	Electrically conductive
UHMWPE/clay	Earthquake-resistance pipes – Yantai Haili Ind. & Commerce of China
Polypropylene/clay	Packaging – Clariant
PEO/LS	Airplane interiors, fuel tanks, brakes and tires, components in electrical and electronic parts
PLA/LS	Lithium battery development
PET/clay	Food packaging application. Specific examples include packaging for processed meats, cheese, confectionery, cereals and boil-in-the-bag foods, fruit juice and dairy products, beer, and carbonated drinks bottles
Thermoplastic olefin/clay	Beverage containers
Polyimide/clay	Automotive step assists (GM Safari and Astra Vans)
Epoxy/MMT	Materials for electronics
SPEEK/laponite	Direct methanol fuel cells
EVA/clay	Wires and cables (Kabelwerk Eupen of Belgium)
Unsaturated polyester/clay	Marine, transportation – Polymeric Supply

Source: Adopted from Henrique Cury Camargo et al. (n.d.); Sanchez et al. (2005).

1.5 TYPES OF INORGANIC NANOFILLERS

There are many types of inorganic nanoparticles (Figures 1.5–1.8):

- nanoclays
- silica
- metallic phosphates
- transition metal oxides
- nanometals
- metal chalcogenides

1.6 SYNTHESIS OF NANOFILLERS

Nanoparticles can be synthesized from many materials with particles differing in their elemental composition, shape, size, and chemical or physical properties using various physical and chemical methods (Masala & Seshadri, 2004). The physical methods depend on the principle of sub-dividing bulk precursor materials into smaller nanoparticles and generally involve vapor deposition. The chemical method is followed by the controlled aggregation of atoms and generally involves the reduction of metal ions into metal atoms in the presence of stabilizing agents (Chen & Liu, 2011). Chemical methods proved to be more effective than the use of physical methods for the synthesis of nanoparticles. The properties of nanoparticles of material completely different from their bulk counterparts. When the size decreases the

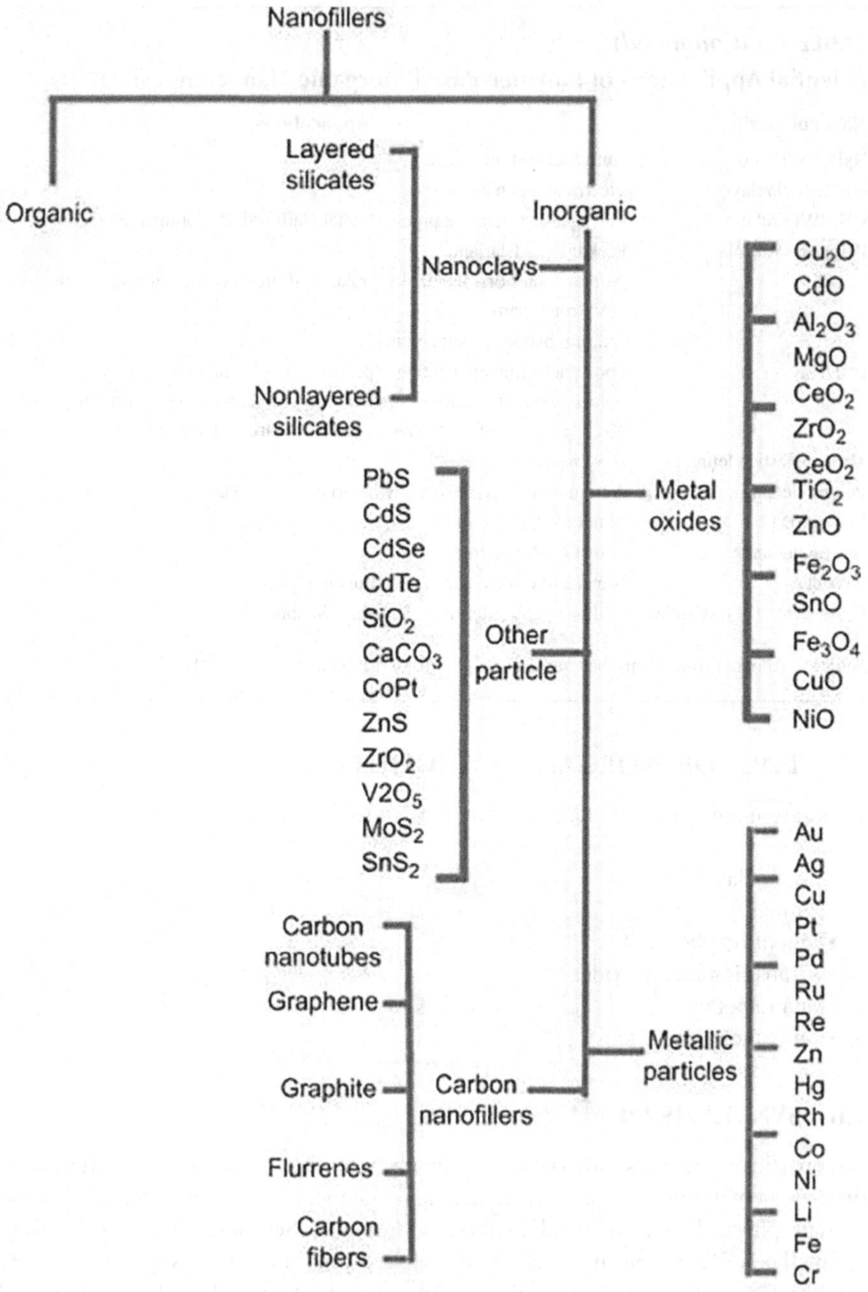

FIGURE 1.5 Types of inorganic nanofillers. (Adopted from Jeon & Baek, 2010.)

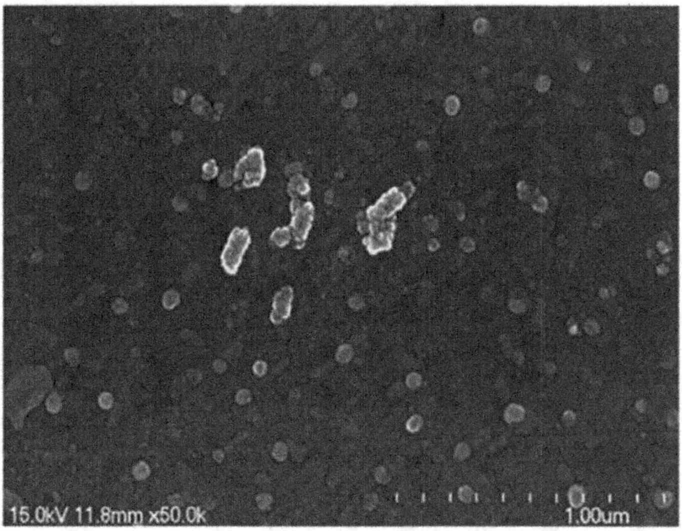

FIGURE 1.6 SEM micrograph of silica nanoparticles prepared at optimal conditions. (Adopted from Luo et al., 2012.)

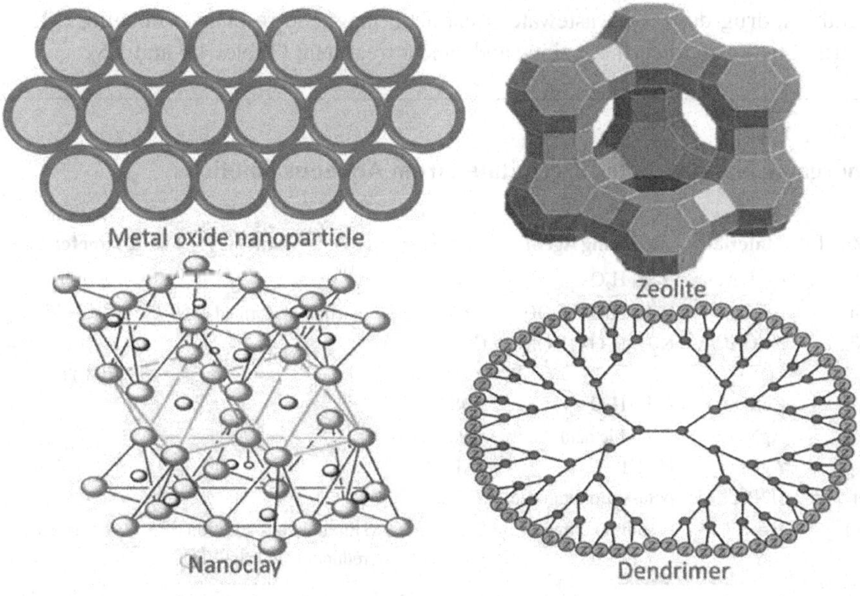

FIGURE 1.7 Some functional materials used in membranes for water purification. (Adopted from Paul & Robeson, 2008.)

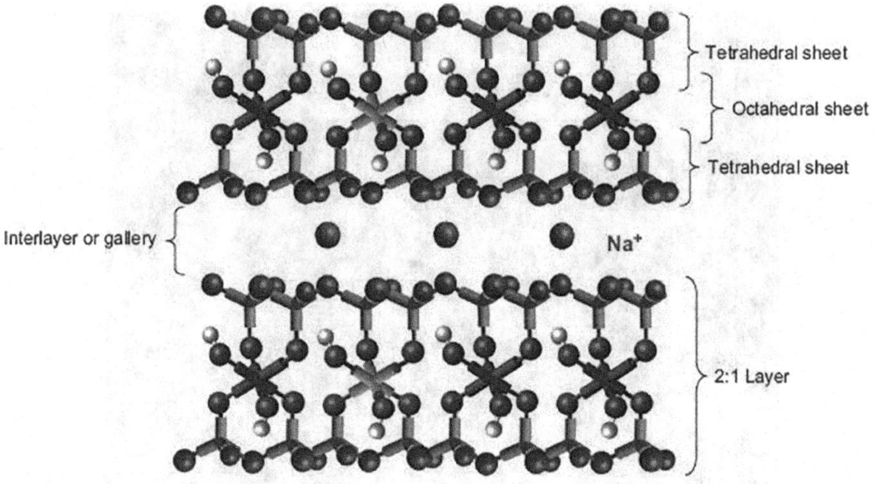

FIGURE 1.8 Structure of sodium montmorillonite. (Adopted from Paul & Robeson, 2008.)

proportion of surface atoms increases and due to that the reactivity makes them highly reactive catalysts (Hanemann & Szabó, 2010) Due to their nanometer-scale size, they possess unique electronic, magnetic, optical, and mechanical properties. With these unique properties, nanoparticles can be used in applications in various fields, such as catalysis, drug delivery, wastewater treatment, magnetic resonance imaging (MRI), textiles, tissue engineering, paints, and cancer treatment (Tables 1.4 and 1.5).

TABLE 1.4
Inorganic Nanoparticles Precipitated from Aqueous Solutions

Metal	Starting Material	Reducing Agent	Stabilizer	Notes	Average Diameter (nm)
Co	$Co(OAc)_2$	$N_2H_4{\cdot}H_2O$	none		~20
Ni	$NiCl_2$	$N_2H_4{\cdot}H_2O$+NaOH	CTAB	reaction performed at 60°C	10–36
Ni	$Ni(OAc)_2$	$N_2H_4{\cdot}H_2O$+NaOH	none		(10–20) × (200–300) rods
Cu	$CuSO_4$	$N_2H_4{\cdot}H_2O$	SDS		~35
Ag	$AgNO_3$	Ascorbic acid	Daxad 19		15–26
Ag	$AgNO_3$	$NaBH_4$	TADDD		3–5
Pt	$H2PtCl_6$	potassium bitartrate	TDPC	60°C	<1.5
Au	$HAuCl_4$	trisodium citrate	S3MP	simultaneous addition of reductant and stabilizer	not stated

Source: Adapted from Cushing et al. (2004).

CTAB, cetyltrimethylammonium bromide; SDS, sodium dodecyl sulfate; Daxad 19, sodium salt of high-molecular-weight naphthalene sulfonate formaldehyde condensate; TADDD, bis(11-trimethylammonium decanoylaminoethyl)-disulfide dibromide; TDPC, 3,3'-thiodipropionic acid; S3MP, sodium 3-mercaptopropionate.

TABLE 1.5
Inorganic Nanoparticles Precipitated by Reduction from Nonaqueous Solutions

Compound	Starting Material	Solvent[a]	Reductant[b]	Stabilizer[c]	Conditions	Product Size[d] (nm)
Fe	$Fe(OEt)_2$	THF	$NaBEt_3H$	THF	16 h at 67°C	10–100
Fe	$Fe(acac)_3$	THF	Mg^+	THF		~8 [e]
$Fe_{20}Ni_{80}$	$Fe(OAc)_2$ $Ni(OAc)_2$	EG	EG	EG	reflux (150–160°C)	6 (A)
Co	$Co(OH)_2$	THF	$NaBEt_3H$	THF	2h at 23°C	10-100
Co	$CoCl_2$	THF	Mg^+	THF		~12
$Co_{20}Ni_{80}$	$Co(OAc)_2$ $Ni(OAc)_2$	EG	EG	EG	reflux (150–160°C)	18–22 (A)
Ni	$Ni(acac)_2$	HDA	$NaBH_4$	HDA	160°C	3.7 (C)
Ni	$NiCl_2$	THF	Mg^+	THF		~94 [e]
Ni	$Ni(OAc)_2$	EG	EG	EG	reflux (150–160°C)	25 (A)
Ru	$RuCl_3$	1,2-PD	1,2-PD	Na(OAc) and DT	170°C	1–6 (C)
Ag	$AgNo_3$	methanol	$NaBH_4$	MSA	room temp	1–6 (C)
Ag	$AgClO_4$	DMF	DMF	3-APTMS	20–156°C	7–20 (C)
Au	$AuCl_3$	THF	$K^+(15C)_2K^-$	THF	-50°C	6–11 (C)
Au	$HAuCl_3$	formamide	Formamide	PVP	30°C	30 (C)

Source: Adapted from Cushing et al. (2004).

[a] EG, ethylene glycol; DMF, dimethylformamide; HAD, hexadecylamine; THF, tetrahydrofuran; 1,2-PD, 1,2-propanediol.

[b] See text for descriptions of reducing agents.

[c] MSA, mercaptosuccinic acid; 3-APTMS, 3-(aminopropyl)trimethoxysilan; PVP, poly(vinylpyrrolidone); DT, dodecanethiol.

[d] (A), agglomerated; (C), colloidal/monodispersed estimated from BET surface area assuming spherical shape.

1.7 SURFACE MODIFICATION OF NANOFILLERS

However, due to the poor compatibility between inorganic filler and polymer matrix, a homogeneous distribution of inorganic fillers in the polymer resin is not always obtained and insufficient dispersal in the polymer matrix can happen. That restricts the application of these nanocomposites in many fields because there is no chemical bonding between the nanoparticles and the polymer matrices. Also, nanoparticles have a strong tendency to undergo agglomeration and to form clusters or agglomerates. Due to aggregation and growth of the size, their properties can decrease because interparticle forces such as van der Waals and electrostatic forces, as well as magnetic attraction, become stronger and need proper chemical treatment to reduce the surface energy (Chandra et al., 2008; Ivanov et al., 2009; Jordan et al., 2005). Need to avoid clustering of particles in polymer nanocomposites by doing modifications to balance the competition between particle-particle, particle-matrix, and matrix-matrix associations. For that, surface modification in inorganic particles is the most popular method because it produces excellent integration and an improved interface between inorganic filler and the polymer matrix.

There are two main approaches for the modification of the surface of the nanoparticles:

i. the surface modification of the inorganic particles by chemical treatment
ii. functional polymeric molecules grafting to the hydroxyl groups existing on the particles.
    - The 'Grafting-from' method
    - The 'Grafting-to' method
    - Plasma surface modification

The surface of the nanoparticles must be modified involving polymer surfactant molecules or other modifiers to take advantage of the properties of nanoparticles by generating a strong repulsion between nanoparticles. One method is to introduce steric stabilization by coating the nanoparticle with a thin layer of the polymer. The van der Waals influence from the nanoparticles can be masked, and the compatibility between the inorganic nanoparticle and hosting organic polymer can be improved by coating the nanoparticle with a thin layer of the polymer (Chandra et al., 2008; Wu & Ke, 2007) and surface functionalization of the nanoparticles (Chang et al., 2003; Guo et al., 2006; Han & Yu, 2006; Shenhar et al., 2005). It will facilitate better nanoparticle dispersion and increased loading amount.

Surface modification improves the inherent characteristics of the nanoparticles and also allows the preparation of composite materials inexistent in nature. With that, excellent combination and an enhanced interface between inorganic filler and the polymer matrix can be obtained (Kalia et al., 2014; Thakur & Thakur, 2014). An additional problem found in nanocomposites is a lower impact strength than that found in the organic precursor alone due to the stiffness of the inorganic material, leading to the use of elastomeric additives to increase the toughness of the composites (Bao & Tjong, 2007) (Figures 1.9 and 1.10).

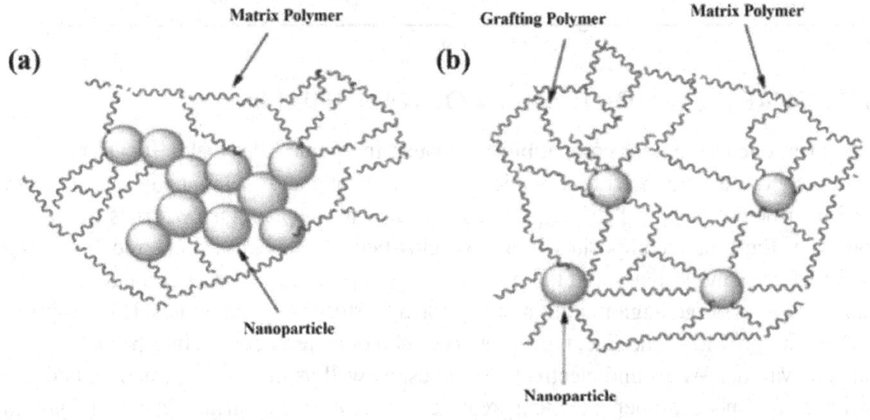

FIGURE 1.9 Schematics of: (a) agglomerated nanoparticles in the matrix polymer in the case without grafting polymer and (b) separation of particles due to the grafting polymer. (Adopted from Rong et al., n.d.)

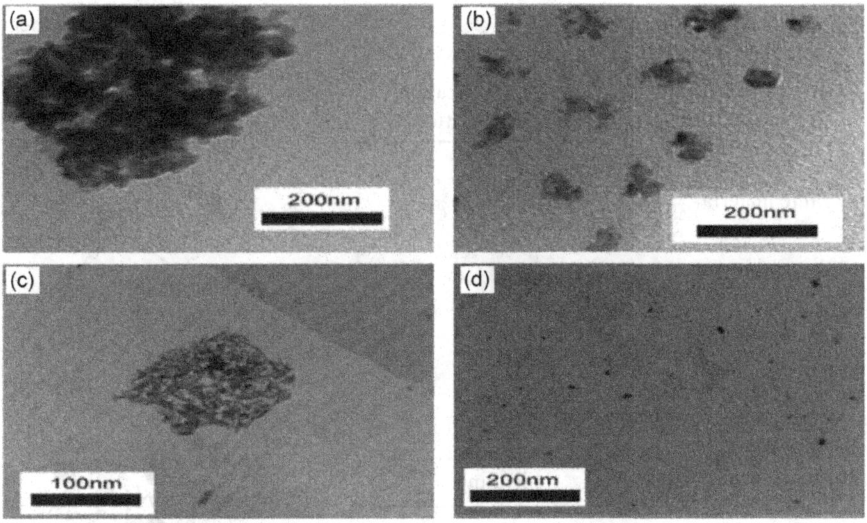

FIGURE 1.10 TEM images of nongrafted Fe_3O_4 (d ¼ 10 nm) particles (a), PS-grafted Fe_3O_4 (b), nongrafted TiO_2 (d ¼ 10 nm) particles (c), and PS-grafted TiO_2 dispersed in chloroform (0.01 mg/mL) (d). (Adapted from Kango et al., 2013.)

When we talk about polymeric membranes. Also, the introduction of organic functional groups to an inorganic filler surface sometimes contributes to better absorption and transportation of penetrants, which results in permeability and favorable selectivity (Patel et al., 2004) together with a better dispersion of the inorganic material in the polymer membrane. Either by the degree of cross-linking of the polymer matrix, or the types of connection bonds between the polymer and inorganic phases in the nanocomposite material, the membrane structure can also be controlled (Saito et al., 1995).

Two types of PINC membranes can be found according to their structure:

a. inorganic and polymer phases connected by covalent bonds and
b. inorganic polymer phases connected by van der Waals force or hydrogen bonds (Jung et al., 2008) (Figure 1.11).

1.8 PREPARATION OF POLYMER-INORGANIC NANOCOMPOSITES

The method of nanocomposite preparation has an impact on the properties of the resulting composite.

The mainly used methods are:

- Sol–gel processing
- Blending
  - Solution blending
  - Emulsion or suspension blending
  - Melt blending
- *in situ* polymerization

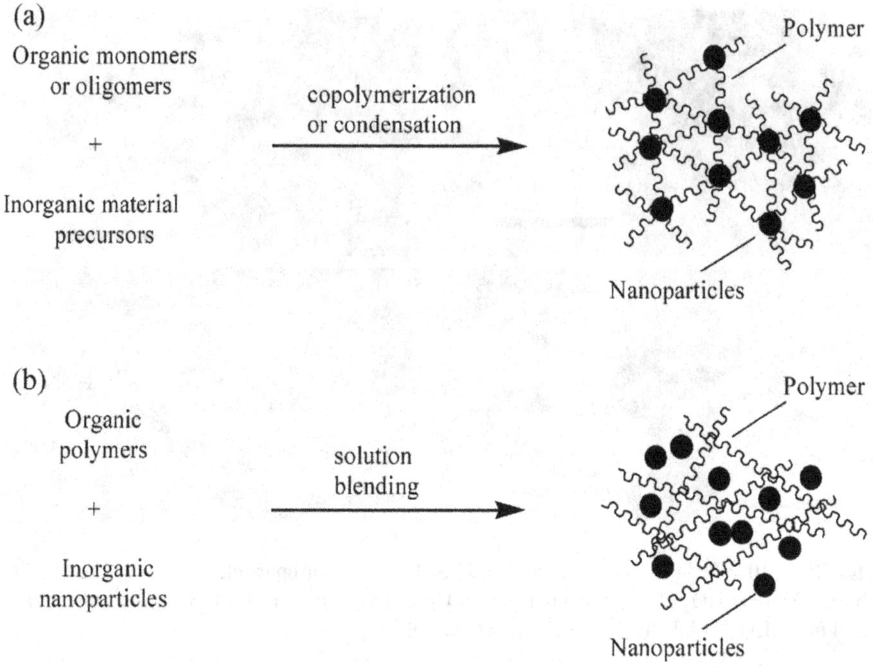

FIGURE 1.11 Illustration of different types of polymer-inorganic nanocomposite membranes. (a) Polymer and inorganic phases connected by covalent bonds and (b) polymer and inorganic phases are connected by van der Waals force or hydrogen bonds. (Adopted from Cong, Radosz, et al., 2007.)

1.8.1 Sol–Gel Processing

This method results in the formation of interpenetrating networks between inorganic and organic moieties at mild temperatures by the processing of nanoparticles inside of a polymer dissolved in nonaqueous or aqueous solutions. This network builds strong interfacial interaction between the two phases and improves the compatibility between constituents. This process is used to prepare nanocomposites with calcium oxide, silica, alumina, and titania in a wide range of polymer matrices.

1.8.2 Blending

This method is done by direct mixing the nanoparticles into the polymer, and it is the most conventional and simple method for the synthesis of PINCs. Generally, the missing can be done by melt blending or solution blending. There is a strong tendency to form agglomerates of nanoparticles and due to that achieving an effective dispersion of the nanoparticles in the polymer matrix is the main difficulty in the mixing process.

1.8.2.1 Solution Blending

Solution blending provides a good level of molecular mixing and is widely used in material preparation and processing, and it is a liquid-state powder processing method. If both the polymer and the nanoparticles are dissolved or dispersed in solution, some of the limitations of melt mixing can be overcome but there is a cost depending on the solvent and its recovery.

1.8.2.2 Emulsion or Suspension Blending

It is quite similar to solution blending; the only difference is that instead of a simple solution here normally uses an emulsion or suspension solution. Emulsion blending is also known as suspension blending. Where polymers are difficult to dissolve, this method is far more effective (Hong et al., 2009).

1.8.2.3 Melt Blending

In this method, particles are dispersed into a polymer melt and PINCs are then obtained by extrusion. With direct mixing of components followed by melt compounding using a twin-screw extruder prior to spinning, polypropylene/silica nanocomposite filaments have been prepared (Erdem et al., 2009).

1.8.3 In Situ Polymerization

This method is a simple and efficient method to incorporate inorganic NPs into polymer composites (Pan et al., 2010; Park & Park, 2006; Sheng et al., 2006; Wang et al., 2000). The particles are generated from their respective particle precursors in the presence of the polymer matrix in this method. The incorporation of precursors into the polymeric matrix can achieve either from the gas or liquid phases, but it is possible, though unusual, to mix the components in the solid phase. This step is followed by the removal of unbound chemical products after the NPs have formed. For the in situ fabrication of NP various pathways can be used including chemical reductions, photoreductions, and thermal decompositions (*5 Chemical Methods of Metal-Polymer Nanocomposite Production*, n.d.; Balan & Burget, 2006) (Figure 1.12).

1.9 CONCLUSION

Combining inorganic nanoparticles and polymers could improve various properties, and advanced new functions can also be generated to the PINCs.

Specifically, inorganic nanoparticles possess outstanding optical, catalytic, electronic, and magnetic properties, and organic polymer-based nanocomposites generally have many advantages such as long-term stability and good processability. By the hybridization of the attractive functionalities of both components, the final product of nanocomposite can be used in various areas such as aerospace and automotive. When developing these nanocomposites, there is a strong tendency to agglomerate the inorganic nanoparticles. To improve the dispersion of nanoparticles in the polymer matrix, surface of those nanofillers need to be modified. This improves the interfacial interactions between the inorganic nanofillers and polymer matrices, which results in unique properties, such as very high mechanical toughness and other electronic, optical, gas-barrier and flame-retardance properties. Thus, the surface modification of inorganic nanofillers is a must to produce high-performance organic-inorganic nanocomposite materials.

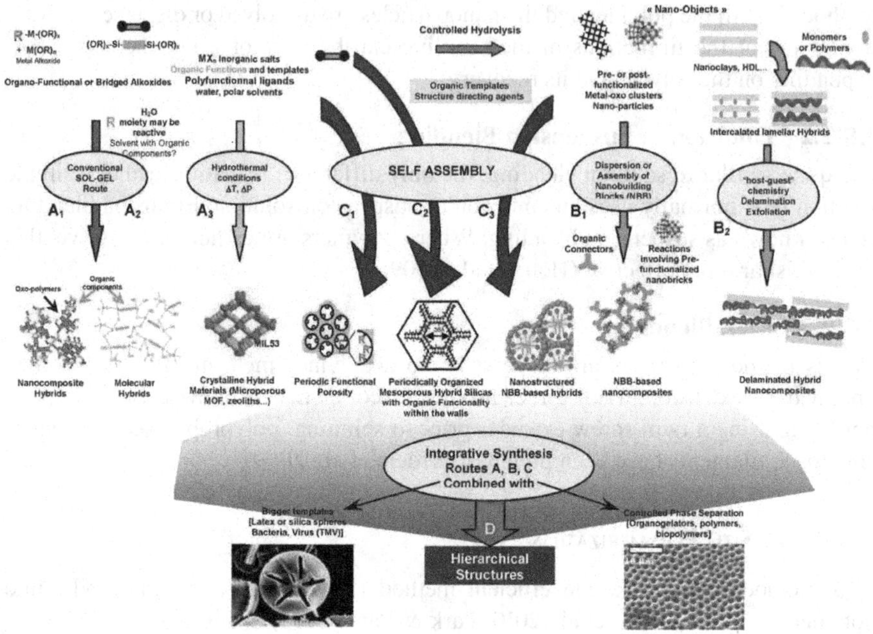

FIGURE 1.12 Illustration of the main chemical routes for the synthesis of polymer-inorganic nanocomposites. (a) sol–gel process; (b) assembly or dispersion; (c) self-assembly procedures; (d) an integrative synthesis. (Adopted from Sanchez et al., 2005.)

REFERENCES

Ajayan, P. M., Schadler, L. S., & Braun, P. V. (2003). *Nanocomposite science and technology*. Weinheim, Germany: Wiley-VCH.

Ajayan, P. M., Schadler, L. S., & Braun, P. V. (2006). *Nanocomposite science and technology*. John Wiley & Sons. Wiley-VCH.

Balan, L., & Burget, D. (2006). Synthesis of metal/polymer nanocomposite by UV-radiation curing. *European Polymer Journal*, *42*(12), 3180–3189. https://doi.org/10.1016/j.eurpolymj.2006.08.016

Bao, S. P., & Tjong, S. C. (2007). Impact essential work of fracture of polypropylene/montmorillonite nanocomposites toughened with SEBS-g-MA elastomer. *Composites Part A: Applied Science and Manufacturing*, *38*(2), 378–387. https://doi.org/10.1016/j.compositesa.2006.03.005

Bhattacharya, M. (2016). Polymer nanocomposites-A comparison between carbon nanotubes, graphene, and clay as nanofillers. *Materials, 9*(4). MDPI AG. https://doi.org/10.3390/ma9040262

Brune, D. A., & Bicerano, J. (n.d.). *Micromechanics of nanocomposites: Comparison of tensile and compressive elastic moduli, and prediction of effects of incomplete exfoliation and imperfect alignment on modulus q*. www.elsevier.com/locate/polymer

Burda, C., Chen, X., Narayanan, R., & El-Sayed, M. A. (2005). Chemistry and properties of nanocrystals of different shapes. *Chemical Reviews, 105*(4), 1025–1102. https://doi.org/10.1021/cr030063a

Caseri, W. (2000). Nanocomposites of polymers and metals or semiconductors: Historical background and optical properties. *Macromolecules Rapid Communication*, 21.

Caseri, W. (2009). Inorganic nanoparticles as optically effective additives for polymers. *Chemical Engineering Communications*, *196*(5), 549–572. https://doi.org/10.1080/00986440802483954

Chan, W. C. W., & Nie, S. (1998). Quantum dot bioconjugates for ultrasensitive nonisotopic detection. *Science*, *281*(5385), 2016–2018. https://doi.org/10.1126/science.281.5385.2016

Chandra, A., Turng, L. S., Gopalan, P., Rowell, R. M., & Gong, S. (2008). Study of utilizing thin polymer surface coating on the nanoparticles for melt compounding of polycarbonate/alumina nanocomposites and their optical properties. *Composites Science and Technology*, *68*(3–4), 768–776. https://doi.org/10.1016/j.compscitech.2007.08.027

Chang, J. H., An, Y. U., Cho, D., & Giannelis, E. P. (2003). Poly(lactic acid) nanocomposites: Comparison of their properties with montmorillonite and synthetic mica (II). *Polymer*, *44*(13), 3715–3720. https://doi.org/10.1016/S0032–3861(03)00276–3

Chen, H. M., & Liu, R. S. (2011). Architecture of metallic nanostructures: Synthesis strategy and specific applications. *Journal of Physical Chemistry C*, *115*(9), 3513–3527. https://doi.org/10.1021/jp108403r

Cho, E. S., Kim, J., Tejerina, B., Hermans, T. M., Jiang, H., Nakanishi, H., Yu, M., Patashinski, A. Z., Glotzer, S. C., Stellacci, F., & Grzybowski, B. A. (2012). Ultrasensitive detection of toxic cations through changes in the tunnelling current across films of striped nanoparticles. *Nature Materials*, *11*(11), 978–985. https://doi.org/10.1038/nmat3406

Cho, J. W., & Paul, D. R. (n.d.). *Nylon 6 nanocomposites by melt compounding*. www.elsevier.nl/locate/polymer

Cong, H., Hu, X., Radosz, M., & Shen, Y. (2007). Brominated poly(2,6-diphenyl-1,4-phenylene oxide) and its silica nanocomposite membranes for gas separation. *Industrial and Engineering Chemistry Research*, *46*(8), 2567–2575. https://doi.org/10.1021/ie061494x

Cong, H., Radosz, M., Towler, B. F., & Shen, Y. (2007). Polymer-inorganic nanocomposite membranes for gas separation. *Separation and Purification Technology*, *55*(3), 281–291. https://doi.org/10.1016/j.seppur.2006.12.017

Cushing, B. L., Kolesnichenko, V. L., & O'Connor, C. J. (2004). Recent advances in the liquid-phase syntheses of inorganic nanoparticles. *Chemical Reviews*, *104*(9), 3893–3946. https://doi.org/10.1021/cr030027b

Daniel, M.-C., & Astruc, D. (2004). *Gold nanoparticles: Assembly, supramolecular chemistry, quantum-size-related properties, and applications toward biology, catalysis, and nanotechnology*. https://pubs.acs.org/doi/full/10.1021/cr030698+

Dong, X., Ji, X., Jing, J., Li, M., Li, J., & Yang, W. (2010). Synthesis of triangular silver nanoprisms by stepwise reduction of sodium borohydride and trisodium citrate. *Journal of Physical Chemistry C*, *114*(5), 2070–2074. https://doi.org/10.1021/jp909964k

Erdem, N., Cireli, A. A., & Erdogan, U. H. (2009). Flame retardancy behaviors and structural properties of polypropylene/nano-SiO2 composite textile filaments. *Journal of Applied Polymer Science*, *111*(4), 2085–2091. https://doi.org/10.1002/app.29052

Fornes, T. D., & Paul, D. R. (2003). Crystallization behavior of nylon 6 nanocomposites. *Polymer*, *44*(14), 3945–3961. https://doi.org/10.1016/S0032-3861(03)00344-6

Gomes, D., Nunes, S. P., & Peinemann, K. V. (2005). Membranes for gas separation based on poly(1-trimethylsilyl-1-propyne)- silica nanocomposites. *Journal of Membrane Science*, *246*(1), 13–25. https://doi.org/10.1016/j.memsci.2004.05.015

Guo, Z., Pereira, T., Choi, O., Wang, Y., & Hahn, H. T. (2006). Surface functionalized alumina nanoparticle filled polymeric nanocomposites with enhanced mechanical properties. *Journal of Materials Chemistry*, *16*(27), 2800–2808. https://doi.org/10.1039/b603020c

Han, K., & Yu, M. (2006). Study of the preparation and properties of UV-blocking fabrics of a PET/TiO_2 nanocomposite prepared by in situ polycondensation. *Journal of Applied Polymer Science*, *100*(2), 1588–1593. https://doi.org/10.1002/app.23312

Hanemann, T., & Szabó, D. V. (2010). Polymer-nanoparticle composites: From synthesis to modern applications. *Materials, 3*(6), 3468–3517). https://doi.org/10.3390/ma3063468

Henrique Cury Camargo, P., Gundappa Satyanarayana, K., & Wypych, F. (n.d.). Nanocomposites: Synthesis, structure, properties and new application opportunities. *Materials Research, 12*(1). Hong, R. Y., & Chen, Q. (2014). Dispersion of inorganic nanoparticles in polymer matrices: Challenges and solutions. *Advances in Polymer Science, 267*, 1–38. https://doi.org/10.1007/12_2014_286

Hong, R. Y., Feng, B., Liu, G., Wang, S., Li, H. Z., Ding, J. M., Zheng, Y., & Wei, D. G. (2009). Preparation and characterization of Fe_3O_4/polystyrene composite particles via inverse emulsion polymerization. *Journal of Alloys and Compounds*, *476*(1–2), 612–618. https://doi.org/10.1016/j.jallcom.2008.09.060

Hu, S. H., Chen, S. Y., & Gao, X. (2012). Multifunctional nanocapsules for simultaneous encapsulation of hydrophilic and hydrophobic compounds and on-demand release. *ACS Nano*, *6*(3), 2558–2565. https://doi.org/10.1021/nn205023w

Isitman, N. A., Dogan, M., Bayramli, E., & Kaynak, C. (2012). The role of nanoparticle geometry in flame retardancy of polylactide nanocomposites containing aluminium phosphinate. *Polymer Degradation and Stability*, *97*(8), 1285–1296. https://doi.org/10.1016/j.polymdegradstab.2012.05.028

Ivanov, M. R., Bednar, H. R., & Haes, A. J. (2009). Investigations of the mechanism of gold nanoparticle stability and surface functionalization in capillary electrophoresis. *ACS Nano*, *3*(2), 386–394. https://doi.org/10.1021/nn8005619

Jeon, I. Y., & Baek, J. B. (2010). Nanocomposites derived from polymers and inorganic nanoparticles. *Materials*, *3*(6), 3654–3674. https://doi.org/10.3390/ma3063654

Jordan, J., Jacob, K. I., Tannenbaum, R., Sharaf, M. A., & Jasiuk, I. (2005). Experimental trends in polymer nanocomposites: A review. *Materials Science and Engineering A, 393*(1–2), 1–11. https://doi.org/10.1016/j.msea.2004.09.044

Jung, S. H., Oh, E., Lee, K. H., Yang, Y., Park, C. G., Park, W., & Jeong, S. H. (2008). Sonochemical preparation of shape-selective ZnO nanostructures. *Crystal Growth and Design*, *8*(1), 265–269. https://doi.org/10.1021/cg070296l

Kalia, S., Boufi, S., Celli, A., & Kango, S. (2014). Nanofibrillated cellulose: Surface modification and potential applications. *Colloid and Polymer Science, 292*(1), 5–31. https://doi.org/10.1007/s00396-013-3112–9

Kango, S., Kalia, S., Celli, A., Njuguna, J., Habibi, Y., & Kumar, R. (2013). Surface modification of inorganic nanoparticles for development of organic-inorganic nanocomposites: A review. *Progress in Polymer Science, 38*(8), 1232–1261. https://doi.org/10.1016/j.progpolymsci.2013.02.003

Khan, W. S., Hamadneh, N. N., & Khan, W. A. (2016). 4 polymer nanocomposites-synthesis techniques, classification and properties. Conference Proceedings.

Kim, K. S., Zhao, Y., Jang, H., Lee, S. Y., Kim, J. M., Kim, K. S., Ahn, J. H., Kim, P., Choi, J. Y., & Hong, B. H. (2009). Large-scale pattern growth of graphene films for stretchable transparent electrodes. *Nature*, *457*(7230), 706–710. https://doi.org/10.1038/nature07719

Kramer, I. J., & Sargent, E. H. (2014). The architecture of colloidal quantum dot solar cells: Materials to devices. *Chemical Reviews, 114*(1), 863–882). https://doi.org/10.1021/cr400299t

Kruenate, J., Tongpool, R., Panyathanmaporn, T., & Kongrat, P. (2004). Optical and mechanical properties of polypropylene modified by metal oxides. *Surface and Interface Analysis*, *36*(8), 1044–1047. https://doi.org/10.1002/sia.1833

Kumar, A. P., Depan, D., Singh Tomer, N., & Singh, R. P. (2009). Nanoscale particles for polymer degradation and stabilization-Trends and future perspectives. *Progress in Polymer Science, 34*(6), 479–515. https://doi.org/10.1016/j.progpolymsci.2009.01.002

Lee, H. T., & Lin, L. H. (2006). Waterborne polyurethane/clay nanocomposites: Novel effects of the clay and its interlayer ions on the morphology and physical and electrical properties. *Macromolecules*, *39*(18), 6133–6141. https://doi.org/10.1021/ma060621y

Li, B. L., Setyawati, M. I., Chen, L., Xie, J., Ariga, K., Lim, C. T., Garaj, S., & Leong, D. T. (2017). Directing assembly and disassembly of 2D MoS2 nanosheets with DNA for drug delivery. *ACS Applied Materials and Interfaces*, *9*(18), 15286–15296. https://doi.org/10.1021/acsami.7b02529

Li, S., Meng Lin, M., Toprak, M. S., Kim, D. K., & Muhammed, M. (2010). Nanocomposites of polymer and inorganic nanoparticles for optical and magnetic applications. *Nano Reviews*, *1*(1), 5214. https://doi.org/10.3402/nano.v1i0.5214

Luo, Z., Cai, X., Hong, R. Y., Wang, L. S., & Feng, W. G. (2012). Preparation of silica nanoparticles using silicon tetrachloride for reinforcement of PU. *Chemical Engineering Journal*, *187*, 357–366. https://doi.org/10.1016/j.cej.2012.01.098

Mann, A. K. P., & Skrabalak, S. E. (2011). Synthesis of single-crystalline nanoplates by spray pyrolysis: A metathesis route to Bi2WO6. *Chemistry of Materials*, *23*(4), 1017–1022. https://doi.org/10.1021/cm103007v

Masala, O., & Seshadri, R. (2004). Synthesis routes for large volumes of nanoparticles. *Annual Review of Materials Research*, *34*, 41–81. https://doi.org/10.1146/annurev.matsci.34.052803.090949

Matos, A. C., Marques, C. F., Pinto, R. v., Ribeiro, I. A. C., Gonçalves, L. M., Vaz, M. A., Ferreira, J. M. F., Almeida, A. J., & Bettencourt, A. F. (2015). Novel doped calcium phosphate-PMMA bone cement composites as levofloxacin delivery systems. *International Journal of Pharmaceutics*, *490*(1–2), 200–208. https://doi.org/10.1016/j.ijpharm.2015.05.038

Maximous, N., Nakhla, G., Wan, W., & Wong, K. (2009). Preparation, characterization and performance of Al2O3/PES membrane for wastewater filtration. *Journal of Membrane Science*, *341*(1–2), 67–75. https://doi.org/10.1016/j.memsci.2009.05.040

Min, K. D., Kim, M. Y., Choi, K. Y., Lee, J. H., & Lee, S. G. (2006). Effect of layered silicates on the crystallinity and mechanical properties of HDPE/MMT nanocomposite blown films. *Polymer Bulletin*, *57*(1), 101–108. https://doi.org/10.1007/s00289-006-0537–z

Nie, G., Li, G., Wang, L., & Zhang, X. (2016). Nanocomposites of polymer brush and inorganic nanoparticles: Preparation, characterization and application. *Polymer Chemistry*, *7*(4), 753–769. https://doi.org/10.1039/c5py01333j

Okada, A., & Usuki, A. (2006). Twenty years of polymer-clay nanocomposites. *Macromolecular Materials and Engineering*, *291*(12), 1449–1476. https://doi.org/10.1002/mame.200600260

Olad, A., & Nosrati, R. (2013). Preparation and corrosion resistance of nanostructured PVC/ZnO-polyaniline hybrid coating. *Progress in Organic Coatings*, *76*(1), 113–118. https://doi.org/10.1016/j.porgcoat.2012.08.017

Pan, F., Cheng, Q., Jia, H., & Jiang, Z. (2010). Facile approach to polymer-inorganic nanocomposite membrane through a biomineralization-inspired process. *Journal of Membrane Science*, *357*(1–2), 171–177. https://doi.org/10.1016/j.memsci.2010.04.017

Park, J. H., & Park, O. O. (2006). Photorefractive properties in poly (N-vinylcarbazole)/CdSe nanocomposites through chemical hybridization. *Applied Physics Letters*, *89*(19). https://doi.org/10.1063/1.2374804

Patel, N. P., Zielinski, J. M., Samseth, J., & Spontak, R. J. (2004). Effects of pressure and nanoparticle functionality on CO_2-selective nanocomposites derived from crosslinked poly(ethylene glycol). *Macromolecular Chemistry and Physics*, *205*(18), 2409–2419. https://doi.org/10.1002/macp.200400356

Pattanayak, A., & Jana, S. C. (2005). Properties of bulk-polymerized thermoplastic polyurethane nanocomposites. *Polymer*, *46*(10), 3394–3406. https://doi.org/10.1016/j.polymer.2005.03.021

Paul, D. R., & Robeson, L. M. (2008). Polymer nanotechnology: Nanocomposites. *Polymer, 49*(15), 3187–3204. https://doi.org/10.1016/j.polymer.2008.04.017

Pradhan, S., Lach, R., Le, H. H., Grellmann, W., Radusch, H.-J., & Adhikari, R. (2013). Effect of filler dimensionality on mechanical properties of nanofiller reinforced polyolefin elastomers. *ISRN Polymer Science, 2013*, 1–9. https://doi.org/10.1155/2013/284504

Ramanathan, T., Stankovich, S., Dikin, D. A., Liu, H., Shen, H., Nguyen, S. T., & Brinson, L. C. (2007). Graphitic nanofillers in PMMA nanocomposites: An investigation of particle size and dispersion and their influence on nanocomposite properties. *Journal of Polymer Science, Part B: Polymer Physics, 45*(15), 2097–2112. https://doi.org/10.1002/polb.21187

Rong, M. Z., Zhang, M. Q., Zheng, Y. X., Zeng, H. M., Walter, R., & Friedrich, K. (n.d.). *Structure-property relationships of irradiation grafted nano-inorganic particle filled polypropylene composites*. www.elsevier.nl/locate/polymer

Saito, Y. C., Okubo, T., & Sadakata, M. (1995). Gas permeation of porous organic/inorganic hybrid membranes. *Journal of Sol-Gel Science and Technology, 5,* 127–134.

Sanchez, C., Julián, B., Belleville, P., & Popall, M. (2005). Applications of hybrid organic-inorganic nanocomposites. *Journal of Materials Chemistry, 15*(35–36), 3559–3592. https://doi.org/10.1039/b509097k

Sanjeeva Rao, K., Senthilnathan, J., Ting, J. M., & Yoshimura, M. (2014). Continuous production of nitrogen-functionalized graphene nanosheets for catalysis applications. *Nanoscale, 6*(21), 12758–12768. https://doi.org/10.1039/c4nr02824d

Shahjamali, M. M., Salvador, M., Bosman, M., Ginger, D. S., & Xue, C. (2014). Edge-gold-coated silver nanoprisms: Enhanced stability and applications in organic photovoltaics and chemical sensing. *Journal of Physical Chemistry C, 118*(23), 12459–12468. https://doi.org/10.1021/jp501884s

Sheng, W., Kim, S., Lee, J., Kim, S. W., Jensen, K., & Bawendi, M. G. (2006). In-situ encapsulation of quantum dots into polymer microspheres. *Langmuir, 22*(8), 3782–3790. https://doi.org/10.1021/la051973l

Shenhar, R., Norsten, T. B., & Rotello, V. M. (2005). Polymer-mediated nanoparticle assembly: Structural control and applications. *Advanced Materials, 17*(6), 657–669. https://doi.org/10.1002/adma.200401291

Siril, P. F., Ramos, L., Beaunier, P., Archirel, P., Etcheberry, A., & Remita, H. (2009). Synthesis of ultrathin hexagonal palladium nanosheets. *Chemistry of Materials, 21*(21), 5170–5175. https://doi.org/10.1021/cm9021134

Stratakis, E., & Kymakis, E. (2013). Nanoparticle-based plasmonic organic photovoltaic devices. *Materials Today, 16*(4), 133–146. https://doi.org/10.1016/j.mattod.2013.04.006

Thakur, V. K., & Thakur, M. K. (2014). Processing and characterization of natural cellulose fibers/thermoset polymer composites. *Carbohydrate Polymers, 109*, 102–117. https://doi.org/10.1016/j.carbpol.2014.03.039

Tiwari, J. N., Tiwari, R. N., & Kim, K. S. (2012). Zero-dimensional, one-dimensional, two-dimensional and three-dimensional nanostructured materials for advanced electrochemical energy devices. *Progress in Materials Science, 57*(4), 724–803. https://doi.org/10.1016/j.pmatsci.2011.08.003

Vaisman, L., Wachtel, E., Wagner, H. D., & Marom, G. (2007). Polymer-nanoinclusion interactions in carbon nanotube based polyacrylonitrile extruded and electrospun fibers. *Polymer, 48*(23), 6843–6854. https://doi.org/10.1016/j.polymer.2007.09.032

Verdejo, R., Mar Bernal, M., Romasanta, L. J., Tapiador, F. J., & Lopez-Manchado, M. A. (2011). Reactive nanocomposite foams. *Cellular Polymers, 30*(2), 45–62.

Vizireanu, S., Stoica, S. D., Luculescu, C., Nistor, L. C., Mitu, B., & Dinescu, G. (2010). Plasma techniques for nanostructured carbon materials synthesis. A case study: Carbon nanowall growth by low pressure expanding RF plasma. *Plasma Sources Science and Technology, 19*(3). https://iopscience.iop.org/article/10.1088/0963-0252/19/3/034016

Wang, K., Chen, L., Wu, J., Toh, M. L., He, C., & Yee, A. F. (2005). Epoxy nanocomposites with highly exfoliated clay: Mechanical properties and fracture mechanisms. *Macromolecules*, *38*(3), 788–800. https://doi.org/10.1021/ma048465n

Wang, S., Yang, S., Yang, C., Li, Z., Wang, J., & Ge, W. (2000). Poly(N-vinylcarbazole) (PVK) photoconductivity enhancement induced by doping with CdS nanocrystals through chemical hybridization. *Journal of Physical Chemistry B*, *104*(50), 11853–11858. https://doi.org/10.1021/jp0005064

Wu, T., & Ke, Y. (2007). Melting, crystallization and optical behaviors of poly (ethylene terephthalate)-silica/polystyrene nanocomposite films. *Thin Solid Films*, *515*(13), 5220–5226. https://doi.org/10.1016/j.tsf.2006.12.029

Yang, H., Zhang, Q., Guo, M., Wang, C., Du, R., & Fu, Q. (2006). Study on the phase structures and toughening mechanism in PP/EPDM/SiO 2 ternary composites. *Polymer*, *47*(6), 2106–2115. https://doi.org/10.1016/j.polymer.2006.01.076

Zhang, H., Liu, Y., Yao, D., & Yang, B. (2012). Hybridization of inorganic nanoparticles and polymers to create regular and reversible self-assembly architectures. *Chemical Society Reviews*, *41*(18), 6066–6088. https://doi.org/10.1039/c2cs35038f

2 A Compendium of Metallic Inorganic Fillers' Properties and Applications Employed in Polymers

Qazi Adfar
Vishwa Bharti Degree College

Mohammad Aslam and Shrikant S. Maktedar
National Institute of Technology

CONTENTS

DOI: 10.1201/9781003279389-2

2.1 INTRODUCTION

Polymer nanocomposites are probably the most versatile materials today. They do not only possess excellent properties but can also be made to exhibit combination of properties for multifunctional applications. They are useful in almost every area of life including sports, automotive, aerospace, electronics, electrical, marine, energy, biomedical sciences, agriculture, pollution control, etc. It also involves determination of the best production process, considering the end-use requirements, and finally fabrication of the material. The effects such as quantum confinement, quantization of energy, molecular motion, and electromagnetic forces become very active at the nanoscale (Figure 2.1). Their immense effects give rise to increased intermolecular bonding, hydrogen bonding, van der Waals, hydrophobic effect, catalysis, magnetism, surface energy, etc. This prevalent increase in effects forms the basis for nanotechnology and nanostructured materials.

Even though elastomeric composites with nanosized fillers have been in use since 1959, proper research on polymer nanocomposites began in 1990 when Toyota attempted to exfoliate clay nanofillers in nylon six. The study demonstrated significant improvement in a wide range of mechanical properties and heat deflection temperature resulting in the use of the material in the fabrication of automobile tires. Following this study, extensive research in polymer nanocomposites and applications has been carried out globally. Today, polymer nanocomposites are used in several applications including packaging, electronic, electrical, structural, energy storage, automotive, aerospace, etc. The Tribology section of the Institute for Composite Materials, Kaiserslautern, Germany, has also developed various kinds of tribology materials for automotive and calendar roll applications from polymer nanocomposites. In polymer nanocomposites, some concepts are very fundamental and must be taken into consideration as they have direct effect on the properties of the resulting

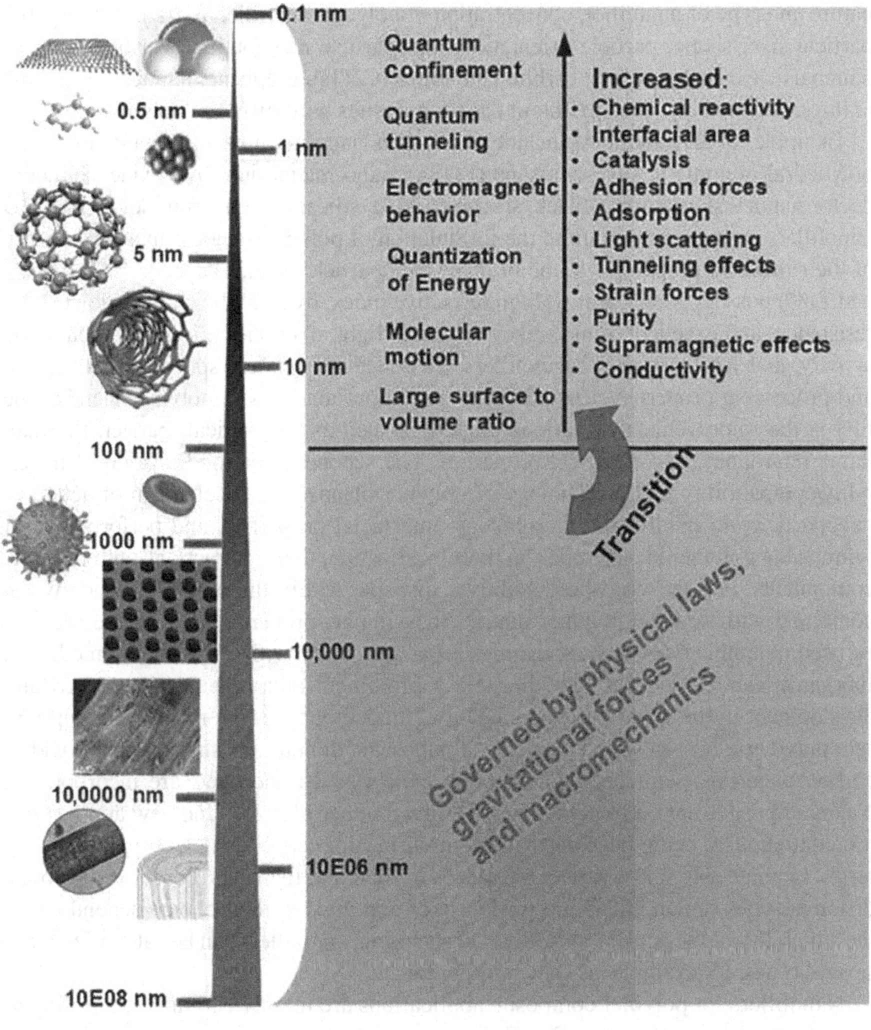

FIGURE 2.1 Change in various effects at nanoscale. (Adapted from ScienceDirect.)

materials. These include size, shape, volume fraction, and state of dispersion of the nanofillers. These factors affect most properties of the nanocomposites.

When a polymer matrix is combined with reinforcement materials, i.e., nanoparticles that are the altered form of basic elements with different atomic and molecular properties, having at least one of its dimensions in the nanoscale, a 3D structure is formed called polymer nanocomposite material. The ideal design of a nanocomposite involves individual nanoparticles homogeneously dispersed in a matrix polymer. The dispersion state of nanoparticles is the key challenge in order to obtain the full potential of properties enhancement. This uniform dispersion of nanofillers can lead to a large interfacial area between the constituents of the nanocomposites. The reinforcing effect of filler is attributed to several factors, such as properties of the polymer matrix, nature and type of nanofiller, concentration of polymer and filler, particle aspect ratio, particle size, shape, particle orientation, and particle distribution. Various types of nanoparticles, such as clays, carbon nanotubes (CNTs), graphene, nanocellulose, and halloysite, have been used to obtain nanocomposites with different polymers.

Examples of 3D nanofillers include nano-silica, nano-alumina, nano-titanium oxide, polyhedral oligomeric silsesquioxane (POSS), nano-magnesium hydroxide, semiconductor nanoclusters, carbon black, silicon carbide, silica, and quantum dots (QDs). 3D nanofillers are very important in the formulation of polymer nanocomposites because of their inherent properties. Some of these nanoparticles (e.g., TiO_2, Ag, SiO_2, Fe_3O_4, and ZnO) possess good stability, high refractive index, hydrophilicity, ultraviolet (UV) resistance, and excellent transparency to visible light, nontoxicity, high photocatalytic activity, and low cost. Since nanofillers are added to augment specific performance and processing properties, they can play two important roles in polymer blends. The first is the improvement of various properties such as mechanical, barrier, thermal, flame retardancy, and electrical properties. The second is the modification of miscibility/compatibility and morphology of polymer blends. The mechanism of action of nanoparticles to modify the morphology, interfacial properties, and performance of immiscible polymer blends relies on their localization, their interactions with polymer components, and the way these additives disperse within the polymer blend. When combined with suitable polymer matrix, they impart properties that enable them to be used in applications such as coatings separation and purification, and biomedicine. Inorganic nanocomposites have become a prominent area of current research and development in the field of nanotechnology. Nanocomposites are materials composed of a polymeric host in which particles of nanoscale dimensions such as metal oxides, carbon materials, semiconductor metallic nanocrystals, and clays are incorporated. Nanocomposites are currently being used in a number of fields, and new applications are continuously being developed including thin-film capacitors, electrolytes for batteries, biomaterials, and a variety of devices in solar and fuel cells. For these purposes, different types of nanofillers are made use of and making their choice depending on the suitability of the applications. Broadly speaking, nanofillers can be categorized into several types, for example, as shown in Figure 2.2.

Nanofillers for polymer composite applications are mainly classified according to their dimensions as shown in Figure 2.3.

These include the following:

1. One-dimensional nanofillers;
2. Two-dimensional nanofillers; and
3. Three-dimensional nanofillers.

For example (Figure 2.4):

2.1.1 One-Dimensional Nanofillers

One-dimensional (1D) nanofillers are those fillers that have one of their dimensions less than 100 nm. They are usually in the form of sheets of one to a few nanometers thick to hundreds and thousands of nanometers long. Montmorillonite clay and

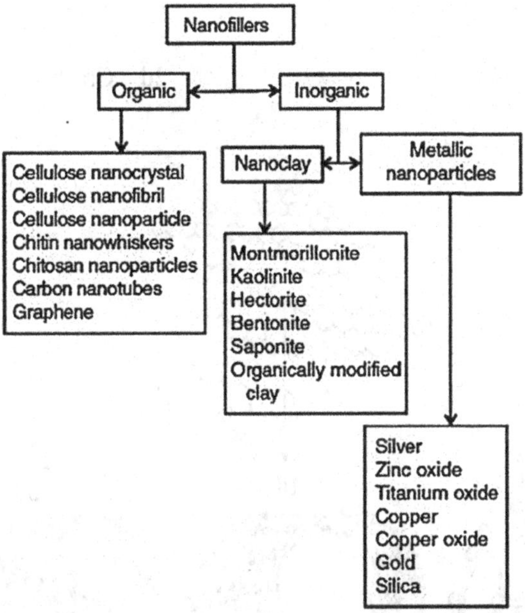

FIGURE 2.2 Different types of nanofillers. (Adapted from Academic library.)

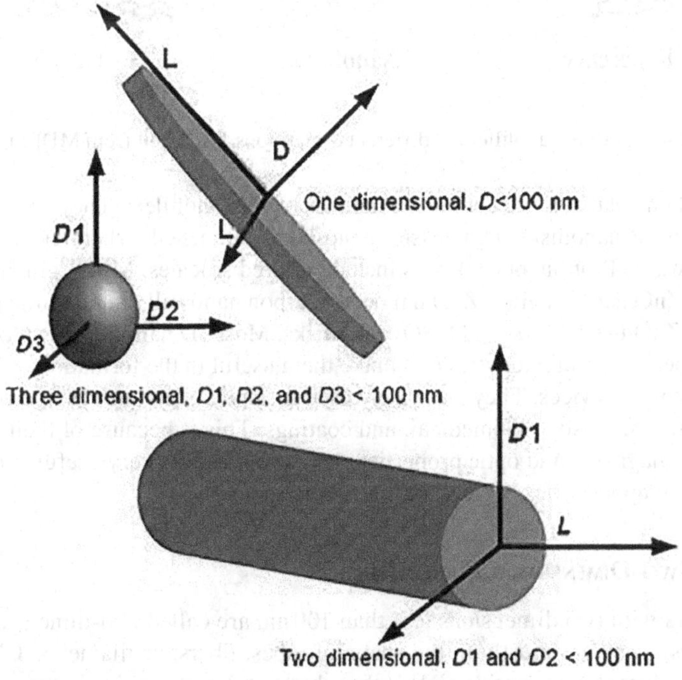

FIGURE 2.3 Classification of nanofillers. (Adapted from ScienceDirect.)

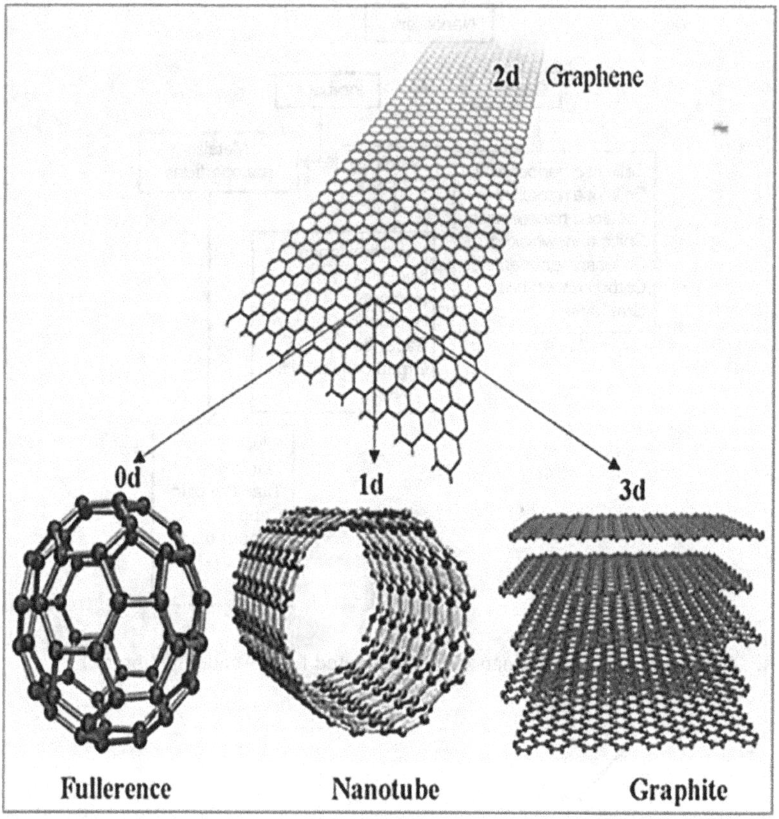

FIGURE 2.4 Carbon nanofillers of different dimensions. (Adapted from MDPI.)

nanographene platelets are common examples of 1D nanofillers. They are also found in the form of nanodisks, nanoprism, nanosheets, branched structures, nanoplates, and nanowalls. Prominent examples include layered silicates, MMT, graphite nanoplatelets, ZnO nanoplatelets, ZnO nanodisk, carbon nanowall, amphiphilic graphene platelets, ZnO nanosheets, and Fe_3O_4 nanodisks. Most 1D nanofillers exhibit unique shape-dependent characteristics that make them useful in the formation of key components of nanodevices. They are widely used in applications such as microelectronics, biosensors, sensors, biomedical, and coatings. This is because of their excellent electrical, magnetic, and optic properties. 1D nanofillers are very useful in electrical and thermal applications because of their dimensionality.

2.1.2 Two-Dimensional Nanofillers

Nanofillers with two dimensions less than 100 nm are called two-dimensional (2D) nanofillers. They are mostly in the shape of tubes, fibers, or filaments. CNTs, cellulose whiskers, boron nitride (BN) tubes, boron carbon nitride tubes, gold or silver nanotubes, 2D graphene, black phosphorus, and clay nanotubes are the most common examples of 2D nanofillers. Some 2D fillers used in the formulation of polymer nanocomposites are natural sepiolite clay fibers, cellulose fibers, nanotubes, sisal fibers, carbon fibers, gold nanowires, zinc oxide, titanium dioxide, silica, cerium dioxide, and copper oxide. Others include molybdenum disulfide (MoS_2), hexagonal boron nitride (h-BN), graphene, MoS_2, and graphene oxide. 2D nanofillers are very useful in energy, sensors, catalysis, photocatalysts, nanocontainers, nanoreactors, electronics, and optoelectronics. They are found to impart better flame-retardant property compared with 3D and 1D nanofillers. They are also found to provide higher degree of reinforcement compared with 3D fillers.

2.1.3 Three-Dimensional Nanofillers

Three-dimensional (3D) nanofillers are relatively equiaxed particles having all three dimensions in the nanoscopic (nanometer) scale. They are usually in spherical and cubical shapes, and are generally referred to as nanoparticles or isodimensional or zero-dimensional nanoparticles. 3D nanofillers are also commonly called nanospheres, nanogranules, and nanocrystals. Examples of 3D nanofillers include nano-silica, nano-alumina, nano-titanium oxide, POSS, nano-magnesium hydroxide, semiconductor nanoclusters, carbon black, silicon carbide, silica, and quantum dots (QDs). 3D nanofillers are very important in the formulation of polymer nanocomposites because of their inherent properties. Some of these nanoparticles (e.g., TiO_2, Ag, SiO_2, Fe_3O_4, and ZnO) possess good stability, high refractive index, hydrophilicity, ultraviolet (UV) resistance and excellent transparency to visible light, nontoxicity, high photocatalytic activity, and low cost. When combined with suitable polymer matrix, they impart properties that enable them to be used in applications such as coatings, separation and purification, and biomedicine.

2.2 PROPERTIES AND APPLICATIONS OF METALLIC INORGANIC FILLERS

The development of polymer nanocomposites has been an area of high scientific and industrial interest in the recent years, due to several improvements achieved in these materials, as a result of the combination of a polymeric matrix and, usually, an inorganic nanomaterial. The improved performance of those materials can include mechanical strength, toughness and stiffness, electrical and thermal conductivity, superior flame retardancy, and higher barrier to moisture and gases. Nanocomposites can also show unique design possibilities, which offer excellent advantages in creating functional materials with desired properties for specific applications. The possibility of using natural resources and the fact of being environmentally friendly have also offered new opportunities for applications. This chapter aims at presenting the prominent properties enhanced by metallic inorganic fillers, and their applications, that are employed in polymers. Metal nanoparticles (NPs) are the focus of scientific interest and play an important role across a range of fields, including (i) catalysis, (ii) sensing, (iii) biomedicine, (iv) optics, and (v) electronics, etc.

2.2.1 Silver (Ag) NPs

Silver NPs (Ag NPs) are especially attractive because of their high electrical conductivity, remarkable optical properties, such as surface-enhanced Raman scattering (SERS), and effective biocidal ability while remaining innocuous to humans. These properties are highly dependent on the size and shape of NPs. Silver nanoparticles have attracted increasing attention for the wide range of applications in biomedicine. Silver nanoparticles, generally smaller than 100 nm, contain 20–15,000 silver atoms, and have distinct physical, chemical, and biological properties compared to their bulk parent materials. The optical, thermal, and catalytic properties of silver nanoparticles are strongly influenced by their size and shape. Additionally, owning to their broad-spectrum antimicrobial ability, silver nanoparticles have also become the most widely used sterilizing nanomaterials in consumable and medical products, for instance, textiles, food storage bags, refrigerator surfaces, and personal care products.

2.2.1.1 Properties

2.2.1.1.1 Optical Property

When silver nanoparticles are exposed to a specific wavelength of light, the oscillating electromagnetic field of the light induces a collective coherent oscillation of the free electrons, which causes a charge separation with respect to the ionic lattice, forming a dipole oscillation along the direction of the electric field of the light. The amplitude of the oscillation reaches maximum at a specific frequency, called surface plasmon resonance (SPR).

The absorption and scattering properties of silver nanoparticles can be changed by controlling the particle size, shape, and refractive index near the particle surface. For example, smaller nanoparticles mostly absorb light and have peaks near 400 nm, while larger nanoparticles exhibit increased scattering and have peaks that broaden and shift toward longer wavelengths. Besides, the optical properties of silver nanoparticles can also change when particles aggregate and the conduction electrons near each particle surface become delocalized (Figure 2.5).

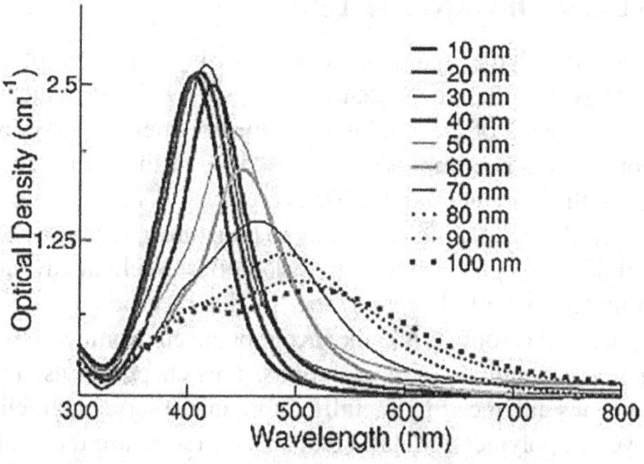

FIGURE 2.5 Extinction spectra of ten sizes of silver nanoparticles. (Adapted from Fortis Life Science/nanocomposite several research works are conducted while using silver NPs, e.g.)

The effect of silver nanomaterials morphology on electrical conductivity properties was carried out (Fang and Lafdi, 2021). Silver nanowires (AgNW)-based nanocomposites have shown a lower percolation threshold. Silver nanoparticles are considered to apply as a silver paste for electrode because of their high conductivity. However, the dispersion of silver nanoparticles in electronically conductive adhesives (ECAs) restricts them to be used as conductive films (Chen et al., 2009). The volume electrical resistivity tests of ECA determine that their electrical resistivity closely correlates with the various sintering temperatures and ECA could achieve the volume electrical resistivity of (3–4) × 9 10^{-5} omega after sintering at 160C for 20min. These are also shown ECA also showed excellent electrical, thermal, and mechanical properties. Nanometal particles such as Ag, Cu, Zn, and Au are particularly useful for electrical circuitry development because nanosized metal particles can be shielded from inks and can also be used to boost electrical conductivity.

1. A novel composite-based, functionalized, multiwalled carbon nanotubes (FMWCNTs) reduced with N, N-dimethyl formamide and cross-linked with silver nitrate, i.e., AgMWCNTs, were fabricated for effective removal of Cu (11) and CD (11) ions. This study accounted for surface, kinetic, equilibrium, and thermal absorption properties. Optimum adsorption of Cu (11) and CD (11) was observed at pH 6.0. Kinetic studies revealed that the adsorption data fitted well to the pseudo-second-order kinetic model and Langmuir model gave a better fit than the other models. The thermodynamic properties, i.e., ^G, ^H, and ^S, showed that adsorption of Cu (11) and Cd (11) onto Ag-MWCNTs was endothermic spontaneous and feasible in the temperature range of 293–313 K.
2. Experimental antimicrobial composite adhesives (ECAs) containing silica nanofillers and silver nanoparticles have been prepared for orthodontic purposes. As shown by the comparison of the effect of salivary pellicles on the adhesion, silver nanoparticles were capable of penetrating the saliva coating. This study suggests that ECAs containing silver NPs and nanofillers can contribute to preventing enamel demineralization around brackets without significant effects on the physical properties (Sug-Joon Ahn et al.—Seoul National University).

2.2.1.2 Applications

2.2.1.2.1 Sensors

Peptide-capped silver nanoparticle for colorimetric sensing has been mostly studied in past years, which focus on the nature of the peptide and silver interaction and the effect of the peptide on the formation of the silver nanoparticles. Besides, the efficiency of silver nanoparticles-based fluorescent sensors can be very high and overcome the detection limits.

Silver nanoparticles are also used as sensors in water treatments. In fact, the pollutants present in the water disturb the spontaneity of life-related mechanisms, such as the synthesis of cellular constituents and the transport of nutrients into cells, and this causes long-/short-term diseases. For this reason, research continuously tends to develop always innovative, selective, and efficient processes/technologies to remove pollutants from water. The silver nanoparticles are used as colorimetric sensors for water pollutants (Figure 2.6).

2.2.1.2.2 Optical Probes

Silver nanoparticles are widely used as probes for SERS and metal-enhanced fluorescence (MEF). Compared to other noble metal nanoparticles, silver nanoparticles exhibit more advantages for probe, such as higher extinction coefficients, sharper extinction bands, and high field enhancements.

2.2.1.2.3 Antibacterial Effects

The antibacterial effects of silver nanoparticles have been used to control bacterial growth in a variety of applications, including dental work, surgery applications, wounds and burns treatment, and biomedical devices. It is well known that silver ions and silver-based compounds are highly toxic to microorganisms. Introduction of silver nanoparticles into bacterial cells can induce a high degree of structural and morphological changes, which can lead to cell death. Scientists have demonstrated that the antibacterial effect of silver nanoparticles is mostly due to the sustained release of free silver ions from the nanoparticles, which serve as a vehicle for silver ions (Figure 2.7).

2.2.1.3 As Catalyst

Silver nanoparticles have been demonstrated to present catalytic redox properties for biological agents such as dyes, as well as chemical agents such as benzene. The chemical environment of the nanoparticle plays an important role in their catalytic properties. In addition, it is important to know that complicated catalysis takes place by adsorption of the reactant species to the catalytic substrate. When polymers, complex ligands, or surfactants are used as the stabilizer or to prevent coalescence of the nanoparticles, the catalytic ability is usually decreased due to reduced adsorption ability. In general, silver nanoparticles are mostly used with titanium dioxide as the catalyst for chemical reactions (Figure 2.8).

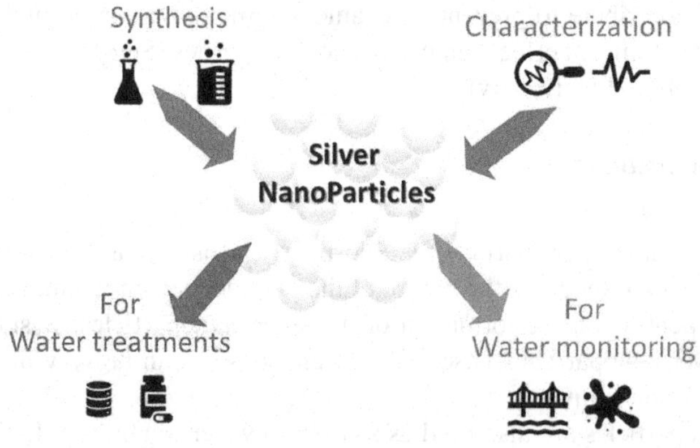

FIGURE 2.6 Silver nanoparticles as colorimetric sensors for water pollutants. (Adapted from MDPI.)

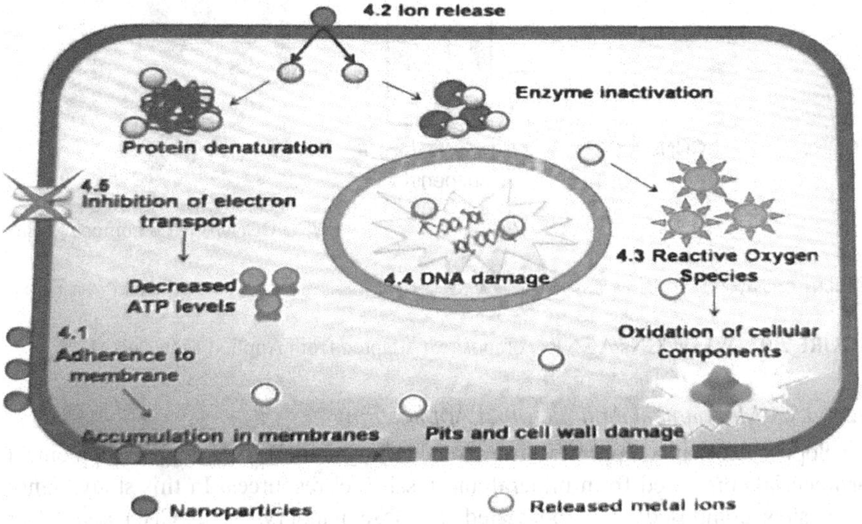

FIGURE 2.7 Two different mechanisms of antimicrobial action of silver nanoparticle. (Adapted from ResearchGate.)

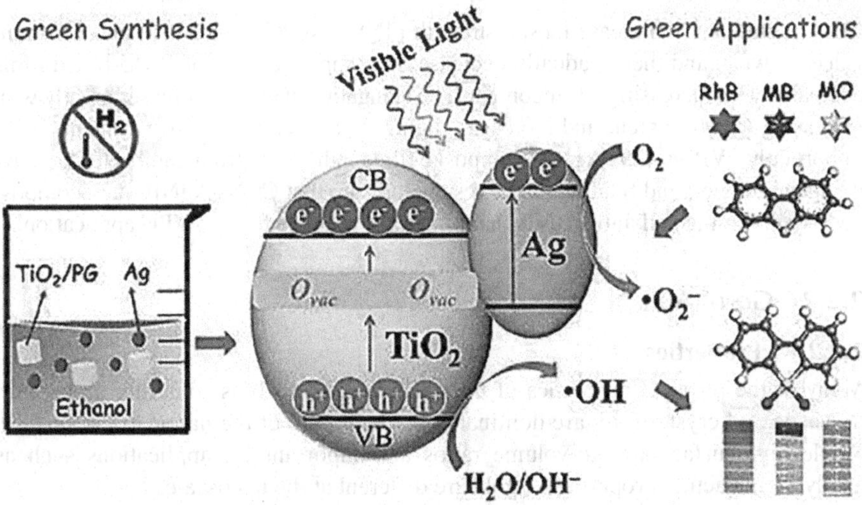

FIGURE 2.8 Green synthesis of Ag–TiO_2 supported on porous glass with enhanced photocatalytic performance for oxidative desulfurization and removal of dyes under visible light. (Adapted from ACS Publications.)

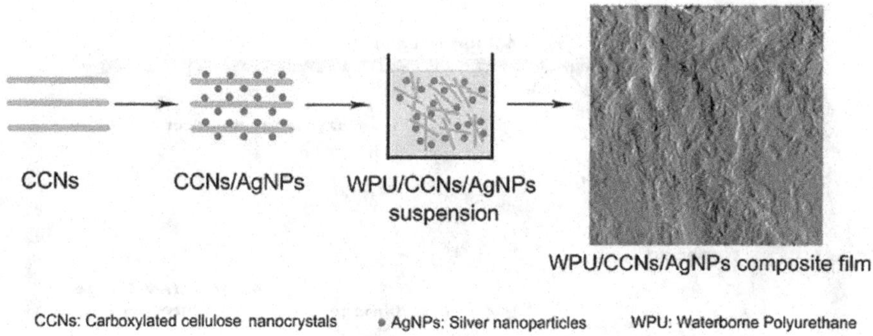

FIGURE 2.9 WPU/CCNs/Ag NPs composite. (Adapted from Applied Materials.)

2.2.1.3.1 Mechanical/Antimicrobial Applications

Developing bio-nanocomposites from renewable biomass is a viable supplement for materials produced from mineral and fossil fuel resources. In this study, nanocomposites composed of carboxylated cellulose nanocrystals (CCNs) and silver nanoparticles (AgNPs) were prepared and used as bifunctional nanofillers to improve the mechanical and antimicrobial properties of waterborne polyurethane (WPU). Morphology, structure, and performance of the CCNs/AgNPs nanocomposites and WPU-based films were investigated. WPU-based composite films were homogeneous and reinforced. The WPU/CCNs/AgNPs composite showed excellent antimicrobial properties in killing both Gram-negative E. coli and Gram-positive S. aureus. The CCNs/AgNPs nanocomposites (Figure 2.9).

2.2.1.3.2 CCNs

CCNs significantly increase tensile strength (TS) of WPU-based films and optimum value (10 wt%) and then gradually decrease. In comparison, TS of WPU-based films decreases with increasing silver content. The elongation at break decreases greatly with increasing CCNs content and increases slightly with increasing silver content. More importantly, WPU/CCNs/AgNPs composite films indicate a strong antibacterial activity against E. coli and S. aureus. The results indicate that CCNs/AgNPs nanocomposites as reinforcing and antibacterial nanofiller are valuable for the WPU application.

2.2.2 Gold (Au)

2.2.2.1 Properties

Many of the physical properties of the nanoparticles such as solubility and stability shape and crystallinity are dominated by the nature of the nanoparticle surface. While high-surface-area-to-volume ratios are important for applications such as catalysis, the actual properties of gold are different at the nanoscale.

2.2.2.1.1 Shape and Crystallinity

Gold nanoparticles can be produced with various sizes and shapes depending on the fabrication method. Typically, anisotropic shapes are formed in the presence of a

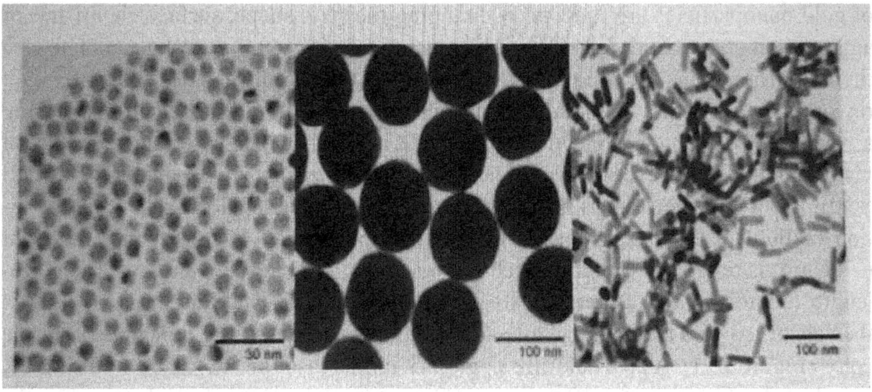

FIGURE 2.10 TEM image of 7-nm BioPure gold nanospheres, image of 100-nm BioPure gold nanospheres, and image of gold nanorods fabricated at nanocomposites (Left to Right). (Adapted from nanocomposites.)

stabilizing polymer that preferentially binds to one crystal face and results in one crystal direction growing faster than others. The size of the crystalline domains within a nanoparticle is dependent on the fabrication method. At nanocomposite, we have developed methods for minimizing the size of crystalline domains within our spherical nanoparticles to yield less variation in their spectral properties (Figure 2.10).

2.2.2.1.2 Particle Stability

Preventing nanoparticle aggregation can be very challenging depending on the application. Nanoparticles are either charge-stabilized or sterically stabilized. For charge-stabilized particles, the zeta potential is a measure of the particle's stability. Typically, nanoparticles with zeta potentials greater than 20mV or less than –20 mV have sufficient electrostatic repulsion to remain stable in solution. However, it is important to note that the surface-bound molecules on nanocomposite NanoXact and BioPure gold 36 nanoparticles are easily displaced, and the zeta potential is highly responsive to other molecules or contaminants in solution. An unwashed pipet tip may introduce sufficient material to displace the ionically bound stabilizing molecules and destabilize the particles. Highly acidic or basic solutions can also increase the dissolution rate of the nanoparticles into an ionic form that can re-deposit onto existing nanoparticles changing the average diameter and size distribution. Particle stability can be accurately tracked using UV/visible spectroscopy or dark-field microscopy as well as dynamic light scattering.

2.2.2.1.3 Optical and Electronic Properties of Gold Nanoparticles

The plasmon resonance of spherical gold nanoparticles results in the particle's exceptional ability to scatter visible light. Colloidal gold nanoparticles have been utilized for centuries by artists due to the vibrant colors produced by their interaction with visible light. More recently, these unique optoelectronic properties have been researched and utilized in high-technology applications such as organic photovoltaics, sensory probes, therapeutic agents, drug delivery in biological and medical applications, electronic conductors, and catalysis. The optical and electronic properties of gold nanoparticles are tunable by changing the size, shape, surface chemistry, or aggregation state. Gold nanoparticles' interaction with light is strongly dictated by their environment, size, and physical dimensions. Oscillating electric fields of a light ray propagating near a colloidal nanoparticle interact with the free electrons causing a concerted oscillation of electron charge that is in resonance with the frequency of visible light. These resonant oscillations are known as surface plasmons. For small (~30 nm) monodisperse gold nanoparticles, the SPR phenomenon causes an absorption of light in the blue-green portion of the spectrum (~450 nm) while red light (~700 nm) is reflected, yielding a rich red color. As particle size increases, the wavelength of SPR-related absorption shifts to longer, redder wavelengths. Red light is then absorbed, and blue light is reflected, yielding solutions with a pale blue or purple color (Figure 2.11). As particle size continues to increase toward the bulk limit, SPR wavelengths move into the IR portion of the spectrum and most visible wavelengths are reflected, giving the nanoparticles clear or translucent color. The SPR can be tuned by varying the size or shape of the nanoparticles, leading to particles with tailored optical properties for different applications.

This phenomenon is also seen when excess salt is added to the gold solution. The surface charge of the gold nanoparticle becomes neutral, causing nanoparticles to aggregate. As a result, the solution color changes from red to blue. To minimize aggregation, the versatile surface chemistry of gold nanoparticles allows them to be coated with polymers, small molecules, and biological recognition molecules. This surface modification enables gold nanoparticles to be used extensively in chemical, biological, engineering, and medical applications. Typical properties of gold nanoparticles are presented in Figure 2.12.

2.2.2.2 Gold NanoUrchins

Gold NanoUrchins have unique optical properties compared to spherical gold nanoparticles of the same core diameter. The spiky uneven surface causes a red shift in the surface plasmon peak and a larger enhancement of the electromagnetic field

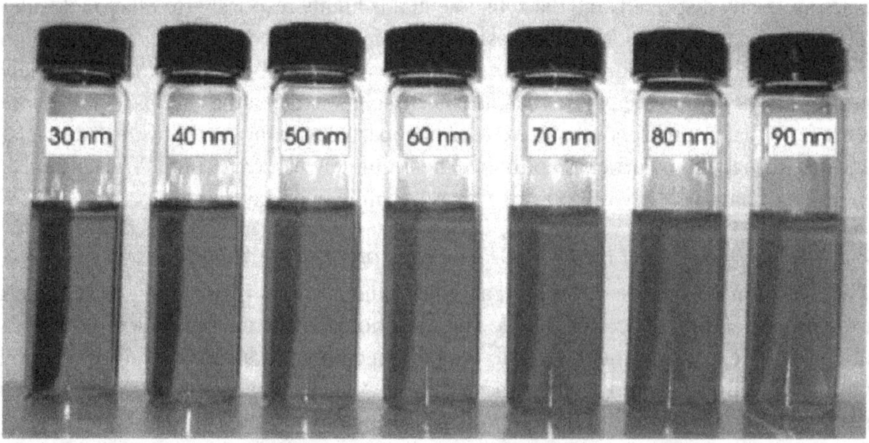

FIGURE 2.11 Colors of various-sized monodispersed gold nanoparticles.

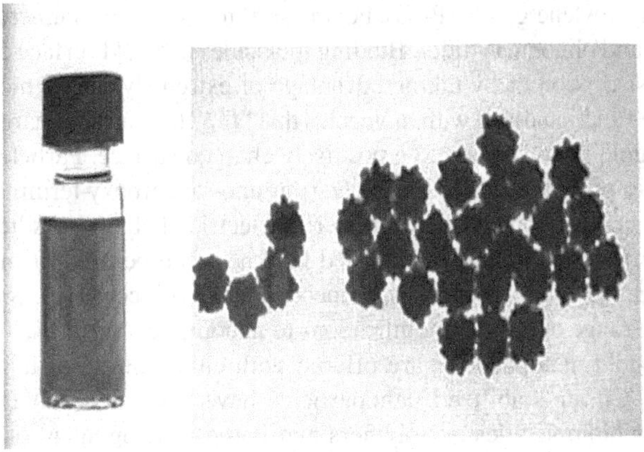

FIGURE 2.12 TEM of 100-nm Gold NanoUrchins. (Adapted from Sigma Aldrich.)

at the tips of the Gold NanoUrchin spikes compared to spherical particles. As an example, 100-nm spherical gold nanoparticles have an SPR peak at 570 nm while 100-nm Gold NanoUrchins have a SPR peak at around 680 nm (Figure 2.13), UV/Vis spectra of 100-nm Gold NanoUrchins (blue), and 100-nm standard gold nanoparticles (green). Note the red shift in the SPR peak. Right—UV/Vis spectra of Gold NanoUrchins ranging in size from 50 to 100 nm in diameter.

2.2.2.3 Surface Chemistry and Functionalization

Gold nanoparticles can be functionalized with a wide range of materials. Polymers such as polyvinylpyrrolidone and tannic acid are capping agents typically used to stabilize gold nanoparticles. Gold nanoparticles used in biological applications are commonly

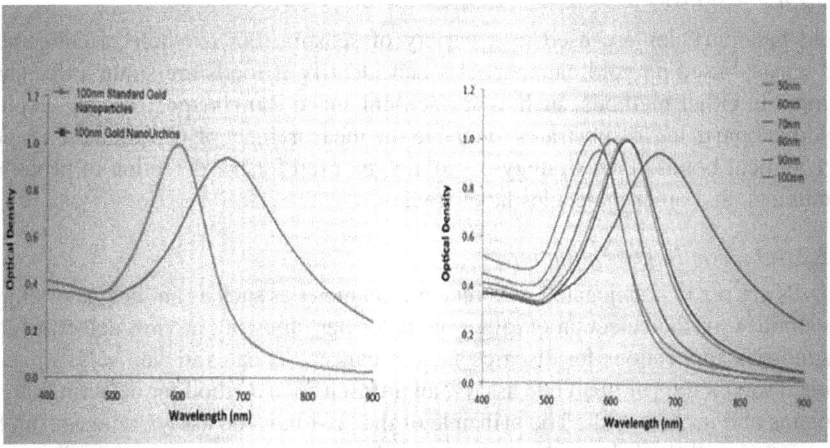

FIGURE 2.13 (Left) UV/Vis spectra of 100-nm Gold NanoUrchins (blue) and 100-nm standard gold nanoparticles (green). Note the red shift in the SPR peak. (Right) UV/Vis spectra of Gold NanoUrchins ranging in size from 50 nm to 100 nm in diameter. (Adapted from Sigma Aldrich.com.)

coated with polyethylene glycol (PEG), bovine serum albumin, or numerous other proteins, peptides, and oligonucleotides. Binding molecules to a gold surface can be accomplished by physisorption or by taking advantage of extremely stable thiol-gold bonds. Particles can be functionalized with molecules that "flip" the surface charge of the negatively charged gold nanoparticles to a positively charged surface. Particles can also be functionalized to provide reactive groups (e.g., amine- or carboxy-terminated surfaces) for subsequent conjugation by the customer. Dielectric shells (e.g., silica, aluminum oxide, and TiO_2) with a precisely controlled thickness can be used to encapsulate the particles, change the optical properties, or incorporate fluorescent dyes. Nanocomposite provides a wide range of custom modifications to nanoparticle surfaces.

Nowadays, gold nanoparticles are offered with either citrate or tannic acid as a capping agent. Citrate-stabilized nanoparticles have a surface that is readily displaced by other biomolecules or polymers and is the capping agent of choice when physisorption is to be used to further modify the nanoparticle surface. Tannic acid is a multidentate capping agent, and it often requires the formation of highly stable thiol-gold bonds to displace tannic acid with other molecules. Tannic acid-coated nanoparticles are more stable in higher ionic strength solutions and have a wider pH tolerance. Both citrate and tannic-capped gold nanoparticles are stable for at least 6 months if stored at 4°C away from light.

2.2.2.4 Applications

2.2.2.4.1 Electronics

Gold nanoparticles are designed for use as conductors from printable inks to electronic chips. As the world of electronics has become smaller, nanoparticles are important components in chip design. Nanoscale gold nanoparticles are being used to connect resistors, conductors, and other elements of an electronic chip.

2.2.2.4.2 Sensors

Gold nanoparticles are used in a variety of sensors. For example, a colorimetric sensor based on gold nanoparticles can identify if foods are suitable for consumption. Other methods, such as surface-enhanced Raman spectroscopy, exploit gold nanoparticles as substrates to enable the measurement of vibrational energies of chemical bonds. This strategy could also be used for the detection of proteins, pollutants, and other molecules label-free.

2.2.2.4.3 For Target Detections

AuNPs are readily conjugated with recognition moieties such as antibodies or oligonucleotides for the detection of target biomolecules, allowing in vitro detection and diagnostics applications for diseases such as cancer. As an example, AuNPs play a critical role in the "bio-barcode assay", an ultrasensitive method for detecting target proteins and nucleic acids. The principle of the "bio-barcode assay" utilizes AuNPs conjugated with both barcode oligonucleotides and target-specific antibodies, and magnetic microparticles functionalized with monoclonal antibodies for the target moiety. These complexes produce a sandwich complex upon detection of the target molecule that releases a large amount of barcode oligonucleotides, providing both identification and quantification of the target (Figure 2.14). As an example of the

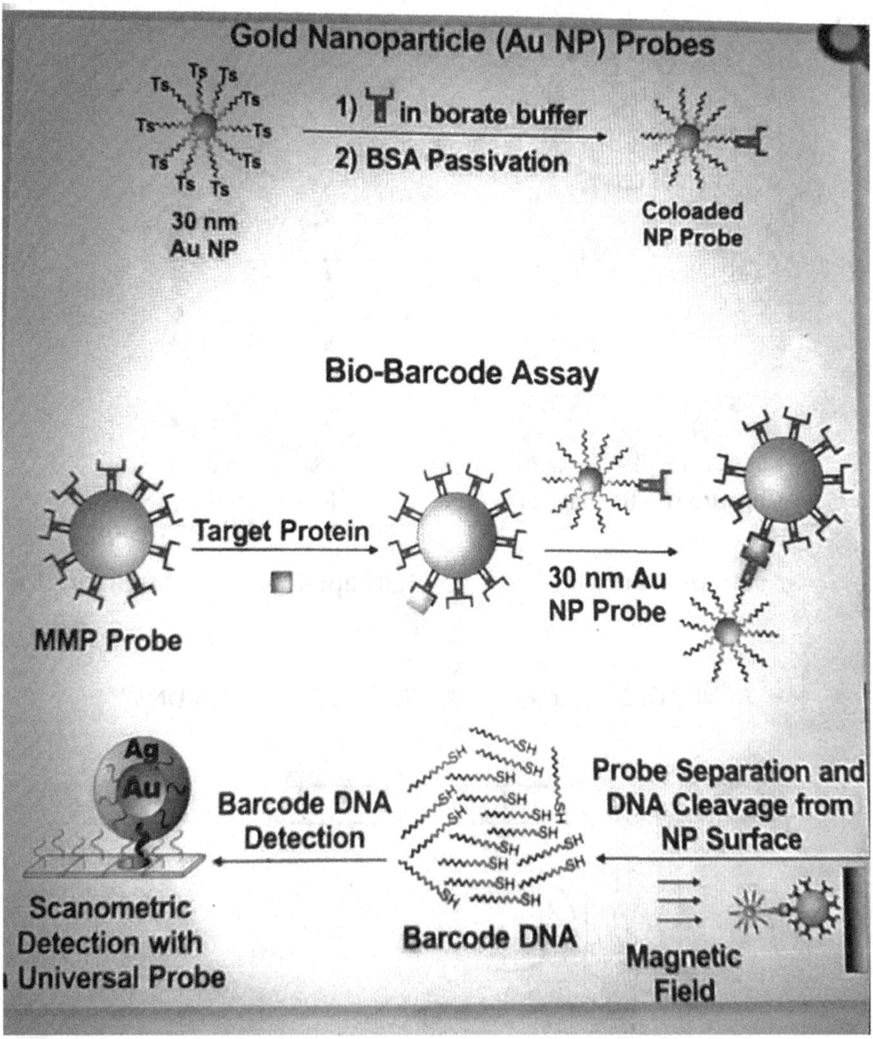

FIGURE 2.14 AuNP-based bio-barcode detection strategy. (Adapted from Nanoscale.)

sensitivity of this method, Mirkin et al. have demonstrated the detection of prostate-specific antigen using this methodology with a limit of detection of 330 fg/mL.

Aptamer-conjugated AuNPs that combine the selectivity and the affinity of aptamers with the spectroscopic properties of AuNPs were utilized to detect small molecule and cancer cells. Zeng et al. have demonstrated an aptamer-nanoparticle strip biosensor (ANSB) system for the detection of Ramos (lymphoma) cells (Figure 2.15). Under optimal conditions, the ANSB showed a detection limit of 4000 Ramos cells using visual detection and 800 Ramos cells with a portable strip reader.

Ramos cells are captured on the test zone through specific aptamer–cell interaction, while excess aptamer-conjugated AuNPs are captured on the control zone through aptamer-DNA hybridization. (A) Schematic illustration of detecting Ramos cells on ANSB. Ramos cells are captured on the test zone through specific

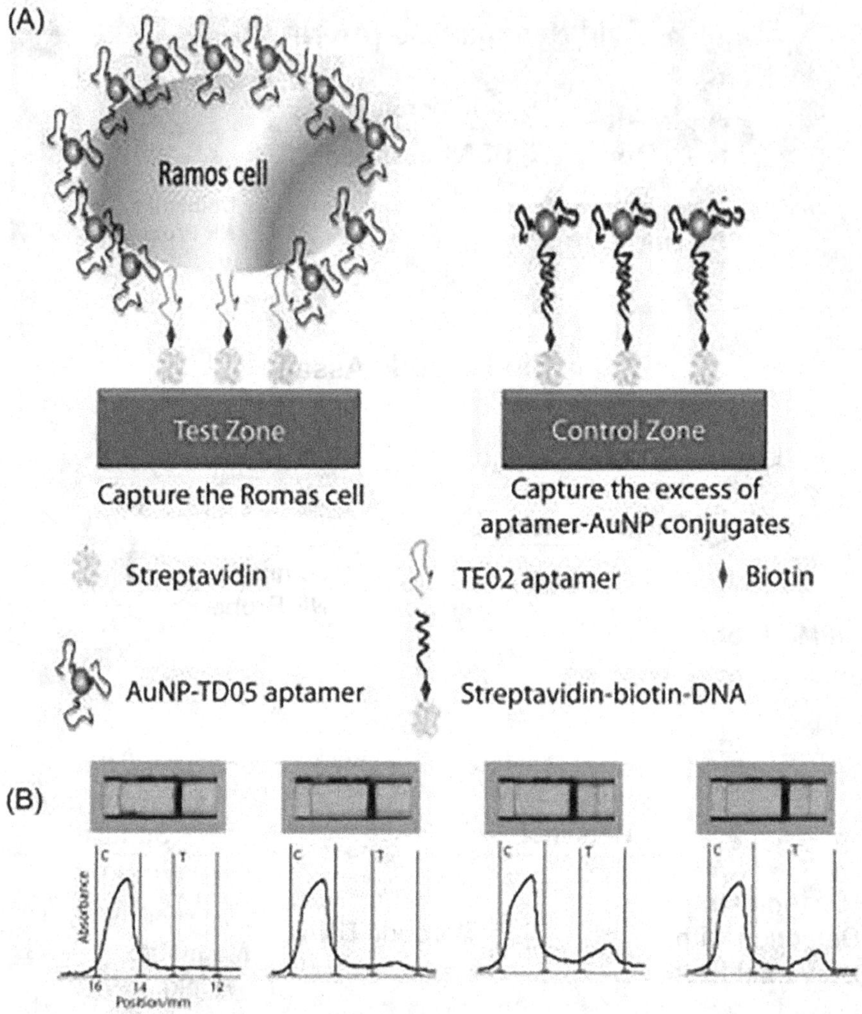

FIGURE 2.15 Detecting Ramos cells on ANSB.

aptamer–cell interaction, while excess aptamer-conjugated AuNPs are captured on the control zone through aptamer-DNA hybridization. A new "chemical nose" methodology using non-covalent conjugates of AuNP and fluorophore was introduced by Rotello and co-workers to provide high-sensitivity sensing of biomolecular targets. Fifty AuNP-fluorophore conjugates provide an alternative biodetection method to "lock and key" specific recognition-based approaches, using an array of selective receptors to generate a pattern that is able to recognize analytes. The initial sensor system was composed of quaternary ammonium-functionalized AuNPs with poly(para-phenylene ethynylene) (PPE) where the PPE serves as a fluorescence transduction element that can be quenched by cationic AuNPs.

Competitive binding of analyte can disrupt the PPE from the complex, resulting in the recovered fluorescence from PPE and producing a readable signal. This method was able to differentiate 12 different species/strains of bacteria with 95% accuracy. Moreover, this strategy was used to differentiate normal, cancerous, and metastatic cells in a rapid and accurate assay (Figure 2.16). Additionally, green fluorescent protein replaced the polymer transducer to provide higher sensitivity (5000 cells relative to the previous 20,000) in mammalian cancer cells sensing.

Recently, the array-based sensing strategy was adapted to an enzyme-amplified array-sensing approach, where the sensitivity is amplified through enzymatic catalysis. The system works by having the analyte protein competitively bind with the AuNP, releasing the β-galactosidase (β-Gal) and restoring its activity. The cleavage of the substrate provides an enzyme-amplified fluorescent readout of the binding event, allowing the identification of proteins even in desalted human urine. A similar approach was used for the construction of a colorimetric enzyme–nanoparticle conjugate system by using chlorophenol red β-D-galactopyranoside (CPRG), a chromogenic substrate, for the detection of bacteria (Figure 2.16). Bacteria sensing was achieved at concentrations of 1×10^2 bacteria/mL in solution and 1×10^4 bacteria/mL in a field-friendly test strip format.

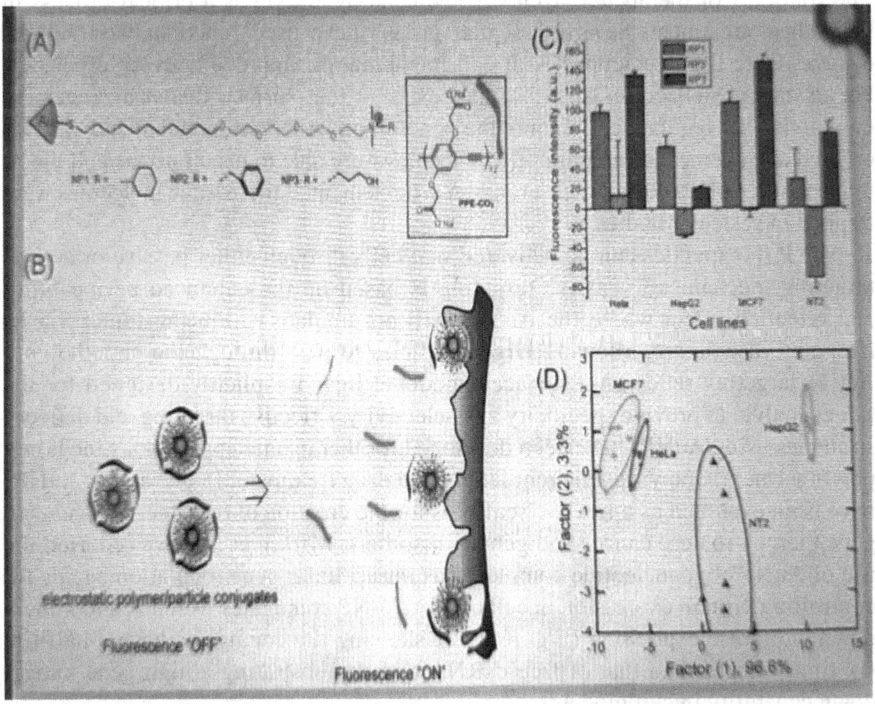

FIGURE 2.16 (a) Molecular structures of the cationic AuNPs and the fluorescent polymer ($PPECO_2$). (b) Displacement of quenched $PPECO_2$ by cell with concomitant restoration of fluorescence. (c) Fluorescence change for four different cancer cell lines using AuNP-$PPECO_2$. (Adapted from Nanoscale.)

1. **As Probes:** Gold nanoparticles also scatter light and can produce an array of interesting colors under dark-field microscopy. The scattered colors of gold nanoparticles are currently used for biological imaging applications. In addition, gold nanoparticles are relatively dense, making them useful as probes for transmission electron microscopy.
2. **In Fuel cells:** Gold nanoparticles are being developed for fuel cell applications. These technologies would be useful in the automotive and display industry.
3. **As Catalysts:** Gold nanoparticles are used as photocatalysts in a number of chemical reactions. Because of the unique SPR property, the surface of gold nanoparticles can be used for selective oxidation or reduce a reaction in certain cases. Normally, gold nanoparticles are raised as photocatalyst with the combination of titanium dioxide, which can be useful in the chemical industry.
4. **In Biomedical Applications:** Therapeutic agents can also be coated onto the surface of gold nanoparticles. The large-surface-area-to-volume ratio of gold nanoparticles enables their surface to be coated with hundreds of molecules (including therapeutics, targeting agents, and anti-fouling agents). Because of their small size, embedded drugs have the ability to spread extensively in tissues.

The transport of therapeutic agents to the cells by AuNPs is a critical process in biomedical treatment. Several research groups have used functionalized AuNPs to investigate the interactions with cell membrane to improve delivery efficiency. For example, Stellacci et al. have demonstrated that surface ligand arrangement on AuNPs can regulate cell membrane penetration. AuNPs functionalized with an ordered arrangement of amphiphilic molecules were able to penetrate the cell membrane while AuNPs coated with a random arrangement of these same molecules were trapped in vesicular bodies.

AuNP therapeutics can be delivered into cells through either passive or active-targeting mechanisms. Passive targeting is based on the enhanced permeability and retention effect where the AuNPs will accumulate within the tumor via its irregular vasculature, allowing larger particles to pass through the endothelium. Active targeting relies on a surface functional ligand explicitly designed for the target analyte to provide specificity and selectivity. Effective targeting and delivery strategies using AuNPs have been developed for therapeutic applications including photothermal therapy, genetic regulation, and drug treatment. In one arena, AuNPs have been exploited as attractive scaffolds for the creation of transfection agents in gene therapy to cure cancer and genetic disorders. Mirkin et al. have reported the use of AuNP/oligonucleotide complexes as intracellular gene regulation agents for controlling protein expression in cells. RNA/AuNP conjugates were used to knock-down luciferase expression (Figure 2.17), showing the conjugates have a half-life six times longer than that of free dsRNA and demonstrating a high gene knock-down capability in cell models.

Rotello et al. also have demonstrated that cationic AuNPs, featuring cationic amino acid-based side chains, can be used for DNA transfection. Lysine-based motif coating AuNPs provided effective nontoxic transfection vectors for DNA delivery that were up to 28 times more effective than polylysine.

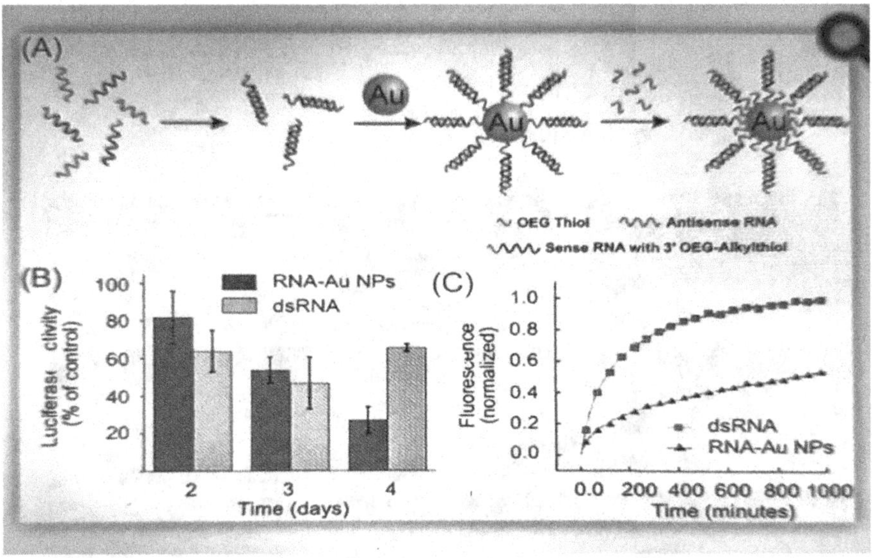

FIGURE 2.17 (a) Preparation of polyvalent RNA/AuNP conjugates. (b) Knockdown of luciferase expression over four days. (c) Stability of RNA/AuNP conjugates, showing the comparison of the stability of dsRNA (red) and RNA/AuNP conjugates (blue) in 10% serum. (Adapted from nanoscale.)

Loading of drugs onto AuNP can be performed through either non-covalent interactions or covalent conjugation. Drug encapsulation with AuNPs has been demonstrated using hydrophobic or hydrophilic pockets generated by the monolayer. In a recent example, Burda et al. utilized PEG-coated AuNPs to provide an amphiphilic environment to capture the hydrophobic silicon phthalocyanine 4 (Pc 4), a photodynamic therapy (PDT) drug (Figure 2.6a and b). They found that the drug release mechanism was through passive accumulation, and the non-covalent Pc 4-AuNP conjugates released their drugs quickly and penetrated deeply into tumors within hours (Figure 2.18).

Drugs covalently conjugated with AuNPs can be released by glutathione (GSH) displacement or through cleavable linkers. Rotello et al. have demonstrated GSH-mediated release using AuNPs featured a mixed monolayer composed of cationic ligands and fluorogenic ligands. The cationic surface of the nanoparticles facilitated their penetration through cell membranes, and the payload release was triggered by intracellular GSH. Kotov et al. applied the GSH-mediated release strategy using 6-mercaptopurine-9-b-D-ribofuranoside functionalized AuNPs to enhance the anti-proliferative effect against K-562 leukemia cells compared to the free drug. Recently, Forbes and Rotello have applied the GSH-mediated approach to investigate the movement of AuNPs carrying either fluorescein or doxorubicin molecules in a tumor model. The results indicate that cationic AuNPs may be more effective in delivering payloads to the majority of tumor cells while anionic AuNPs are able to deliver drug deep into tissues. Alternatively, Rotello et al. have used a light-controlled

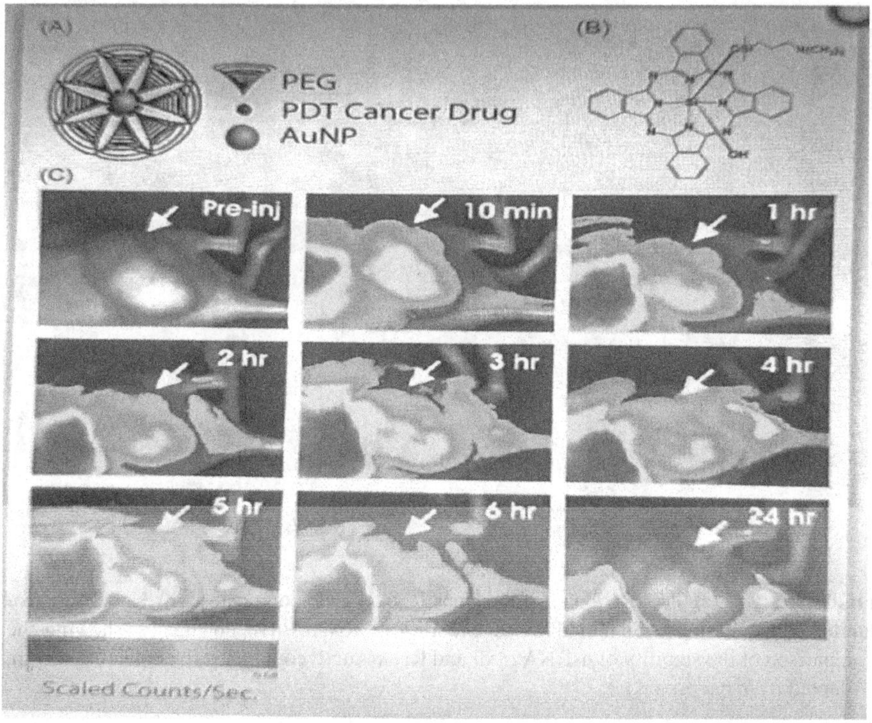

FIGURE 2.18 (a) PEG-functionalized AuNPs with loading of PDT cancer drug. (b) Chemical structure of the PDT drug Pc 4. (c) In vivo fluorescence imaging of AuNP-Pc 4 conjugates injected mouse at various time points within 24 hr. Arrows indicate the tumor location. (Adapted from nanoscale.)

external release strategy to deliver the anticancer drug 5-fluorouracil into cells using AuNPs featured a mixed monolayer of zwitterionic and photocleavable ligands on the surface. Other strategies for delivering covalently attached drugs using AuNPs include the reduction of disulfide bonds and pH-mediated release. Using AuNPs as therapeutic moieties in their own right is another potential approach for medical treatment. For instance, Feldheim et al. have synthesized mixed thiol monolayer-coated AuNPs for bacterial growth inhibition. Mukherjee et al. have utilized naked AuNPs for treatment studies of multiple myeloma, a plasma cell disorder. AuNPs have been shown to inhibit the proliferation of multiple myeloma cells through cell-cycle arrest in the G1 phase via the upregulation of cell-cycle proteins p21 and p27. Recently, Rotello et al. demonstrated a host/guest system to mediate the activation of a therapeutic diamino hexane-terminated AuNP (AuNP-NH2) and modulate its cytotoxicity. The results show that the threading of cucurbit[7]uril (CB[7]) onto a particle surface reduces the high toxicity of the AuNP-NH2 through sequestration of the particle in the endosomes. When treated with 1-adamantylamine (ADA), the CB [7] is displaced from the nanoparticle surface, releasing the toxic particle from the endosome and killing the cancer cell.

2.2.2.4.4 *Diagnostics*

Gold nanoparticles are also used to detect biomarkers in the diagnosis of heart diseases, cancers, and infectious agents. They are also common in lateral flow immunoassays, a common household example being the home pregnancy test. Gold nanoparticles (AuNPs) have several biomedical applications in diagnosis and treating of disease such as targeted chemotherapy and in pharmaceutical drug delivery due to their multifunctionality and unique characteristics. AuNPs can be conjugated with ligands, imaging labels, therapeutic drugs, and other functional moieties for site-specific drug delivery application (Figure 2.19).

Gold nanoparticles (AuNPs) have been widely studied and applied in the field of tumor diagnosis and treatment because of their special fundamental properties. Among the physical properties of AuNPs, localized SPR (LSPR), radioactivity, and high X-ray absorption coefficient are widely used in the diagnosis and treatment of tumors. As an advantage over many other nanoparticles in chemicals, AuNPs can form stable chemical bonds with S- and N-containing groups. This allows AuNPs to attach to a wide variety of organic ligands or polymers with a specific function. These surface modifications endow AuNPs with outstanding biocompatibility, targeting, and drug delivery capabilities. These applications for diagnosis and treatment of tumors include image agent, phototherapy, radiotherapy, targeting, nano-enzyme, and drug delivery (Figure 2.20).

2.2.3 AuNPs Application in Cancer Treatment

AuNPs provide nontoxic carrier system for pharmaceutical drug and gene delivery applications. Currently, various anticancer drugs are available, but these cause the necrosis of cancerous cell as well as normal cells. AuNPs cause the necrosis of only cancer cells; therefore, we can utilize it as a delivery vehicle as well as anticancer agent.

PDT—Near-IR absorbing gold nanoparticles (including gold nanoshells and nanorods) produce heat when excited by light at wavelengths from 700 to 800 nm.

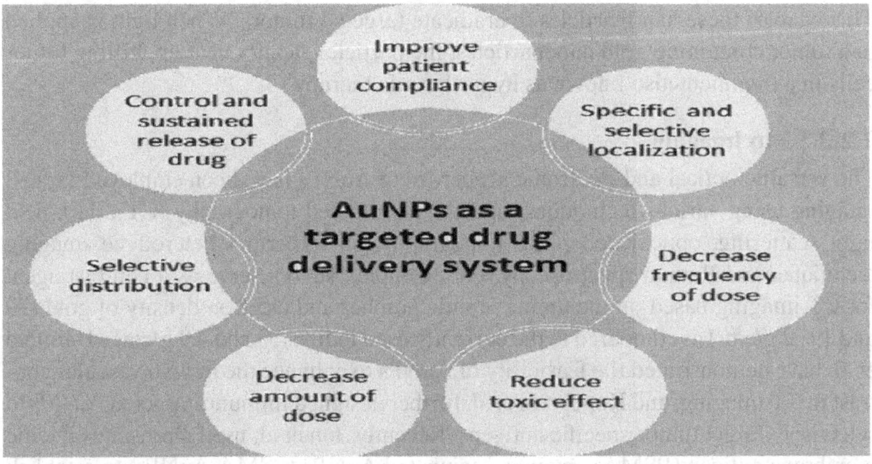

FIGURE 2.19 Advantages of AuNPs as a targeted drug delivery system. (Adapted from ResearchGate.)

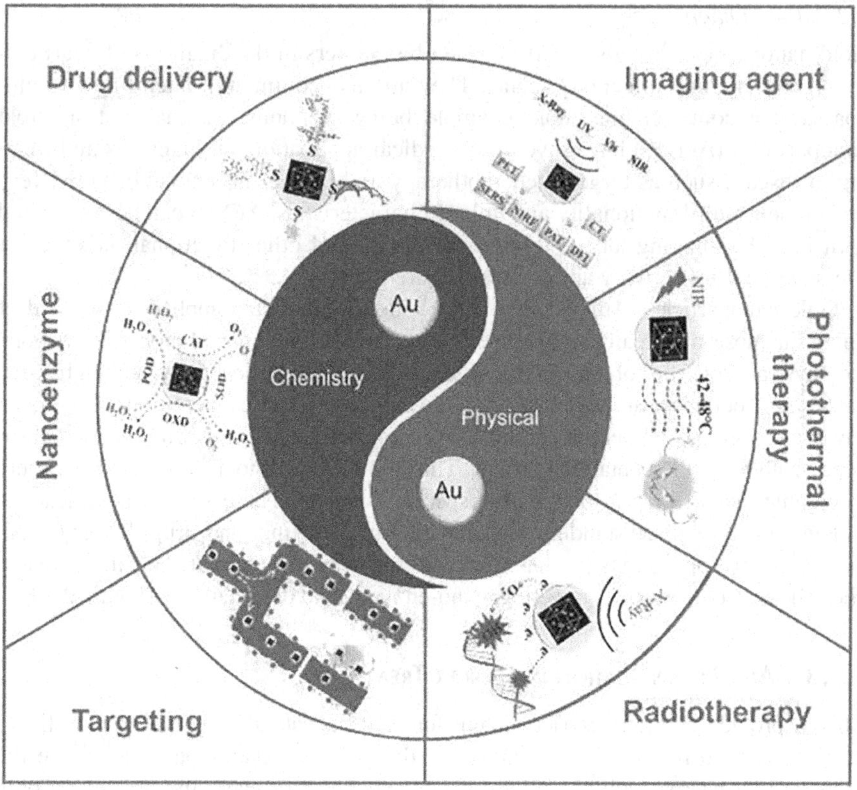

FIGURE 2.20 Applications for tumor diagnosis and treatment based on the basic physical and chemical properties of gold nanoparticles (AuNPs). (Adapted from MDPI.)

This enables these nanoparticles to eradicate targeted tumors. When light is applied to a tumor containing gold nanoparticles, the particles rapidly heat up, killing tumor cells in a treatment also known as hyperthermia therapy.

2.2.3.1 In Imaging

The versatile optical and electronic properties of AuNPs have been employed for cell imaging using various techniques, including computed tomography (CT), dark-field light scattering, optical coherence tomography, photothermal heterodyne imaging technique, and Raman spectroscopy. For example, AuNPs serve as a contrast agent for CT imaging based on the higher atomic number and electron density of gold (79 and 19.32 g/cm^3) as compared to the currently used iodine (53 and 4.9 g/cm^3). Hainfeld et al. have demonstrated the feasibility of AuNPs to enhance the in vivo vascular contrast in CT imaging, and Kopelman et al. further designed immuno-targeted AuNPs to selectively target tumor-specific antigens. Recently, Jon et al. used a prostate-specific membrane antigen (PSMA) aptamer-conjugated AuNPs (PSMA-AuNPs) to establish a molecular CT image for the specific imaging of prostate cancer cells (Figure 2.21). These results showed that PSMA-AuNPs had a 4-fold greater CT intensity for a

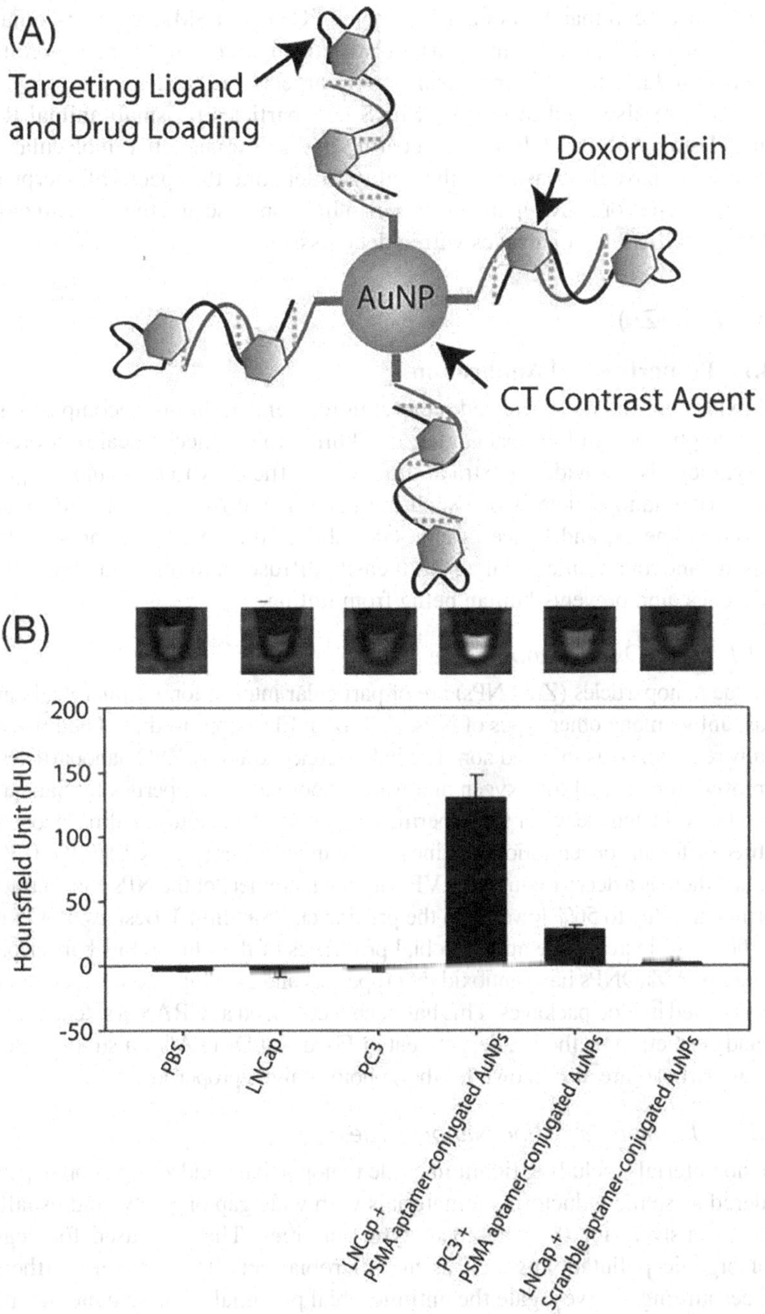

FIGURE 2.21 (a) Schematic of the method for preparing doxorubicin-loaded aptamer-conjugated AuNPs. (b) CT images and (c) HU values of PBS, LNCaP, PC3 cells, LNCaP, and PC3 cells treated with PSMA aptamer-conjugated AuNPs (5 nM) or scramble aptamer-conjugated AuNPs (5 nM) for 6 h. (Adapted from Nanoscale PMC.)

targeted LNCaP cell than that of a non-targeted PC3 cell. PSMA aptamer-conjugated AuNPs loaded with the anticancer drug doxorubicin were significantly more potent against targeted LNCaP cells than against non-targeted PC3 cells.

AuNPs were also used to prepare SERS nanoparticles for small-animal Raman imaging. Using AuNPs with a silica coating and a Raman-active molecular layer, Gambhir et al. have demonstrated the ability to separate the spectral fingerprints of ten different types of SERS nanoparticles in a living mouse and the colocalization of five different SERS nanoparticles within deep tissues.

2.2.4 Zinc (Zn)

2.2.4.1 Properties and Applications

Metal oxides are the most referred nanoreinforcements in bio-nanocomposite structures for application in food packaging in addition to the mechanical reinforcement capacity; they also provide an extra antimicrobial efficiency to bio-nanocomposites.

Zinc oxide nanoparticle is one such inorganic metal oxide which fulfills all the above requirements, and hence, it can safely be used as medicine, preservative in packaging, and an antimicrobial agent. It easily diffuses into the food material, kills the microbes, and prevents human being from falling.

2.2.4.1.1 Zinc Oxide Nanoparticle

Zinc oxide nanoparticles (ZnO NPs) are of particular interest for antimicrobial applications as, unlike many other types of NPs, ZnO is an FDA-approved food additive and is generally recognized as safe and nontoxic in low concentrations. ZnO nanoparticles have been proved to be suited for oxygen and water vapor barrier properties by Sanuja et al. (2015) who evaluated the water vapor permeability (WVP) of chitosan films incorporated with three different concentrations of zinc oxide nanoparticles (ZnONPs) (0.1%, 0.3% and 0.5%), and there is a decrease in the WVP with the increment of the NPs concentration in the formulation (up to 56% lower than the pristine chitosan films). Besides, ZnONPs are very efficient to increase the antimicrobial properties of the film and its barrier properties. Moreover, ZnONPs have antioxidant properties and are nontoxic when added to the film being used in food packages. This has been recognized as GRAS (a safe tag for food, drug, and cosmetics) by the US department of Food and Drug Administration. Besides, ZnO nanoparticles are also known for their photocatalytic properties.

2.2.4.1.2 Titanium and Zinc Nanoparticles

The nanomaterials including titanium oxide nanoparticles and zinc nanoparticles are considered as semiconductor nanomaterials with wide gap of 3.2 eV, and usually are studied extensively for their photocatalytic activities. They are used for degradation of organic pollutants as well as antimicrobial activities. Therefore, there are evidences aiming to investigate the antimicrobial potential of these nanomaterials.

2.2.5 Cobalt (Co) NPs

Cobalt is a Block D, Period 4 element. The morphology of cobalt nanoparticles is spherical, and their appearance is a gray or black powder. Cobalt nanoparticles possess magnetic properties, which leads to applications in imaging, sensors, and many other areas. Cobalt nanoparticles are graded as harmful with chances of allergic skin reactions. They have also been tested to be causing asthma symptoms and breathing difficulties if inhaled.

2.2.5.1 Properties

Cobalt nanoparticles (CoNPs) exhibit quite unique magnetic, catalytic, and optical properties. Bishoy Morcos studied magnetic, structural, and chemical properties of cobalt nanoparticles, synthesized in ionic liquid, in 2018. In this work, imidazolium-based ionic liquids (ILs) are successfully used to elaborate magnetically responsive suspensions of quite monodisperse CoNPs with diameters below 5 nm. The as-synthesized CoNPs adopt the noncompact and metastable structure of ϵ-Co that progressively evolves at room temperature toward the stable hexagonal close-packed allotrope of Co. Accordingly, magnetization curves are consistent with zero-valent Co. As expected in this size range, the CoNPs are superparamagnetic at room temperature. Their blocking temperature is found to depend on the size of the IL cation. The CoNPs produced in an IL with a large cation exhibit a very high anisotropy, attributed to an enhanced dipolar coupling of the NPs, even though a larger interparticle distance is observed in this IL. Finally, the presence of surface hydrides on the CoNPs is assessed and paves the way toward the synthesis for Co-based bimetallic NPs.

2.2.5.1.1 Applications

Cobalt compounds have been in use for hundreds of years as a dye. It is also an essential human nutrient as part of vitamin B12.

The following are the areas of application of cobalt nanoparticles:

1. **In Alloys:** Cobalt nanoparticles along with small amounts of iron and nickel are used as super alloys for jet engine parts. Super alloys are resistant to corrosion and retain their properties at high temperature.
2. Cobalt NPs with chromium and molybdenum alloys are used for artificial joints, stents, and wire in strumpets for surgery metals as they have good mechanical properties and biocompatibility (Figure 2.22).
3. **In Cobalt Batteries:** CoNPs batteries are electrochemical devices that convert chemical energy into electrical energy.
4. **In Electromagnetic Wave Absorption:** CoNPs are used in electromagnetic wave absorption which has attracted great interest due to rapid increases of electromagnetic interference (EMI) pollution by communication devices like mobile phones, local area network, radar systems, etc.
5. **In Bioimaging (MRI) Magnetic Resonance:** CoNPs are also used in hyperthermia applications and for MRI with great efficiency comparable to T2 contrast agents. These CoNPs are also biocompatible.
6. **In CoNPs Inks:** CoNPs are utilized as developed inks in various printed electronic applications such as antenna miniaturization substrates and magnetic sensors. These nanomaterials enhance permeability values in devices that are used in manipulation of electromagnetic waves. These devices include resonator antennas, filter phase shifters, and radio

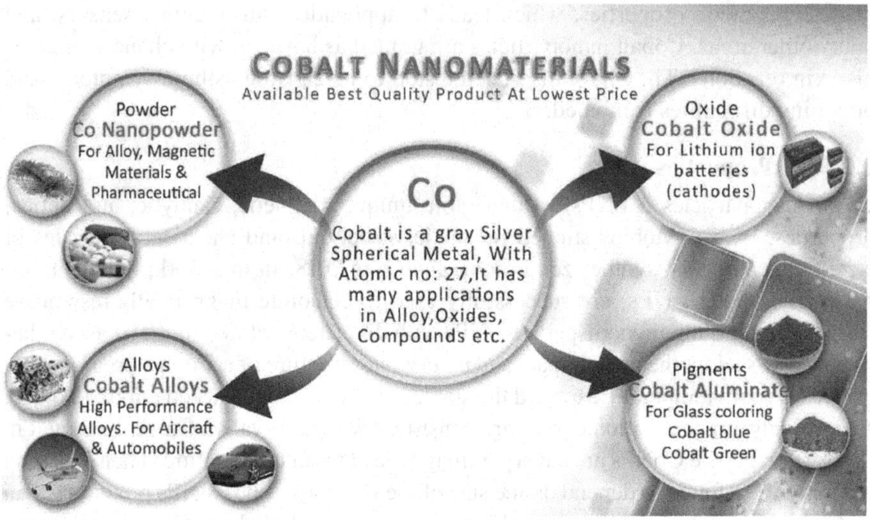

FIGURE 2.22 CoNPs used in various applications. (Adapted from nanoshel.com.)

frequency absorbents for flexible materials with good electrical and magnetic properties.

7. They are used in coatings, ferrofluids, plastics, nanofibers, textiles, for coloring glass as cobalt blue and cobalt green.
8. As drug delivery agents for cancer therapies.
9. The cytotoxicity studies demonstrate that CoNPs exhibit the mild anti-proliferative character against the cancer cells (cisplatin-resistant ovarian cancer and safe nature toward the normal cells). Hemolytic behavior of human red blood cells (RBCs) revealed (<5%) hemolysis signifying the compatibility of CoNPs with human RBC which is an essential feature in vivo biomedical application without creating any harmful effects in the human blood stream.
10. **Medical Sensors:** Biomedicine as a contrast enhancement agent for magnetic resonance imaging (MRI) and site-specific drug delivery agents for cancer therapies.
11. Coatings, plastics, nanofibers, nanowires, textiles, and high-performance magnetic recording materials.
12. As a magnetic fluid made of iron, cobalt, nickel, and its alloy nanoparticles.
13. Microwave-absorption materials.
14. Cobalt oxide particles can also be used in several military applications such as high-performance invisible materials for absorbing extremely high-frequency millimeter wave, visible light, and infrared.
15. The synthesis of silver-doped cobalt oxide nanoparticles by microwave-assisted method and their structural, optical, antibacterial activities were studied. The doping concentrations were chosen as 5, 10, 15, and 20 wt percentages. The sample was undergone powder X-ray diffraction studies, and the result shows the good crystalline nature of the sample. Also, the average crystallite size increases from 13.95 nm, 21.26 nm, 26.13 nm,

to 28.35 nm with different doping concentrations. The transmission electron microscopy image shows cubic and spherical morphology. The optical properties were tested by UV–Vis–NIR absorption spectrum. It indicates the decrease of band gap value. From the antibacterial activity studies, the 20 wt % Ag-doped nanoparticles exhibit better activity.

2.2.6 Iron (Fe)

2.2.6.1 Properties and Applications

Other types of metal oxides such as iron oxide (Fe_3O_4) and magnesium oxide (MgO) have equally been used in interaction with chitosan but more specifically in applications related to biomedical and effluent remediation. However, two quality studies applied MgO in a chitosan packaging system. Silva et al. (2017) fabricated Ch/MgO thin films with improved physical properties for potential packaging applications. Between the two MgO concentrations added (5% and 10%, w/w Ch), the TS and Young's modulus were better for the lower amount improving the values by 86% and 38%, respectively. More recently, Wang et al. (2020) produced carboxy methyl chitosan (CMCS) and MgO for food packaging. In this case, the amount of MgO explored was much lower (0.5% and 1%, w/w Ch) than in the previous study. Uncommonly, the authors perceived a decrease in TS for both composites compared to chitosan film. In contrast, a greater increase was observed in the elastic modulus (EM) and elongation at break. It was concluded that the blend of CMCs and MgO might be used as a novel food packaging when ductility and elasticity are intended. In another study, Yadav et al. (2014) added 0.5% (w/W) Fe_3O_4 to chitosan/graphene oxide (GO) composite. In comparison with the Ch/GO composite, the addition of iron oxide showed improvement by 10% and 22% in accordance with the TS and EM. It was concluded from above studies that in case of metal oxide used as nanofillers, only smaller amounts are sufficient to reach the optimal reinforcement values. The small optional percentage allows achieving optimal nanofillers dispersion, while above these values nanoparticles tend to aggregate which makes the nanocomposite films more brittle and more prone to gas diffusion pollution control applications of iron oxide.

In recent years, many studies have focused on the functionalization of magnetic iron oxide nanoparticles to obtain efficient adsorbents for oil removal from water as the oil pollution has posed a great threat to marine ecosystems and human health. Recently, carbon-coated magnetic nanoparticles have received considerable attention because magnetic carbon nanoparticles have many advantages such as a high chemical and thermal stability and biocompatibility. Banerjee et al. synthesized a carbon–Fe_3O nanocomposite by pyrolysis of an iron-containing metal organic framework and tested the oil adsorption capacity of the material. The water contact angle of the nanoparticles was 143°, indicating a near super hydrophobic character and the particles could absorb an amount of oil of more than 40 times its own weight.

Zhu et al. obtained core shell Fe_2O_3 nanoparticles by thermal decomposition of the precursor material. The specific surface area of Fe_2O_3 was 94.04 m^2/g, and the water contact angle was approximately 162°; these magnetic nanoparticles were unsinkable, hydrophobic, and super oleophilic, and could absorb up to 3.8 times their own weight. After absorbance, the magnetic nanoparticles could be quickly and easily collected and regenerated. The magnetic nanoparticles still maintained a large water contact angle, which was greater than 150° after six cycles.

Another widely employed approach to synthesize efficient magnetic oil sorbents is to attach iron nanoparticles to the pore walls and streets of certain porous materials by different techniques. Three-dimensional porous materials in the form of aerogels, sponges, and foams are considered oil sorbents due to their low density, high porosity, and specific surface area. Fe_3O_4 nanoparticles can be modified with SiO_2 shell in advance to make them compatible with the organo-silames and to obtain a uniform magnetic porous material; otherwise, phase separation is inevitable. The magnetic particles increased the roughness, hydrophobicity, and oleophilicity of the porous material.

1. In another study, Wu et al. inserted Fe_3O_4 nanoparticles into a sponge by polymerization-induced macrophage separation. This approach provided magnetic sponges with a high hydrophobicity (water contact angle of 140°) and oil adsorption capacity from 9.9 to 20.3 g/g, besides the adsorption process was fast and tested for only 10 minutes at which point the adsorption saturation was reached.
2. In yet another study, a remarkable performance was obtained by Li et al. in preparing a magnetic carbon fiber aerogel through one-step pyrolysis of Fe_3O_3-coated cotton in an argon atmosphere. The magnetic carbon fiber aerogel could absorb oil selectively and rapidly due to its super hydrophobicity and superoleophobicity.

2.2.7 Nickel (Ni)

2.2.7.1 Properties and Applications

1. **Nickel NPs Are Used as High-Performance Electrode Material:** If the micron nickel powder is replaced by nano-nickel powder, the nickel-hydrogen reaction is significantly increased in the specific surface area, which makes the corresponding nickel metal hydride, battery power increasing several times, and greatly improves charge and discharge efficiency.
2. **Efficient Catalysts:** Due to the large and highly active surface area, nano-nickel powder has a great enhancement catalytic efficiency. Instead of conventional micro-nickel powder, it can be used for hydrogenation of organic compounds. In the vehicle tail gas treatment, nano-nickel can replace the precious metals such as platinum and rhodium, so that the cost can be greatly reduced.
3. **Magnetic Fluid:** Nickel and its alloy powders can be prepared as nanomagnetic fluid production which is used in sealing, medical equipment, sound regulation, mechanical control, and other fields.
4. **Conductive Paste:** Electronic paste is widely used in microelectronics industry packaging, connectivity, and miniaturization of microelectronic devices. Nickel powder with nanoelectronic paste has superior performance as micro-nickel powder. It is widely used in MLCC.

5. **Sintering Additives:** Nickel nanopowder has large volume ratio of surface atoms, which has high energy state. In the powder metallurgical industry, it can be used as sintering additive to reduce sintering temperature in ceramic and diamond tools production.
6. **Non-Metallic Conductive Coatings:** The high surface energy of nano-nickel can be utilized in the form of sprays at lower temperature than its bulk for non-metallic device surfaces to enhance the electric conductivity under anaerobic conditions. The coating can improve the oxidation resistance, corrosion, etc.
7. **Magnetic Recording:** Nano-nickel powder can be used as high-performance magnetic recording materials. It can improve the density of storage information several times for tapes, and hard and soft disks.
8. **Combustion Efficiency:** The nano-nickel powder added to the solid fuel rocket propellant can significantly improve the efficiency of combustion of fuel, and improves combustion stability.
9. **Fuel Cell:** Nanometer nickel is irreplaceable as fuel cell catalysts. As a catalyst for fuel cells, it can replace the expensive platinum, and this can greatly reduce the manufacturing costs of fuel cells. Assisted with nano-nickel powders with appropriate technology, the electrode surface area is larger than micro-nickel particles. It can greatly improve the discharge efficiency of fuel cell.
10. **Stealth Materials:** Nano-nickel can be used as stealth material, because of its special electromagnetic properties.
11. **Lubricant Additive:** Nano-nickel can be dispersed into lubricant to decrease the surface friction and repair microdefect of the friction surface.

2.2.8 Copper (Cu) NPs

2.2.8.1 Properties and Applications

1. Copper is a ductile metal with very high thermal and electrical conductivity. The morphology of copper nanoparticles is round, and they appear as a brown to black powder. Copper is found to be too soft for some applications, and hence, it is often combined with other metals to form numerous alloys such as brass, which is a copper-zinc alloy. Copper nanoparticles are graded as highly flammable solids; therefore, they must be stored away from sources of ignition. They are also known to be very toxic to aquatic life.
2. The effects of cylindrical and spherical copper nanoparticles on the mechanical and thermal properties of polyethylene matrix using molecular dynamics and finite element methods were studied (Ali Shahrokh et al., 2021); in this, the three-dimensional finite element modeling was done to obtain Young's modulus and thermal conductivity of the studied nanocomposite for different geometries, orientations, and volume fractions of nanofillers. The results revealed that reinforcing polymer with copper nanofillers improves its thermal conductivity and EM, remarkably. Besides, nanoparticles with cylindrical geometry and axial crystallographic orientation have a greater

effect on the enhancement of properties than other orientations or spherical nanoparticles. Furthermore, the orientation of the nanoparticles inside the matrix in the direction of applied force and heat flux has greater impact on improving the properties of polymers compared to the other directions (Figures 2.23–2.25).

3. Earlier, Eid et al. (2019) experimentally measured the effects of copper nanoparticles with different volume fraction on the thermal and electrical conductivity of polypropylene and found out that reinforcing polypropylene by copper nanoparticles increased conductivity by about 1.5%. Prior to that, Boudenne et al. (2005) experimentally analyzed the effects of copper particles (with two different sizes) on the thermal and electrical properties of polypropylene composite such as conductivity, permeability, and specific heat. Their results revealed that smaller particles with higher volume fractions enhance the heat transfer properties, more effectively, and in contrast to thermal conductivity, larger copper particles result in composites with

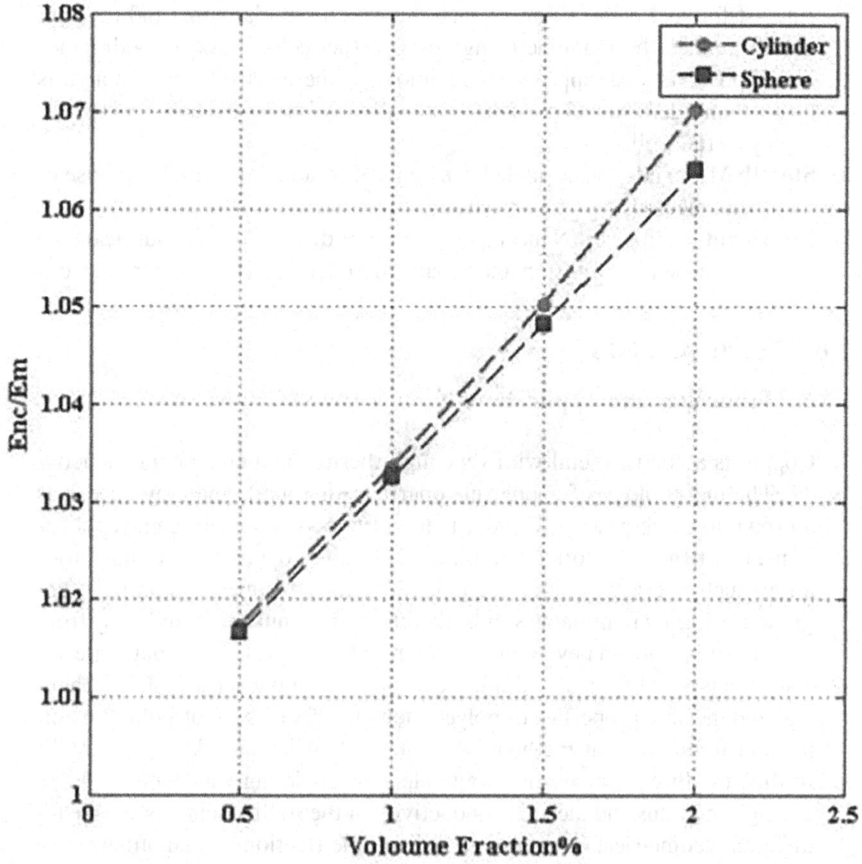

FIGURE 2.23 Effects of volume fraction and copper nanoparticles shape on EM ratio. (Adapted from. ScienceDirect.)

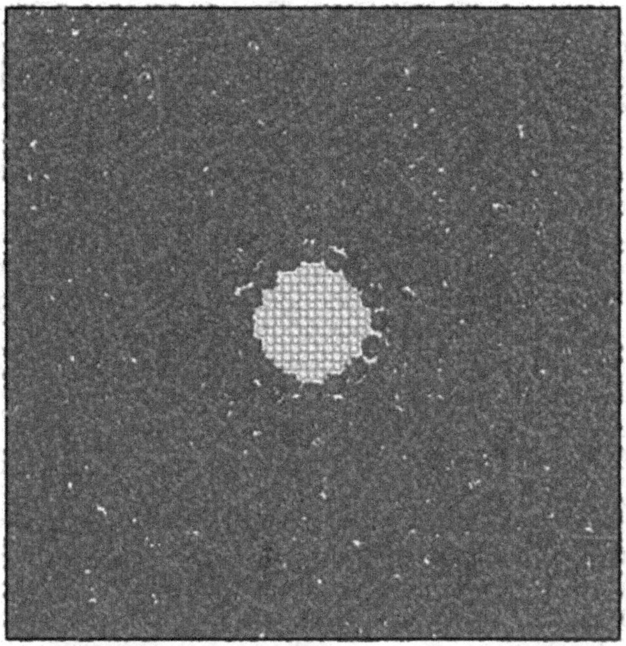

FIGURE 2.24 Structure of polyethylene reinforced by copper nanoparticle: copper, carbon, and hydrogen atoms are represented by yellow, purple, and red colors, respectively. (Adapted from ScienceDirect.)

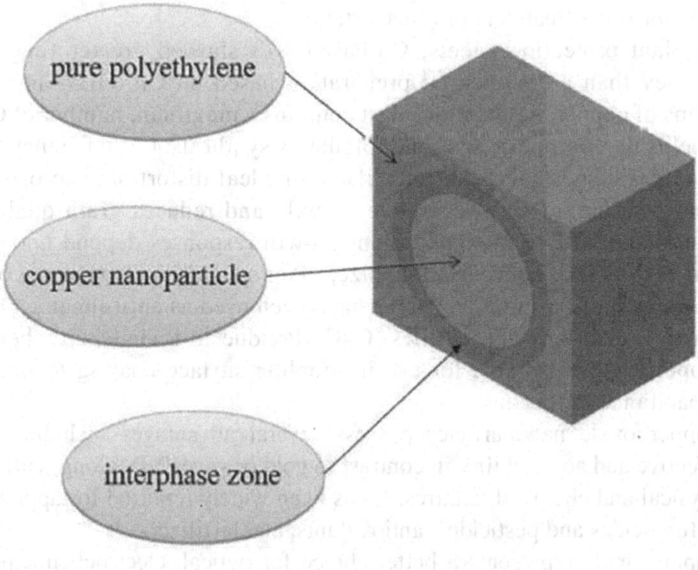

FIGURE 2.25 Three different zones of representative volume element (RVE). (Adapted from ScienceDirect.)

higher electrical conductivity. In another study, Hu et al. (2016) experimentally investigated the thermal conductivity of an epoxy composite reinforced by copper nanowires with a diameter of about 20 nm and a length of about 40 nm. Due to better interaction between the nanofillers and polymer matrix, catalysts and solvents such as nickel and oleylamine were used, to speed up synthesis process and increased the quality of the thermal interphase between copper and epoxy.

4. Again, Li et al. (2016) experimentally investigated the thermal and mechanical properties of polymer nanocomposites. They used multiwalled carbon nanotubes and copper nanoparticles to ameliorate the properties of polypropylene and high-density polyethylene. The thermal conductivity of nanocomposites showed a satisfactory improvement compared to the thermal conductivity of the pure polymer.

2.2.8.2 Applications of Copper NPs

1. Copper has long been known to have antimicrobial activity and is used in drinking water treatment and transportation. It has been recognized by the American Environmental Protection Agency as the first metallic antimicrobial agent in 2008. With ongoing waterborne hospital-acquired infections and antibiotic resistance, research on copper as an antimicrobial agent is gaining importance. Many studies have shown that the use of copper surface and copper particles could significantly reduce the environmental bioburden. In the ancient Egypt, copper was used for the preservation of water and food as well as for medical applications (Vincent et al., 2016; Konieczny and Rdzawski, 2012), and, since this period, copper is widely used for water treatment or transportation.
2. As plant protection agents, Cu-based NPs showed greater fungicidal efficacy than a commercial preparation based on Cu $(OH)_2$ and ionic forms of copper. Plant chloroplast contains a maximum number of Cu as it helps in chloroplast and other pigments synthesis. Cu deficiency leads to several abnormal conditions like young leaf distortion, necrosis, and stem bending, affects vegetative growth, and reduces grain quality in crop plants. CuNP-mediated plant growth responses depend upon several factors like concentration, size, plant species, and structure of the particles. Furthermore, CuONPs have been used as antifungal agents in plastics, coatings, and textiles. CuONPs, due to their electrochemical properties, are suitable for use in graphite surface coating to improve capacitance properties.
3. Copper oxide nanoparticles possess several advantages including cost-effective and accessibility in contrast to gold or silver NPs along with other physical and chemical features. It has been widely reported for application as fungicides and pesticides, antioxidants, and fertilizers.
4. Copper oxide represents a better choice for optical, electrochemical, and sensing attributes due to the low-cost production, easy availability, and optimum physical and chemical characteristics. Cupric oxide has a monoclinic

structure at space group C2/c, and it is a p-type semiconductor having a narrow energy gap of 1.2–2.4.

5. CuONPs are also used in a variety of other products, including intrauterine contraceptive devices, heat transfer fluids, semiconductors, and electronic chips.

2.2.9 Platinum (Pt) NPs

2.2.9.1 Properties

Platinum is one of the rarest and most expensive metals. It has high corrosion resistance and numerous catalytic applications including automotive catalytic converters and petrochemical cracking catalysts. Platinum nanoparticles are usually used in the form of colloid or suspension in a fluid. They are the subject of extensive research due to their antioxidant properties. The chemical and physical properties of platinum nanoparticles (NPs) make them applicable for a wide variety of research applications (Figure 2.26).

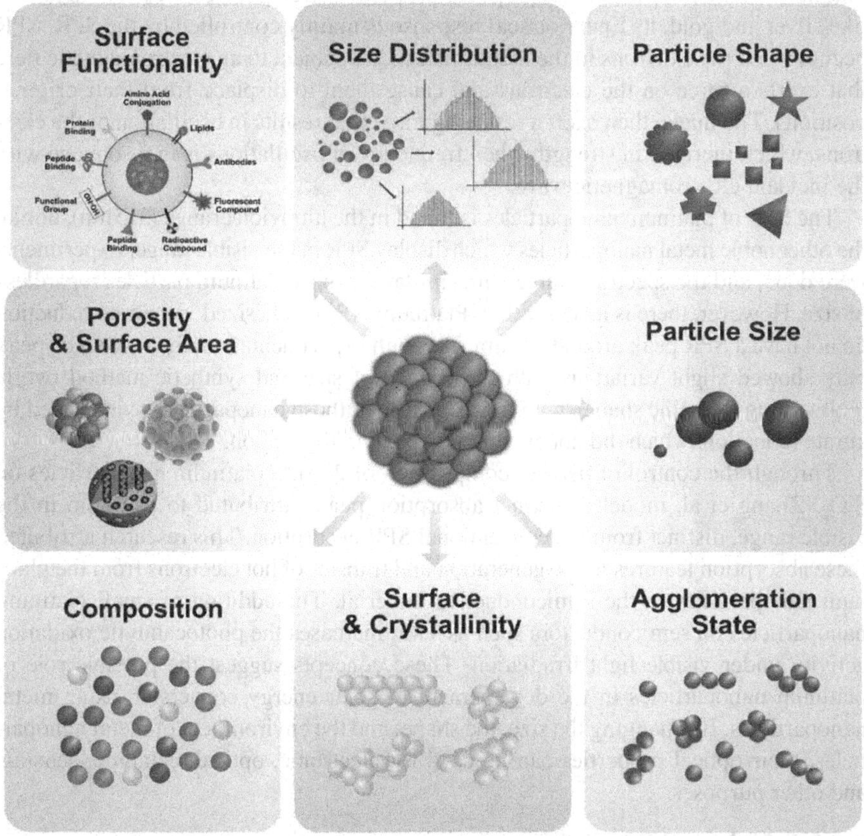

FIGURE 2.26 Physicochemical properties of platinum nanoparticles. (Adapted from encyclopedia.pub.)

2.2.9.2 Catalytic Properties

Platinum NPs are used as catalysts for proton exchange membrane fuel cell, for industrial synthesis of nitric acid, for reduction of exhaust gases from vehicles, and as catalytic nucleating agents for synthesis of magnetic NPs. These NPs can also act as catalysts in homogeneous colloidal solution or as gas-phase catalysts while supported on solid-state material. The catalytic action of the NPs is dependent on the shape, the size, and the morphology of the particle. One type of platinum NPs that have been researched on are colloidal platinum NPs. Monometallic and bimetallic colloids have been used as catalysts in a wide range of organic chemistry, including oxidation of carbon monoxide in aqueous solutions, hydrogenation of alkenes in organic or biphasic solutions, and hydrosilylation of olefins in organic solutions. Colloidal platinum NPs protected by poly(N-isopropylacrylamide) were synthesized, and their catalytic properties measured. It was determined that they were more active in solution and inactive when phase separated due to its solubility being inversely proportional to temperature.

2.2.9.3 Optical Properties

Platinum NPs exhibit fascinating optical properties. Being a free electron metal NP like silver and gold, its linear optical response is mainly controlled by the SPR. SPR occurs when the electrons in the metal surface are subject to an electromagnetic field that exerts a force on the electrons and cause them to displace from their original positions. The nuclei then exert a restoring force that results in oscillation of the electrons, which increase in strength when frequency of oscillations is in resonance with the incident electromagnetic wave.

The SPR of platinum nanoparticles is found in the ultraviolet range (215 nm), unlike the other noble metal nanoparticles which display SPR in the visible range. Experiments were done, and the spectra obtained are similar for most platinum particles regardless of size. However, there is an exception. Platinum NPs synthesized via citrate reduction do not have a SPR peak around 215 nm. Through experimentation, the resonance peak only showed slight variations with the change of size and synthetic method (while maintaining the same shape), with the exception of those nanoparticles synthesized by citrate reduction, which did not exhibit SPR peak in this region.

Through the control of percent composition of 2–5 nm platinum nanoparticles on SiO_2, Zhang et al. modeled distinct absorption peaks attributed to platinum in the visible range, distinct from the conventional SPR absorption. This research attributed these absorption features to the generation and transfer of hot electrons from the platinum nanoparticles to the semiconductive material. The addition of small platinum nanoparticles on semiconductors such as TiO_2 increases the photocatalytic oxidation activity under visible light irradiation. These concepts suggest the possible role of platinum nanoparticles in the development of solar energy conversion using metal nanoparticles. By changing the size, the shape, and the environment of metal nanoparticles, their optical properties can be used for electronic, optical, catalytic, sensing, and other purposes.

2.2.9.4 Biological Interactions

The increased reactivity of nanoparticles is one of their most useful properties and is leveraged in fields such as catalysis, consumer products, and energy storage. However, this high reactivity also means that a nanoparticle in a biological environment may have unintended impacts. For example, many nanoparticles such as silver, copper, and ceria interact with cells to produce reactive oxygen species or ROS, which can cause premature cell death through apoptosis. Determining the toxicity of a specific nanoparticle requires knowledge of the particle's chemical composition, shape, size, and is a field that is growing alongside advances in nanoparticle research.

2.2.9.5 Applications

2.2.9.5.1 Hydrogen Fuel Cells

Among the precious metals, platinum is the most active toward the hydrogen oxidation reaction that occurs at the anode in hydrogen fuel cells. In order to meet cost reductions of this magnitude, the Pt catalyst loading must be decreased. Two strategies have been investigated for reducing the Pt loading: the binary and the ternary Pt-based alloyed nanomaterials and the dispersion of Pt-based nanomaterials onto high-surface-area substrates.

2.2.9.5.2 Methanol Fuel Cells

The methanol oxidation reaction occurs at the anode in direct methanol fuel cells (DMFCs). Platinum is the most promising candidate among pure metals for application in DMFCs. Platinum has the highest activity toward the dissociative adsorption of methanol. However, pure Pt surfaces are poisoned by carbon monoxide, a by-product of methanol oxidation. Researchers have focused on dispersing nanostructured catalysts on high-surface-area supporting materials and the development of Pt-based nanomaterials with high electrocatalytic activity toward MOR to overcome the poisoning effect of CO.

2.2.9.5.3 Electrochemical Oxidation of Formic Acid

Formic acid is another attractive fuel for use in PEM-based fuel cells. The dehydration pathway produces adsorbed carbon monoxide. A number of binary Pt-based nanomaterial electrocatalysts have been investigated for enhanced electrocatalytic activity toward formic acid oxidation.

2.2.9.5.4 Modifying Conductivity of Zinc Oxide Materials

Platinum NPs can be used to dope zinc oxide (ZnO) materials to improve their conductivity. ZnO has several characteristics that allow it to be used in several novel devices such as development of light-emitting assemblies and solar cells. However, because ZnO is of slightly lower conductivity than metal and indium tin oxide, it can be doped and hybridized with metal NPs like platinum to improve its conductivity. A method to do so would be to synthesize ZnO NPs using methanol reduction and incorporate at 0.25 at. % platinum NPs. This boosts the electrical properties of ZnO films while preserving its transmittance for application in transparent conducting oxides.

2.2.9.5.5 Glucose Detection Applications

Enzymatic glucose sensors have drawbacks that originate from the nature of the enzyme. Nonenzymatic glucose sensors with Pt-based electrocatalysts offer several advantages, including high stability and ease of fabrication. Many novel Pt and binary Pt-based nanomaterials have been developed to overcome the challenges of glucose oxidation on Pt surfaces, such as low selectivity, poor sensitivity, and poisoning from interfering species.

2.2.9.5.6 Drug Delivery

A topic of research within the field of nanoparticles is how to use these small particles for drug delivery. Depending on particle properties, nanoparticle may move throughout the human body. They are promising as site-specific vehicles for the transport of medicine. Current research using platinum nanoparticles in drug delivery uses platinum-based carriers to move antitumor medicine. In one study, platinum nanoparticles of diameter 58.3 nm were used to transport an anticancer drug to human colon carcinoma cells, HT-29. Uptake of the nanoparticles by the cell involves compartmentalization of the nanoparticles within lysosomes. The high acidity environment enables leaching of platinum ions from the nanoparticle, which the researchers identified as causing the increased effectiveness of the drug. In another study, a Pt nanoparticle of diameter 140 nm was encapsulated within a PEG nanoparticle to move an antitumor drug, Cisplatin, within a prostate cancer cell (LNCaP/PC3) population. The use of platinum in drug delivery hinges on its ability not to interact in a harmful manner in healthy portions of the body while also being able to release its contents when in the correct environment.

2.2.9.5.7 Toxicology

Toxicity stemming from platinum nanoparticles can take multiple forms. One possible interaction is cytotoxicity or the ability of the nanoparticle to cause cell death. A nanoparticle can also interact with the cell's DNA or genome to cause genotoxicity. These effects are seen in different levels of gene expression measured through protein levels. Last is the developmental toxicity that can occur as an organism grows. Developmental toxicity looks at the impact the nanoparticle has on the growth of an organism from an embryonic stage to a later set point. Most nanotoxicology research is done on cyto- and genotoxicity as both can easily be done in a cell culture lab. Platinum nanoparticles have the potential to be toxic to living cells. In one case, 2-nm platinum nanoparticles were exposed to two different types of algae in order to understand how these nanoparticles interact with a living system. In both species of algae tested, the platinum nanoparticles inhibited growth, induced small amounts of membrane damage, and created a large amount of oxidative stress. In another study, researcher tested the effects of differently sized platinum nanoparticles on primary human keratinocytes. The authors tested 5.8- and 57.0-nm Pt nanoparticles. The 57-nm nanoparticles had some hazardous effects including decreased cell metabolism, but the effect of the smaller nanoparticles was much more damaging. The 5.8-nm nanoparticles exhibited a more deleterious effect on the DNA stability of the primary keratinocytes than did the larger nanoparticles. The damage to the DNA was measured for individual cells using single-gel electrophoresis via the comet assay.

2.2.9.5.8 Antibacterial Activity of Platinum Nanoparticle

The current scenario shows paramount importance of PtNPs in human health and for the protection from various diseases caused by microorganisms. However, microorganisms are powerful and attain resistance to various antibiotics. Because of the recent increase in bacterial resistance, alternative therapeutic agents that are nontoxic to human beings but toxic to pathogenic microorganisms are urgently required. Therefore, the development of NP-mediated antimicrobial agents is most warranted. Recently, several studies have focused on NP-based therapeutic agents against pathogenic bacteria. Metallic NPs such as Pt, Ag, Pd, Cu, Au, ZnO, and TiO_2 play a vital role in antibacterial activity against pathogens. The antibacterial activity depends on NP morphology, size, shape, and its surface charges. Most metallic NPs like Ag, Pt, Au, Pd, ZnO, and Cu have a negative zeta potential and thus have potential cell damaging properties. Although PtNPs have a more negative zeta potential and cause severe damage to the cells, they show enhanced antibacterial activity. Apigenin-functionalized PtNPs exhibited significant antibacterial activity against Pseudomonas aeruginosa and Staphylococcus aureus (Figure 2.27).

2.2.9.6 Nanomaterials

Effect of apigenin-mediated synthesis of PtNPs on cell survival of P. aeruginosa and S. aureus was analyzed. All the test strains were incubated in the presence of different concentrations of PtNPs (25–150 µg/mL). Bacterial survival was determined at 24 hours by a colony-forming unit (CFU) assay. The results are expressed as the means ± SD of three separate experiments, each of which contained three replicates.

2.2.10 Other Applications

1. **Electrocatalysts and Catalytic Converters:** Platinum catalysts are alternatives of automotive catalytic converters, carbon monoxide gas sensors, and petroleum refining.
2. They are also used in magnetic nanopowders, polymer membranes, coatings, plastics, nanofibers, textiles, etc.

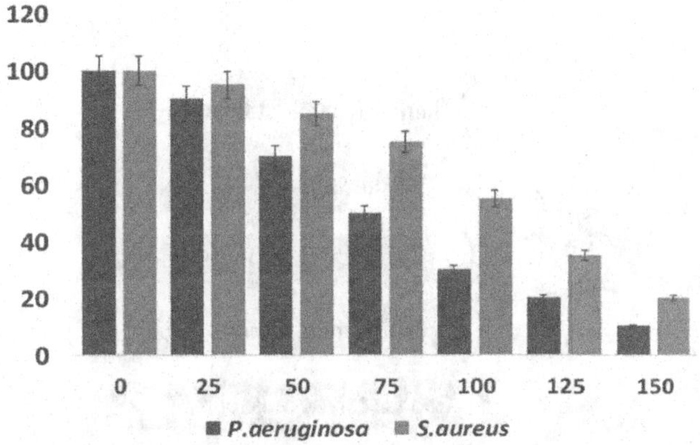

FIGURE 2.27 Antibacterial activity of PtNPs against Gram-negative and Gram-positive bacteria.

2.2.10.1 Chromium Oxide (Cr_2O_3) NPs

2.2.10.1.1 Properties

Transitional metal oxides are regarded as essential materials in industry due to their presence in a wide range of magnetic, thermal, chemical, and electronic properties. Chromic oxide nanoparticles (Cr_2O_3 NPs) are one of the notable inorganic NPs with outstanding features for several branches of modern science and technologies. It is an important component for industrial applications based on its wear resistance and high temperature qualities. Moreover, it can form different stable oxidation states. Chemically, chromium has various stable oxidation states. However, there has been particular attention to chromium (III) oxide due its versatility and intrinsic properties.

Chromium oxide (Cr_2O_3) occurs in nature as a rare mineral called eskolaite. These nanoparticles are predominantly used as dyes. Some of the synonyms of chromium oxide nanoparticles are chromsaeureanhydrid, trioxochromium, monochromium trioxide, chromia, green cinnabar, chrome ochre, and chromtrioxid. Chromium oxide nanoparticles appear in the form of green crystals with a nearly spherical morphology (Figure 2.28).

2.2.10.1.2 Applications

The following are the chief applications of chromium oxide:

The major applications of chromium oxide nanoparticles are in heterogeneous catalysis, liquid crystal displays, wear-resistant and high-temperature materials, solar absorbents as well as in functional pigments with high reflectivity in near-infrared region which is due to their optical properties. In addition, the catalytic applications of chromium nanoparticles have been in numerous reactions, namely toluene oxidation, methanol decomposition, and ethane dehydrogenation. These nanoparticles can lead to ROS with consequent oxidation reactions in biological environments.

Chromium oxide finds use in the area of electronics, chemistry, and medicine. Phytosynthesized Cr_2O_3 NPs have attracted tremendous usage in the fabrication of antibacterial, antifungal, antioxidant, anticancer, antileishmanial, antiviral, antidiabetic, and photocatalytic agents.

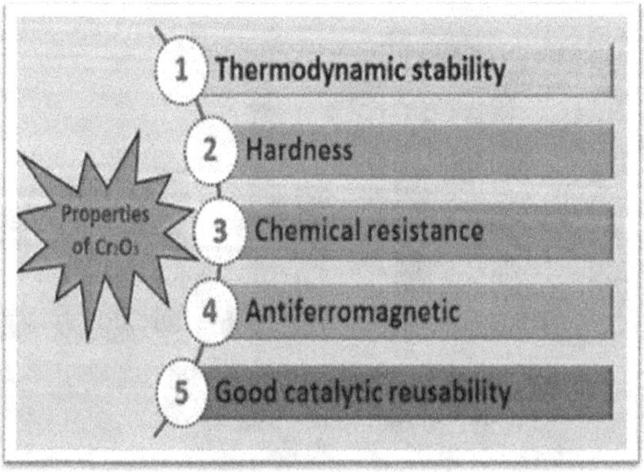

FIGURE 2.28 Properties of Cr_2O_3 NPs.

Hassan et al. synthesized Cr_2O_3 NPs by using flowers extract from Callistemon viminalis, and reported their biological activities. They demonstrated that biosynthesized α-Cr_2O_3 NPs showed impressive anticancer, antimicrobial, and antileishmanial potential with moderate enzyme inhibition and antioxidant properties.

Sharma et al. synthesized Cr_2O_3 NPs using Cannabis sativa and reported their anticancer activity against $HepG_2$ cell line. These green-synthesized Cr_2O_3 NPs possessed promising anticancer activity. Khalil et al. synthesized spherical-shaped Cr_2O_3 NPs using the fruit extract from Hyphaene thebaica and reported their multifunctional biomedical applications. The synthesized Cr_2O_3 NPs displayed a significant fungicidal, bactericidal effect and considerable free radical scavenging potential. In addition, Cr_2O_3 NPs inhibited the protein kinase enzyme and possessed the antiviral nature against polio virus. Kotb et al. revealed the facile green synthesis of Cr_2O_3 NPs utilizing Artemisia herba-alba leaves extract and Melia azedarach fruit extract and evaluated their effective antibacterial activity against Erwinia amylovora by using well diffusion method. Rakesh et al. described the green fabrication of Cr_2O_3 NPs employing Mukia maderaspatana extract and evaluated their $KMnO_4$ decomposition study as well as antibacterial performance against Escherichia coli and Pseudomonas aeruginosa by using disk diffusion method. The synthesized Cr_2O_3 NPs exhibited effective bactericidal effect. Yuvakkumar et al. reported environmentally friendly synthesis of spherical-shaped Cr_2O_3 NPs using fruits peels extract from Nephelium lappaceum and reported their cell viability and cytotoxicity study using MTT assay of human breast cancer cell line.

Iqbal et al. mentioned the facile biogenic fabrication of cubic-shaped Cr_2O_3 NPs from Rhamnus virgata leaves extract and evaluated their anticancer, antileishmanial, antifungal, antioxidant, antibacterial activity, biocompatibility, protein kinase, and α-amylase inhibition study. From the results, Cr_2O_3 NPs are biocompatible, nontoxic, and will be useful in diverse biomedical therapies as nanotherapeutics devices. Ramesh et al. reported the eco-benign production of Cr_2O_3 NPs using Tridax procumbens leaf extract and evaluated their effective antibacterial performance against Escherichia coli. Shanthi and Kamala-Kannan described the biosynthesis of Cr_2O_3 NPs using biomass of Bacillus subtilis which were in the size of 4–50 nm and antibacterial and cytotoxicity study. They evinced effective antibacterial efficacies against Staphylococcus aureus, Escherichia coli, and significant cytotoxic activity against human embryonic cell line.

Due to the pharmacological performance of Cr_2O_3 NPs, they can also be employed in targeted drug delivery to diminish the dose of drugs, enhance specificity, and mitigate toxic effects.

2.2.10.1.3 Biological Exposure to Chromium Nanoparticles

The alteration in brain and kidneys of rats as a consequence of exposure to chromium oxide nanoparticles has been investigated. Accordingly, ROS production seems to cause a considerable increase in malondialdehyde concentration as well as a significant decline in superoxide dismutase and glutathione levels. The pathological evaluations show deleterious oxidative stress as a consequence of ROS generation 2.

1. Besides biological applications, Cr_2O_3 NPs also find usages in the fabrication of microelectronics, sensors, circuits, solar energy collectors, and fuel cells.

2. Chromium oxide nanoparticles are also used in glasses, inks, and paints, as the colorant in "chrome green" and "institutional green".
3. It is used as precursor for the magnetic pigment (chromium dioxide).
4. It is also used in the process of stropping knives.
5. Another main application of chromium oxide nanoparticles is in green pigment production.

2.2.11 Titanium (Ti) NPs

2.2.11.1 Properties and Applications

Titanium nanoparticles have improved strength and radiation resistance. They have high transparency to visible light and high UV absorption.

2.2.11.1.1 Applications

The key applications of titanium nanoparticles are in anti-microbials, antibiotics, and antifungal agents, plastics and soaps, aerospace materials, optical filters, microsensors, coatings, nanofibers, bandages, nanowires, and textiles.

2.2.11.2 Titanium Oxides

Titanium oxide (TiO_2) is available in the form of nanocrystals or nanodots having a high surface area. They exhibit magnetic properties. Titanium oxide is also known as flamenco, rutile, titanium dioxide, and dioxotitanium. Titanium oxide nanoparticles are known for their ability to inhibit bacterial growth and prevent further formation of cell structures. Titanium dioxide (TiO_2) is an inert, nontoxic, and inorganic material, because its high refractive index and high capability to absorb UV light make it promising and environmentally friendly. Titanium oxide nanoparticles appear in the form of black hexagonal crystals. Among the metallic inorganic reinforcements, titanium dioxide (TiO_2) has captured significant attention because of its reasonable price, good stability, photocatalytic activity, UV resistance, antibacterial properties, and nanotoxicity. Kaewkalin et al., (2018) implemented four different concentrations of TiO_2 (0.25%, 0.5%, 1%, and 2% (w/s Ch)) to reinforce chitosan films, and the mechanical properties were evaluated. The TS efficiency increased with the nanoparticle's introduction, with the most significant value being reached for 1% TiO_2.

2.2.11.2.1 Applications

As titanium oxide exhibits good photo catalytic properties, hence is used in antiseptic and antibacterial compositions. Degrading organic contaminants and germs, as a UV-resistant material, manufacture printing ink, self-cleaning ceramics, glass, coating, etc. Making of cosmetic products such as sunscreen creams, whitening creams, morning and night creams, and skin milks is also used in the paper industry for improving the opacity of paper.

2.3 CONCLUSION

Inorganic metallic nanofillers are used for reinforcement of different kinds of polymer composites. Due to their sizes in nanoscale, they exhibit enhanced physicochemical, mechanical, thermal, optical, electrical, and catalytic properties. They find numerous applications in different kinds of industries, viz. electronics, semiconductors, coatings, printings, plastics, textiles, and various polymer composites. The metallic nanofillers of Ag and Au have tremendously contributed in the field of biomedical sciences especially in respect of serving as targeted drug delivery systems in treatment of cancers and tumors. Ni and Pt NPs have contributed greatly in catalytic and photocatalytic processes. Chromium and titanium oxide NPs have worked effectively as antibacterial, antimicrobial, and antifungal agents. Copper NPs have exhibited extraordinary thermal and electrical conductivity enhancements in various polymeric nanocomposites. Iron NPs have played vital role in petroleum pollution control whereas cobalt NPs have shown extraordinary virtues of absorptions of electromagnetic waves, which has attracted great interest due to rapid increases of EMI pollution by communication devices like mobile phones, local area network, radar systems, etc. The reinforcement of inorganic metallic nanofillers has contributed a great deal in various aspects of industries, biomedical applications, and in controlling environmental pollution. Besides, the inclusion of metallic nanofillers like Ag, Co, Fe, Ti, NiO, Al_2O_3, etc., in various functionally graded material composites has made their areas of applications even more diverse. The versatility of their functioning has brought them to the center stage of further investigations. This has broadened the scope and opened the opportunities with new targets set for the young researchers.

REFERENCES

Bai, X., Wang, Y., Song, Z., Feng, Y., Chen, Y., Zhang, D., & Feng, L. (2020). The basic properties of gold nanoparticles and their applications in tumor diagnosis and treatment. *International Journal of Molecular Sciences*, 21(7), 2480.

Boudenne, A., Mamunya, Y., Levchenko, V., Garnier, B., & Lebedev, E. (2015). Improvement of thermal and electrical properties of Silicone–Ni composites using magnetic field. *European Polymer Journal*, 63, 11–19.

Chen, D., Qiao, X., Qiu, X., & Chen, J. (2009). Synthesis and electrical properties of uniform silver nanoparticles for electronic applications. *Journal of Materials Science*, 44(4), 1076–1081.

Fang, Q., & Lafdi, K. (2021). Effect of nanofiller morphology on the electrical conductivity of polymer nanocomposites. *Nano Express*, 2(1), 010019.

Hu, Y., Du, G., & Chen, N. (2016). A novel approach for Al2O3/epoxy composites with high strength and thermal conductivity. *Composites Science and Technology*, 124, 36–43.

Kaewklin, P., Siripatrawan, U., Suwanagul, A., & Lee, Y. S. (2018). Active packaging from chitosan-titanium dioxide nanocomposite film for prolonging storage life of tomato fruit. *International Journal of Biological Macromolecules*, 112, 523–529.

Konieczny, J., & Rdzawski, Z. (2012). Antibacterial properties of copper and its alloys. *Archives of Materials Science and Engineering*, 56(2), 53–60.

Li, W., Li, S., Liu, J., Zhang, A., Zhou, Y., Wei, Q., ... & Shi, Y. (2016). Effect of heat treatment on AlSi10Mg alloy fabricated by selective laser melting: Microstructure evolution, mechanical properties and fracture mechanism. *Materials Science and Engineering: A*, 663, 116–125.

Mohamed, L. Z., Eid, A. I., Eessaa, A. K., & Esmail, S. A. (2019). Studying of physico-mechanical and electrical properties of polypropylene/nano-copper composites for industrial applications. *Egyptian Journal of Chemistry*, 62(5), 913–920.

Sanuja, S., Agalya, A., & Umapathy, M. J. (2015). Synthesis and characterization of zinc oxide–neem oil–chitosan bionanocomposite for food packaging application. *International Journal of Biological Macromolecules*, 74, 76–84.

Shahrokh, A., & Fakhrabadi, M. M. S. (2021). Effects of copper nanoparticles on elastic and thermal properties of conductive polymer nanocomposites. *Mechanics of Materials*, 160, 103958.

De Silva, R. T., Mantilaka, M. M. M. G. P. G., Ratnayake, S. P., Amaratunga, G. A. J., & de Silva, K. N. (2017). Nano-MgO reinforced chitosan nanocomposites for high performance packaging applications with improved mechanical, thermal and barrier properties. *Carbohydrate polymers*, 157, 739–747.

Vincent, M., Hartemann, P., & Engels-Deutsch, M. (2016). Antimicrobial applications of copper. *International Journal of Hygiene and Environmental Health*, 219(7), 585–591.

Yadav, M., Rhee, K. Y., Park, S. J., & Hui, D. (2014). Mechanical properties of Fe3O4/GO/chitosan composites. *Composites Part B: Engineering*, 66, 89–96.

3 An Exploration of Some Essential Inorganic Nanofillers

Dhananjay Singh
Government Polytechnic

Deepak Singh
Institute of Engineering & Technology

Suresh Kumar Patel
Government Polytechnic

G. L. Devnani
Harcourt Butler Technical University

CONTENTS

DOI: 10.1201/9781003279389-3

3.1 INTRODUCTION

Polymers are now become an integral part of the recent advancements in the material science. We have explored specific types of polymers and polymers composites for specific purposes. The fillers play important role in the properties of composites. Nanofillers of specific characters are key components for such specific products (Remanan et al. 2019). Our key target of this chapter is to explore various inorganic nanofillers, which can show specific properties. Every inorganic nanofiller has its distinct property and structure, so we can derive the required composition as per our required application. The main properties are hardness, thermal stability, density, and electric, thermal, and optical properties. Sizes of nanofillers also play a key role in surface chemistry (Sani et al. 2012).

Inorganic nanofillers such as nitrides, carbides, nanoclays, and sulfide have attracted recently in the area of nanomaterial science. Fillers with particle size less than 100 nm are termed as nanofillers. According to their molecular structure, it may be termed as nanoplates, nanofibers, or nanoparticles. Efforts have been made in the area of synthetic materials advancements with respect to the targeted application in last two decades (Qu et al. 2017). The discussed inorganic nanofibers and nanoparticles have different cost, color, and compositions, so they can be picked according to the need (Rallini et al. 2017). Dispersion of these inorganic nanofillers within the polymer matrices also plays an important role, especially the dispersion of nanoclays. Broad classifications of inorganic nanofillers have been shown in Figure 3.1. The detailed applications of each discussed nanofillers have been dictated in the concerned sections.

3.2 NANO-CARBIDES

Nano-carbides-based materials are increasingly used to replace conventional materials in many applications such as environmental remediation, gas sensing, photocatalysis, etc. (Rasaki et al. 2018). Some different types of nano-carbides have been discussed below:

3.2.1 Zirconium Carbide (ZrC)

Zirconium carbide is used as an inorganic nanofiller. It has hard crystalline structure having gray color. Zirconium-based coating has some peculiar significant properties like resistance to corrosion, erosion, wear, and oxidation. It has been found stable at high temperature and has significantly high hardness (Sani et al. 2012). In the initial stage in

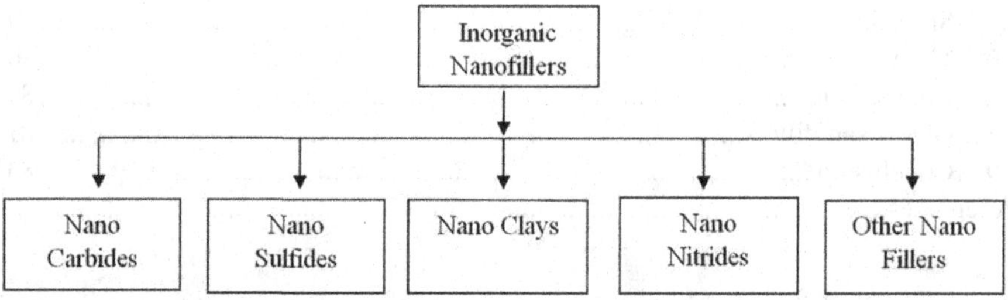

FIGURE 3.1 Classification of inorganic nanofillers.

year 1975, it has been observed that ZrC-coated tungsten surface exhibits high-temperature stability in comparison with carbide of silicon (SiC). ZrC coatings are also effective in biomedical instrumentation to enhance the resistance toward corrosion, in electron microscopes as emission stabilizer because it has very low work function. On the other hand, it has slow significant enhancement in hardness value (Rallini et al. 2017). ZrC may also be used as vehicle starters and circuit breakers. This has been found as inert toward chemicals but highly conductive toward electricity.

3.2.2 Boron Carbide (B_4C)

Boron Carbide has layered crystalline rhombohedral lattice unit structure. Molecules are joined with covalent bond. It is a lightweight compound having low specific gravity 2.52 g/cm^2, high melting point (2427°C), and high boiling point (3500°C). It has significant mechanical properties, especially outstanding hardness (3760 Kg/mm^2) (Sharma et al. 2016). This compound was first developed (1899) by a French scientist Henri Moissan. Oxides of boron and carbon were used for this synthesis, in the presence of electric arc within the furnaces. The preparation steps of the boron carbide are shown in Figure 3.2.

It is used in abrasive coating due to extraordinary abrasion resistance. Due to low density and high hardness, it can be used efficiently in ballistic applications. It is thermally stable at high temperature, so it is efficiently applied in refractory purposes (Remanan et al. 2019). Boron carbide is used extensively in the various applications apart from nanofiller applications as shown in Figure 3.3.

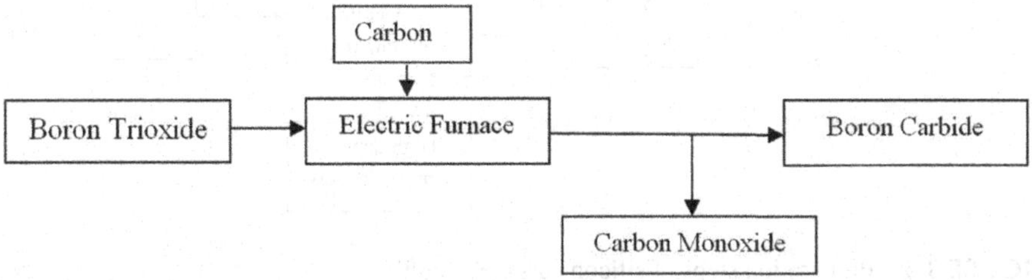

FIGURE 3.2 Preparation steps of boron carbide nanofiller.

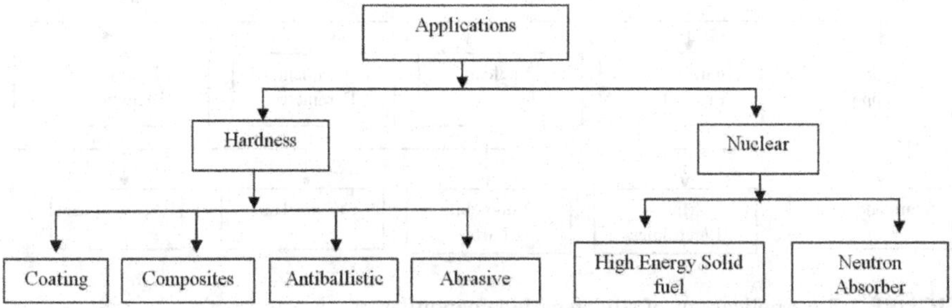

FIGURE 3.3 Applications of boron carbide nanofiller.

3.2.3 Silicon Carbide (SiC)

More than hundred varieties of SiC have been found in the crystalline phase. Si atoms and carbon atom are joined (tetrahedral) by a strong bonding in crystal lattice structure (Naeimirad et al. 2016). SiC has been prepared by the following steps as shown in Figure 3.4.

Silicon carbides have high thermal conductivity. Due to this property, it can be used as heat-sensitive material. Thermal expansion of the SiC has been found negligible. Its strength is significantly high, while its density has been found very low. Chemical resistance of silicon carbide is very low. This material can survive at high temperature without sacrificing its strength; that is, its thermal fluctuations' resistance is quite high. The hardness of SiC is excellent (near to diamond), and wear resistance is quite good. Aluminum and nitrogen impurities can affect the electric conductivity of SiC (Qu et al. 2017). Silicon carbide is used significantly in numerous applications apart from nanofillers in composites as shown in Figure 3.5.

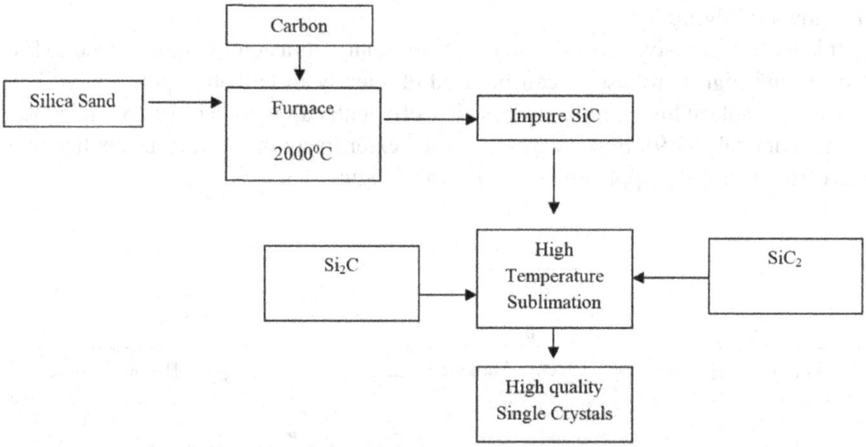

FIGURE 3.4 Preparation steps of silicon carbide nanofiller.

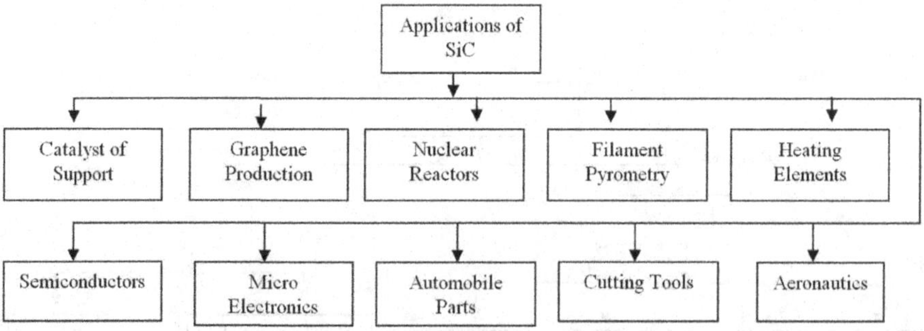

FIGURE 3.5 Applications of silicon carbide nanofiller.

3.2.4 Chromium Carbide (Cr_3C_2)

There are three varieties of chromium carbide nanofillers. Its different structures may be cubic ($Cr_{23}C_6$), hexagonal (Cr_7C_3), or orthorhombic (Cr_3C_2), but the most durable form of chromium carbide is Cr_3C_2, which has been found in orthorhombic crystal structure. This structure shows the strongest covalent character (Cho et al. 2012). For the preparation of Cr_3C_2, chromium oxide, pure aluminum, and graphite have been used. This has been prepared as per the following steps shown in Figure 3.6.

Hard ceramic Cr_3C_2 nanoparticles have been used in nickel–chromium matrix; that is, it is useful in the coating of metal surfaces. Since chromium is itself a corrosion-resistant metal, Cr_3C_2 has shown significant corrosion-resistant property (Cho et al. 2012). Chromium carbide has been used significantly in numerous applications, shown in the following Figure 3.7.

The comparison of all listed nano-carbides has been given in Table 3.1.

3.3 NANO-SULFIDE

Nanosulfides have attracted great attention due to their excellent properties and applications such as manufacturing of electronic, optical, and optoelectronic devices (Njuguna et al. 2014). Some useful nanosulfides are discussed below:

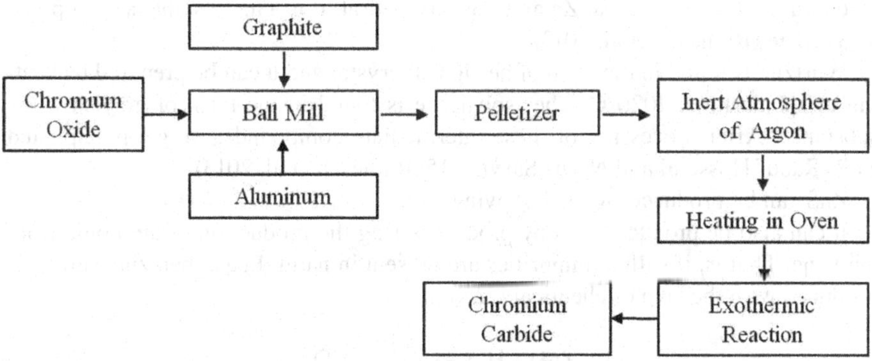

FIGURE 3.6 Preparation steps of Cr_3C_2 nanofiller.

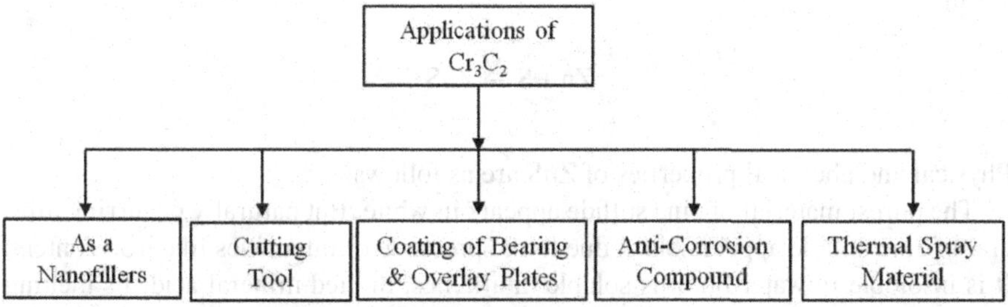

FIGURE 3.7 Applications of Cr_3C_2 nanofiller.

TABLE 3.1
Comparative Study of Different Types of Nano-Carbides

Types of Nano-Carbide	Formula	Molar Mass (g/mol)	Density (g/cm³)	Melting Point (°C)	Boiling Point (°C)	Solubility in Water
Zirconium carbide (ZrC)	ZrC	103.23	6.73	3532	5100	Insoluble
Boron carbide (B_4C)	B_4C	55.25	2.25	2350	3500	Insoluble
Silicon carbide (SiC)	SiC	40.11	3.21	2830	4180	Insoluble
Chromium carbide (Cr_3C_2)	Cr_3C_2	180.01	6.68	1895	3800	Insoluble

3.3.1 Zinc Sulfide (ZnS)

Zinc Sulfide is an inorganic nanoparticle/nanofiller having chemical formula ZnS. Zinc sulfide is a naturally occurring salt and its main source is zinc compound. It is a natural mineral, which is also known as sphalerite. The purest form of ZnS is white, and it is commonly used as a pigment especially in paints and plastic industries, despite the fact that it is usually in black form due to numerous impurities like presence of iron content (Raouf Hosseini and Nasiri Sarvi 2015).

The zinc sulfide is available in two common crystalline polymorphous forms such as sphalerite and wurtzite forms. Sphalerite is found in a cubic crystal structure, and its coordination geometry at Zn and S is tetrahedral. It is a more stable as compared to wurtzite (Ramesan et al. 2013).

Wurtzite is found in the form of hexagonal crystal and it can be prepared by heating of sphalerite at 1020°C. The sphalerite is found in the form of grayish-white substance. After processing of these intermediate compounds, we get precipitated ZnS (Raouf Hosseini and Nasiri Sarvi 2015; Ramesan et al. 2013).

ZnS can be produced by the following:

It can also be produced as a by-product during the production of ammonia from methane. That is, if sulfide impurities are present in natural gas, then zinc sulfide is produced with the help of chemical reaction.

$$ZnO + H_2S \rightarrow ZnS + H_2O$$

In laboratory scale, zinc sulfide can be produced by igniting a mixture of zinc and sulfur.

$$Zn + S \rightarrow ZnS$$

Physical and chemical properties of ZnS are as follows:

The purest material of zinc sulfide appears in white. But naturally occurring mineral of zinc sulfide appears black due to the presence of impurities like iron content. It is insoluble in water but it is soluble in alkalies, diluted mineral acid. Its melting point and density have been found as 1700–1900°C and 4.09 mg/cm³, respectively. Activated form of ZnS can exhibit luminescence, phosphorescence, and fluorescence.

In the presence of oxygen, it produces zinc oxide and sulfur dioxide. When ZnS reacted with diluted hydrochloric acid, zinc chloride and hydrogen sulfide are produced. In the case of sulfuric acid, it produces zinc sulfate and hydrogen sulfide.

$$ZnS + H_2SO_4 \rightarrow ZnSO_4 + H_2S$$

The several uses of zinc sulfides as nanofillers are listed in the following Figure 3.8 (Raouf Hosseini and Nasiri Sarvi 2015).

3.3.2 Lead Sulfide (PbS)

Lead sulfide is an inorganic nanomaterial, and its chemical formula is PbS. PbS is also called as galena, which is the main source of ore of lead compound. It is the most common naturally occurring to be used as a semiconductor (Iram et al. 2020). Lead sulfide containing nanoparticles and quantum dots forms are generally made by combination of lead salt with addition of different sulfide sources. Recently, lead sulfide nanoparticles are mostly used in solar cells for enhancing the absorption rate of solar radiation.

The IUPAC name of PbS is sulfanylidenelead, and its molar mass is 239.3 g/mol. Lead sulfide has a cubic crystal structure with a unit cell formed by one cation surrounding with six anions and vice versa. The structure of PbS is formed by some ionic bonding, mixture of ionic and covalent, and a few only covalent bonding (Seshadri et al. 2011).

Crystalline size can be determined by electron microscopy with range of 2–5 nm. A quantitative analysis can be done by atomic absorption spectrometry (Seshadri et al. 2011). The various chemical techniques are available for the production of PbS nanoparticles which are discussed as follows:

- Lead oxide can be extracted from galena ores by chemical methods to produce this inorganic compound. It can be prepared by the reduction through substitution between hydrogen sulfide gas and lead mixture solution.

$$H_2S + Pb(NO_3)_2 \rightarrow PbS + 2HNO_3$$

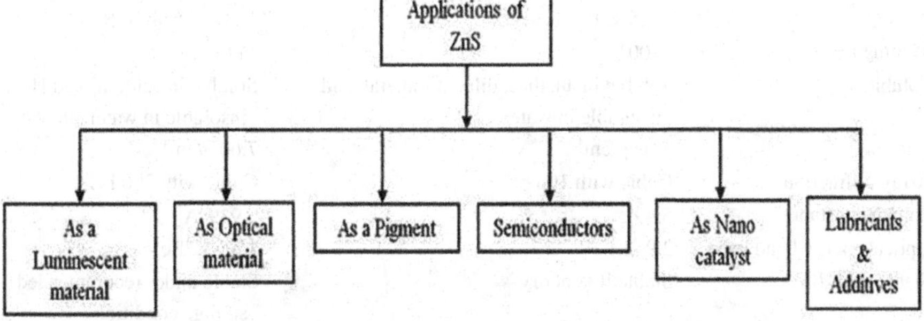

FIGURE 3.8 Applications of ZnS nanofiller.

- Lead sulfide can be produced by addition of hydrogen sulfide to lead soil solution such as $PbCl_2$, and black precipitate of lead sulfide is formed

$$H_2S + Pb^{2+} \rightarrow PbS \downarrow + 2H^+$$

Lead sulfide is found in a form of black crystalline solid or in a silver powder. Its boiling and melting points have been found as 1749°C & 1114°C, respectively. Its vapor pressure has been reported as 1 kPa at 953°C, while density is 7.60 g/cm³. The refractive index and band gap have been reported as 3.921 and 0.37 eV, respectively. Its water & alcohol solubility is negligible but it is highly soluble in hot nitric acid and diluted hydrochloric acid. It has cubic chemical structure having heat of fusion as 72 J/g, heat of formation as 435 kJ/mol, and thermal conductivity as 2.30 w/mK (Iram et al. 2020). Its metallic property is weak, and it has some characteristics of metalloids like semiconductivity or photoconductivity.

The various industrial applications of the lead sulfide are shown in Figure 3.9.

The comparative features of ZnS and PbS are listed in Table 3.2.

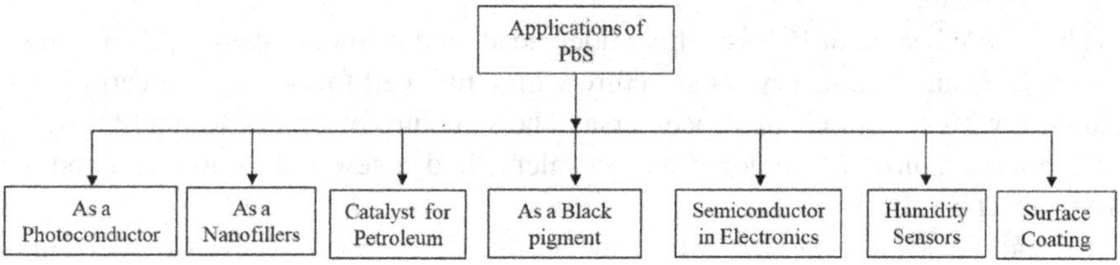

FIGURE 3.9 Applications of lead sulfide.

TABLE 3.2
Comparative Study of Different Types of Nanosulfides

Property	ZnS	PbS
Molar mass	97.474 g/mol	239.3 g/mol
IUPAC Name	Sulfanylidenezine	Sulfanylidenelead
Appearance	White to grayish-white (yellowish powder colorless cubic crystal)	Black to silvery powder or crystalline solid
Boiling point	1935°C (sublimes)	1281°C (sublimes)
Melting point	1700°C	1118°C
Solubility	Soluble in alkalies, diluted mineral acid insoluble in water	Soluble in acid, diluted HCl, insoluble in water, alcohol
Density	4.09 g/cm³	7.60 g/cm³
X-ray diffraction	Cubic with 100 hkl	Cubic with 200 hkl
Lattice constant	5.49°A	5.939°A
Optical energy band gap	2.9–3.1 eV	1.58-1.78 eV
Stable shelf life	Stable if kept dry	Stable under recommended storage condition
Refractive index	2.368	3.921

3.4 NANOCLAY

Recently, many natural-occurring clays have been used for intense academic and commercial activities. Clays are very complex minerals. An important clay rock, which is found in nature, is known as bentonite (Mueller et al. 2014). It is an important source of montmorillonite in nature.

3.4.1 Montmorillonite (MMT)

Montmorillonite (MMT) is natural clay mineral, present in the smectite group. MMT is derived from the bentonite. It contains high amount of sodium or calcium and 10%–20% of other minerals such as calcite, silica, gypsum, etc. (Osman et al. 2017). The molecular structure of the MMT consists of isomorphic substitutions in the tetrahedral sheet of Si^{4+} by Al^{3+} and Al^{3+} by Mg^{2+} in the octahedral ones. The physical structure of montmorillonite is in the form of layers and sheets. The chemical composition of the MMT varies with variation in chemical formula. Therefore, researchers never found the exact theoretical formula of the MMT in nature (Vilarinho et al. 2019).

Montmorillonite can be obtained within the cave environment. The natural atmosphere in the cave formed the aluminosilicates in the bedrocks. After sometime, the slow formation of MMT takes place in the solutions of aluminosilicates. The process of MMT formation is aided naturally by addition of HCO_3. Some physical properties of the MMT have been listed in Table 3.3.

For non-saline soils, MMT produces an increased bulk electrical conductivity. MMT has good water sorption capacity; that is, it absorbs or loses water to change the humidity in the ambient environment. Recently, researchers have been attracted toward the MMT due to its special qualities such as high-surface area, ease of availability, high cation exchange capacity (CEC), low cost, high degree of intercalation, and capability of swelling and expanding (Osman et al. 2017).

Some useful applications of the MMT have been listed below:

- MMT is used for biological purpose.
- MMT is a good heat insulator.

TABLE 3.3
Physical Properties of the MMT

Properties	Description
Crystal system	Monoclinic
Hardness	Soft
Density	2–3 g cm^{-3}
Cleavage	Perfect
Luster	Earthy, dull
Fracture	Irregular
Color	White, yellow, green
Transparency	Translucent

- It provides resistance to nausea and diarrhea.
- MMT provides resistance to tooth decay.
- Montmorillonite is seen in the removal of toxic heavy metals from aqueous solution.
- MMT is also used to improve the thermophysical properties of the biopolymers.

3.4.2 Nano-Calcium Carbonate

Nano-calcium carbonate (Nano-$CaCO_3$) is a chemical compound and found in rocks as mineral calcite and aragonite. These are cubic or hexagonal high-surface area particles. The particles of the nano-calcium carbonate exist in the range of 10–80 nm with specific surface area of 30–60 m^2/g (Liu et al. 2012).

Nano-$CaCO_3$ has been obtained from the phosphogypsum. It is the waste calcium-rich by-product material from the process of phosphoric acid. The aragonite form of the nano-$CaCO_3$ can be prepared by the precipitation at the temperature of 85°C or above, and vaterite form can be obtained at a specific temperature of 60°C There are some chemical and physical properties of the nano-$CaCO_3$ shown in Table 3.4 (Assaedi et al. 2020).

The use of nano-$CaCO_3$ is increasing day by day in the chemical, biological, and industrial applications. The growth in the demand for nano-calcium carbonate from various end-use applications such as paints, sealants, and adhesives is expected to drive the global nano-calcium carbonate market over the forecast period (Lu et al. 2016). There are some other applications of the nano-$CaCO_3$:

- Nano-$CaCO_3$ could be efficacious to improve all of the mechanical properties of geopolymer.
- It is also used to improve the mechanical properties of the cementitious materials.
- It is also used as additives in polymers and adhesive to improve the final product quality.
- Nano-$CaCO_3$ is used in printing ink to improve its quality. This calcium carbonate-treated ink can replace the use of varnishes and oils.

TABLE 3.4
Physical Properties of the Nano-$CaCO_3$

Properties	Description
Morphology	Cubic
Average particle size (nm)	15–40
Bulk density (g/ml)	0.68
Whiteness (%)	89
Moisture content (wt %)	0.5
pH	8.0–9.0
$CaCO_3$ dry basis (%)	97.5
MgO (%)	0.4

3.5 SMECTITE

Smectite is used for a group of phyllosilicate mineral species. These are several species differentiated by the variations in the chemical composition. It involves Al for Si in tetrahedral cation sites and Fe, Mg, and Li in octahedral cation sites. It is divided into three categories: Na smectite, Ca-Mg smectite, and acid earth (Odom 1984). Smectite clays form at surface and shallow subsurface environment. It is ubiquitous components of semi-arid or arid region's soils. Smectite clay is related to the various modes of origin fluids containing Si^{-}-, Al^{-}-, Fe^{-}-, Mg-bearing minerals and volcanic glass.

Physicochemical properties of smectite are very important for industrial and chemical uses. The viscosity properties of the smectite clay are uniform or high variable for single deposit (Odom 1984). Some properties are listed in the Table 3.5.

The smectite clay has been used in industrial, chemical, and commercial purposes. The primary uses of the smectite are divided into two main categories based on volume, that is, large volume and small volume shown in Table 3.6.

TABLE 3.5
Properties of the Smectite

Properties	Description
Crystal structure	Lattice
Surface area	0.2–2.0
Crystal size	Hexagonal
Exchangeable ions	Cation

TABLE 3.6
Applications of the Smectite

Large Volume	Small Volume
Oil well drilling	Building – bricks, sewer pipe
Filtering	Gypsum product
Clarifying	Glazes
Waterproofing	Lubricants
Pelleting – iron ore	Fertilizers
Pelleting – animal feed	Detergents
Pesticide carrier	Paper coating
Foundry sand binder	Seed coating
Oil and grease absorbent	Mortar
Water impedance	Medical, pharmaceutical
decolorizing	Paint

3.6 MICA

Mica is hydrous potassium, aluminum silicate minerals. It exhibits two-dimensional sheet or layer structure. The general formula of mica is $XY_{2-3}Z_4O_{10}(OH, F)_2$ with X=K, Na, Ba, Ca, Cs, (H_3O), (NH_4); Y=Al, Mg, Fe^{2+}, Li, Cr, Mn, V, Zn; and Z=Si, Al, Fe^{3+}, Be, Ti. The crystal structure of mica consists of two polymerized sheets of silica (SiO_4) tetrahedrons (Jouyandeh et al. 2021). Mica is found in three varieties of rocks such as igneous, sedimentary, and metamorphic. It is originated from the diverse process under different conditions. It is also formed by crystallization of the consolidated magmas. The stability range of the mica has been investigated in the laboratory and the chemical composition has been identified with the help of different instruments.

Mica is also derived from different rocks such as:

- Biotite
- Phlogopite
- Muscovite
- Paragonite
- Lepidolite
- Glauconite

Most of rock-forming mica can be divided into two categories: light-colored and dark-colored micas. The physical and chemical properties of the all micas are same as glauconite (Fuse and Sato 2008). The physical and chemical properties of the mica are listed in Tables 3.7 and 3.8, respectively.

Mica is generally used as nanofillers to improve the dielectric properties of the materials. Some other useful applications of the mica are given below (Fuse and Sato 2008):

- Mica is also used in electrical condensers as insulation sheet due to its low thermal and electrical conductivity.
- The sheets of mica are also utilized in optical instruments.
- It is also used in asphalt tiles as filler, lubricant, and absorbent.
- It is also used in the manufacture of wallpaper to give it a shiny luster.
- Mica is also used as water softener due to its high base exchange capacity.

TABLE 3.7
Physical Properties of the Mica

Properties	Description
Appearance	Fine white powder
Specific gravity	2.8
Fines (%)	93
Sand (%)	7
Particle diameter (μm)	53.6
Specific surface area	5.3
Natural water content (%)	0.41
Hardness (mohs)	2.5

TABLE 3.8
Chemical Properties of the Mica

Properties	Description
SiO_2 (%)	49.5
Al_2O_3 (%)	29.2
K_2O (%)	8.9
Fe_2O_3 (%)	4.6
TiO_2 (%)	0.8
MgO(%)	0.7
Na_2O (%)	0.5
CaO (%)	0.4
pH value	7.8
Oil absorption (mL/100 g)	36

TABLE 3.9
Comparative Study of Different Type of Nanoclay

Type of Nanoclays	Formula	Molar Mass	Density (g/cm³)	Melting Point	Boiling Point	Solubility in Water
Montmorillonite (MMT)	$(Na,Ca)_{0.33}(Al,Mg)_2(Si_4O_{10})(OH)_2{\cdot}nH_2O$	549.07 g	3.00	-	-	Soluble
Nano-calcium carbonate	$CaCO_3$	100.08 g/mol	2.93	825°C	-	Soluble
Smectite	$M_{0.33}{}^{+}Al_2(Si_{3.67}\,Al_{0.33})\,O_{10}\,(OH)_2$	282.11 g/mol	2.63	-	-	Soluble
Mica	$AB_{2\text{-}3}(X, Si)_4O_{10}(O, F, OH)_2$	398.71 g	2.82	1300°C	-	Insoluble

The comparative study of all above-listed nanoclays is shown in Table 3.9.

3.7 NANO-NITRIDE

Nano-nitride nanofillers have drawn the attention of researchers worldwide due to their chemical characteristics to make sustainable alternative to noble metals in catalysis (Ashraf et al. 2020). The classification of the nano-nitrides is shown in Figure 3.10.

The comparative study of all above-listed nano-nitrides is shown in Table 3.10.

3.8 OTHER NANOFILLERS

Apart from the natural nanofillers, some synthetic or derived nanofillers have also played important role in chemical and polymer industries (Althues et al. 2007). Some synthetic nanofillers have been listed in Table 3.11 (Liu et al. 2013; Guo et al. 2020).

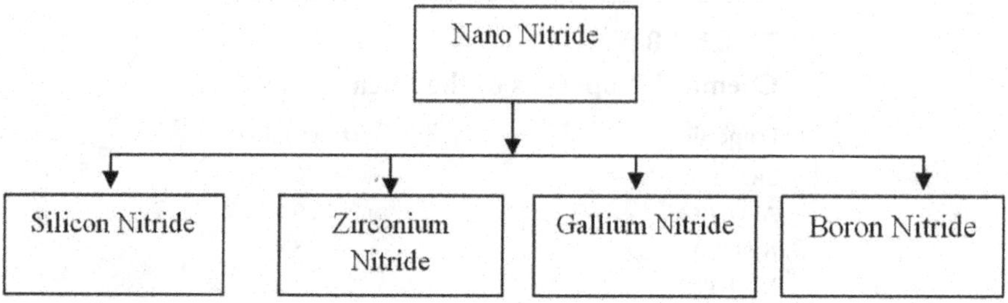

FIGURE 3.10 Different types of nano-nitride nanofillers.

TABLE 3.10
Comparative Study of Different Type of Nano-Nitrides Nanofillers

Type of Nano-Nitride	Formula	Molar Mass (g/mol)	Density (g/cm³)	Melting Point (°C)	Boiling Point	Solubility in Water
Silicon nitride	Si_3N_4	140.28	0.044	1900	-	Insoluble
Zirconium nitride	ZrN	105.23	7.09	2980	3578°C	Insoluble
Gallium nitride	GaN	83.73	6.15	2500	-	Insoluble
Boron nitride	BN	24.80	2.10	3273	-	Insoluble

TABLE 3.11
Properties and Applications of Some Other Synthetic Nanofillers

Nanofillers	Properties	Applications	Remark
ZiF11	Narrow pore size	It is used to improve the separation performance of membranes.	Synthetic
Cadmium telluride (CdTe)	Density: 5.85 g/cm³ Melting point: 1092°C Boiling point: 1130°C Poisson ratio: 0.41	It is used in thin film solar cells, infrared optics, etc.	Synthetic
Polyhedral oligomeric Silsesquioxane (POSS)	Density: 1.39 g/cm³ Refractive index: 1.64 Molecular size: 1–5 nm Form: Colorless	It is used in thermoplastics and thermosetting polymers, drug delivery, and polymer electrolytes.	Synthetic
Titanium boride (TiB_2)	Density: 4.52 g/cm³ Molar mass: 69.48 g/mol	It is used as cathodes in Hall–Heroult cells for primary aluminum smelting. It also finds use as crucibles for handling molten metals and as metal evaporation boats.	Synthetic

3.9 CONCLUSION

The demands for sustainable polymeric materials are increasing day by day. Inorganic nanofillers are basically categorized as follows: nano-carbides, nanosulfides, nano-clays, nano-nitrides, and some other nanofiller types. Among all, montmorillonite (MMT)-, smectite-, and mica-based nanoclay are mostly used for intense academic and commercial activities. Their physical properties, such as hardness, thermal stability, density, and electric, thermal, and optical properties, have also been discussed here.

REFERENCES

Althues H, Henle J, Kaskel S (2007) Functional inorganic nanofillers for transparent polymers. *Chem Soc Rev* 36:1454–1465 https://doi.org/10.1039/b608177k.

Ashraf I, Rizwan S, Iqbal M (2020) A comprehensive review on the synthesis and energy applications of nano-structured metal nitrides. *Front Mater* 7:1–20 https://doi.org/10.3389/fmats.2020.00181.

Assaedi H, Alomayri, T, Kaze, CR, Jindal, BB, Subaer S, Shaikh F, Alraddadi, S. (2020) Characterization and properties of geopolymer nanocomposites with different contents of nano-$CaCO_3$. *Constr Build Mater* 252:119137. https://doi.org/10.1016/j.conbuildmat.2020.119137.

Cho, S., Kikuchi, K., Kawasaki, A., Kwon, H., & Kim, Y. (2012). Effective load transfer by a chromium carbide nanostructure in a multi-walled carbon nanotube/copper matrix composite. Nanotechnology, 23(31), 315705.

Fuse, N., Sato, H., Tanaka, T., & Ohki, Y. (2008, October). Effects of Mica Nanofillers on the Complex Permittivity of Polyamide Nanocomposites. In 2008 Annual Report Conference on Electrical Insulation and Dielectric Phenomena (pp. 540–543). IEEE.

Guo, A., Ban, Y., Yang, K., Zhou, Y., Cao, N., Zhao, M., Yang, W. (2020). Molecular sieving mixed matrix membranes embodying nano-fillers with extremely narrow pore-openings. *J Memb Sci*, 601, 117880.

Iram, S., Mahmood, A., Sitara, E., Batool Bukhari, S. A., Fatima, S. A., Shaheen, R., Azad Malik, M. (2020). Nanostructured lead sulphide depositions by AACVD technique using bis (Isobutyldithiophosphinato) lead (II) complex as single source precursor and its impedance study. *Nanomaterials*, 10(8), 1438.

Jouyandeh, M., Akbari, V., Paran, S. M. R., Livi, S., Lins, L., Vahabi, H., & Saeb, M. R. (2021). Epoxy/ionic liquid-modified mica nanocomposites: network formation–network degradation correlation. *Nanomaterials*, 11(8), 1990.

Liu X, Chen L, Liu A, Wang X (2012) Effect of nano-CaCO3 on properties of cement paste. *Energy Procedia* 16:991–996 https://doi.org/10.1016/j. egypro.2012.01.158.

Liu, Y., Sun, Y., Zeng, F., Liu, J., & Ge, J. (2013). Effect of POSS nanofiller on structure, thermal and mechanical properties of PVDF matrix. *J Nanoparticle Res*, 15, 1–10.

Lu SQ, Lan PQ, Wu SF (2016) Preparation of nano-$CaCO_3$ from phosphogypsum by gas-liquid-solid reaction for CO_2 sorption. *Ind Eng Chem Res* 55:10172–10177 https://doi.org/10.1021/acs.iecr.6b02551.

Mueller KT, Sanders RL, Washton NM (2014) Clay minerals. *eMagRes* 3:13–28 https://doi.org/10.1002/9780470034590.emrstm1332.

Naeimirad M, Zadhoush A, Neisiany RE (2016) Fabrication and characterization of silicon carbide/epoxy nanocomposite using silicon carbide nanowhisker and nanoparticle reinforcements. *J Compos Mater* 50:435–446. https://doi.org/10.1177/0021998315576378.

Njuguna J, Ansari F, Sachse S, (2014) Nanomaterials, nanofillers, and nanocomposites: Types and properties. *Heal Environ Saf Nanomater Polym Nancomposites Other Mater Contain Nanoparticles* 3:27. https://doi.org/10. 1533/9780857096678.1.3.

Odom IE (1984) Smectite clay minerals: properties and uses. *Philos Trans R Soc London* 4:391–409. https://doi.org/10.1098/rsta. 1984.0036.

Osman, A. F., Fitri, T. F. M., Rakibuddin, M., Hashim, F., Johari, S. A. T. T., Ananthakrishnan, R., Ramli, R. (2017). Pre-dispersed organo-montmorillonite (organo-MMT) nanofiller: Morphology, cytocompatibility and impact on flexibility, toughness and biostability of biomedical ethyl vinyl acetate (EVA) copolymer. Mater Sci Eng C, 74, 194–206.

Qu, H., Wang, Y., Ye, Y. S., Zhou, W., Bai, S. P., Zhou, X. P., ... , Mai, Y. W. (2017). A promising nanohybrid of silicon carbide nanowires scrolled by graphene oxide sheets with a synergistic effect for poly (propylene carbonate) nanocomposites. *J Mater Chem A*, 5(42), 22361–22371.

Rallini M, Wu H, Natali M (2017) Nanostructured phenolic matrices: Effect of different nanofillers on the thermal degradation properties and reaction to fire of a resol. *Fire Mater* 41:817–825 https://doi.org/10.1002/fam.2425.

Ramesan MT, Nihmath A, Francis J (2013) Preparation and characterization of zinc sulphide nanocomposites based on acrylonitrile butadiene rubber. *AIP Conf Proc* 1536:255–256 https://doi.org/10.1063/1.4810197.

Raouf Hosseini M, Nasiri Sarvi M (2015) Recent achievements in the microbial synthesis of semiconductor metal sulfide nanoparticles. *Mater Sci Semicond Process* 40:293–301 https://doi.org/10.1016/j.mssp.2015.06.003.

Rasaki, S. A., Zhang, B., Anbalgam, K., Thomas, T., & Yang, M. (2018). Synthesis and application of nano-structured metal nitrides and carbides: A review. *Prog Solid State Chem*, 50, 1–15.

Remanan M, Bhowmik S, Varshney L, Jayanarayanan K (2019) Tungsten carbide, boron carbide, and MWCNT reinforced poly(aryl ether ketone) nanocomposites: Morphology and thermomechanical behavior. *J Appl Polym Sci* 136:1–13 https://doi.org/10.1002/app.47032.

Sani E, Mercatelli L, Sansoni P (2012) Spectrally selective ultra-high temperature ceramic absorbers for high-temperature solar plants. *J Renew Sustain Energy* 4. https://doi.org/10.1063/1.4717515.

Seshadri S, Saranya K, Kowshik M (2011) Green synthesis of lead sulfide nanoparticles by the lead resistant marine yeast, Rhodosporidium diobovatum. *Biotechnol Prog* 27:1464–1469 https://doi.org/10.1002/btpr.651.

Sharma S, Bijwe J, Panier S (2016) Assessment of potential of nano and micro-sized boron carbide particles to enhance the abrasive wear resistance of UHMWPE. *Compos Part B Eng* 99:312–320 https://doi.org/10.1016/j.compositesb.2016.06.003.

Vilarinho F, Vaz MF, Silva AS (2019) The use of montmorillonite (MMT) in food nanocomposites: Methods of incorporation, characterization of MMT/polymer nanocomposites and main consequences in the properties. *Recent Pat Food Nutr Agric* 11:13–26. https://doi.org/10.2174/2212798410666190401160211.

4 An Exploration of Some Essential Inorganic Nanofillers Apart from Metal Oxides, Metallic Particles, and Carbon-Based Nanofillers

Bosely Anne Bose, B. C. Bhadrapriya, and Nandakumar Kalarikkal
Mahatma Gandhi University

CONTENTS

4.1 INTRODUCTION

"Fillers" are separate and distinct materials that form a composite when embedded into the polymer matrix to improve the composite's mechanical, thermal, optical, dielectric, electric, and magnetic properties. The morphological factors like shape and size, state of dispersion of fillers, chemical structure, and the amount of fillers determine the properties of these composites. The filler and polymer matrix interface are also critical for better composite properties. Fillers can be inorganic, organic,

DOI: 10.1201/9781003279389-4

carbonaceous, or metallic, and they can be solid, liquid, or even gaseous phases (Imai, 2020). Carbon black, pyrogenic silica, and diatomite have been exploited as additives in polymers for decades, yielding polymer nanocomposites.

However, at the time, their characterizations and the effect of capabilities induced by the nanometric scale of fillers were not fully recognized. The actual beginning point corresponds to a grasp of how these things work was based on research on polyamide-6 filled with nanoclays that are commonly referred to as fillers, but it was termed as a "hybrid" substance in these works. The pace of research accelerated, and the first application and the term "nanocomposites" was coined in 1994 (Lan et al., 1995; Lan & Pinnavaia, 1994). These new technologies have emerged due to the desire for sustained improvement in the performance of thermoplastic and thermoset polymer materials. The list of nanofillers (nanoclays, nano-oxides, carbon nanotubes, etc.) has increased in recent years, as has the matrix in which they are exploited and interactions with conventional fillers. The research of polymer nanocomposites is currently one of the most active areas of nanomaterial development. The qualities imparted by nanoparticles are diverse, emphasizing optimizing electrical conduction and barrier properties to temperature, gases, and liquids, as well as potential heat release improvement (Mohammed et al., 2011). Nanocomposites are a new class of two-phased materials that blend a fundamental matrix with nanofillers introduced between polymer chains at the molecular scale in the same manner that fibers are reinforced at the macroscopic size. Nanofillers, often in combination with current fillers, can significantly improve or modify the characteristics of the materials into which they are integrated, such as optical, electrical, mechanical, thermal, or heat capabilities. The mixing ratio between the organic matrix and the nanofillers can substantially impact the characteristics of composite materials.

4.2 NANOFILLERS

The term "nanofiller" refers to the relevance of the fillers' dimensional properties. Nanofillers are primary particles with less than 100 nm geometric size at least one dimension. There are three primary forms of nanofillers: one-dimensional (nanoplatelet), two-dimensional (nanofiber), and three-dimensional (nanoparticulate) (Sundarram et al., 2015). The dimensional features of nanofiller distinguish it from conventional fillers by providing a high interfacial interaction with the polymer matrix, enhanced scattering efficiency for optical wavelength, and size-dependent physical properties. The vast surface area of nanofillers can increase the interaction area when these nanofillers are dispersed into the polymer matrix, which alters the polymer characteristics, such as mobility, crystallinity, and orientation from those of the bulk polymer (Ciprari et al., 2006).

Nanofillers, both organic and inorganic, can be incorporated into the polymer matrix. Polymer-inorganic nanocomposites (PINs) are composite materials made up of nanometer-sized inorganic nanoparticles uniformly dispersed in a polymer matrix. These inorganic nanoparticles are also known as "nanofillers" or "nanoinclusions" because they operate as "additives" to improve polymer performance (Vaisman et al., 2007). Nanofillers are added to polymers at a 1%–10% rate (in mass). They are combined with standard fillers and additives, and typical reinforcing fibers (Figure 4.1).

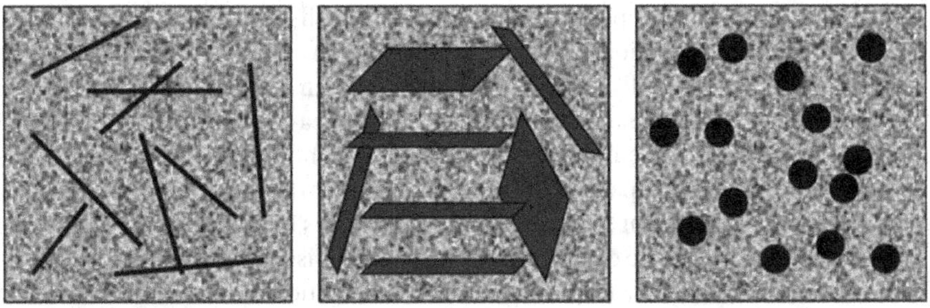

FIGURE 4.1 Primary forms of nanofillers (Gurun, 2010).

4.2.1 Inorganic Nanofillers

Organic–inorganic nanocomposites are composites of organic polymers and inorganic nanoscale building units. The combined benefits of inorganic materials such as rigidity and thermal stability into organic polymers (e.g., flexibility, dielectric, elasticity, and processability) possess specific nanofiller features, resulting in materials with better qualities (Haldorai et al., 2012). Metals and metal alloys (Au, Ag, Cu, Pt, Fe, and CoPt), metal oxides (TiO_2, SiO_2, and ferric oxide), semiconductors (PbS, CdS, CdSe, CdTe, and ZnO), clay minerals (montmorillonite, vermiculite, hectorite, and $CaCO_3$), and carbon-based materials (e.g., carbon nanotube (CNT), graphite, and carbon nanofiber) are commonly used inorganic fillers in the polymer matrix (S. Li et al., 2010). In addition, depending on applications, the choice of polymer also varies as conducting polymers (polyaniline (PANI), polypyrrole), industrial plastics (nylon 6, nylon MXD6, polypropylene (PP), polyimide), and transparent polymers (polymethyl methacrylate (PMMA), polystyrene (PS)). Popular applications of PINs include optical applications, magnetic applications, mechanical applications, catalysis, electrochemical applications, electrical and thermal applications, and biomedical applications.

4.2.2 Factors Affecting Nanofillers

Depending on the product, there are numerous parameters to consider while developing polymer nanocomposites. Regardless of the type of design, certain parameters such as synthesis technique, temperature, pressure, and time must be addressed when designing a novel polymer nanocomposite material. Furthermore, matrix factors such as type of polymer, surface nature and chemistry of the polymer, volume or weight fraction, and structure are all taken into account, as well as filler shape, size, and type; filler surface nature; filler volume or weight fraction. All of these elements fall into three categories: aspect ratio (AR), interface, and orientation.

A critical criterion for nanoparticles used as fillers in transparent polymers is their size. Particle sizes of fewer than 40 nm are necessary for transparent nanocomposites. This is because when particle size rises, the intensity of scattered light increases. As a result, the polymer nanocomposite will have a turbid appearance, limiting its optical applications (Ajayan et al., 2003). Furthermore, when particle size decreases, a homogenous distribution of nanomaterial in the polymer matrix can be formed, increasing interfacial area. A large surface area of small particles can also cause aggregation, which reduces particle distribution uniformity. Many applications require a homogeneous dispersion of the inorganic filler on the nanoscale, especially when it comes to structuring, such as thin films (Althues et al., 2007).

Surface area is one of the characteristics that distinguishes nanofillers from microfillers. It is this region that is responsible for the "interaction" between the matrix and the filler. Nanofillers have a disproportionately large surface area. The entire region in contact with the polymer (interfacial area) is a consequence of the total surface area when they are dispersed in polymers. This indicates that the filler's impact on the polymer will improve as the surface area increases (Akpan et al., 2018).

Another property of nanofillers that are critical for the creation of polymer nanocomposite is the AR. Nanofillers exist in one of three sizes: 1D, 2D, or 3D. These fillers, on the other hand, appear in a variety of shapes and sizes, including rods, fibers, tubes, platelets, wires, disks, flakes, spheroids, and ellipsoids. This geometric dimension is generally defined by the AR. It is defined as the ratio of filler's longest dimension by its shortest dimension. All viable shapes' dimensions can be simplified to AR as it influences the interfacial area per volume of filler to a large extent. It is also important for determining the physical and mechanical properties of polymer nanocomposites, as well as manufacturing polymer nanocomposites with the appropriate qualities (Gulotty et al., 2013).

4.3 INORGANIC NANOFILLERS APART FROM METAL, METAL OXIDES, AND CARBON-BASED NANOFILLERS

Inorganic nanofillers are highly potentially active fillers that are primarily oriented to structural, electrical and electronic properties and to enhance physical, rheological, thermal, and structural properties of polymers. In polymer-based nanocomposites, inorganic nanofillers are essential. The first type of nanofillers is the pure form of metal-based nanoparticles, also known as metal nanoparticles (e.g., silver, copper, gold, titanium, platinum, zinc, magnesium, iron, and alginate nanoparticles). Other inorganic fillers used in the polymer matrix include metal oxide nanoparticles (nanoparticles) such as titanium dioxide, silver oxide, zinc oxide, and so on. Nanofillers that are doped metal/metal oxide/metal nanoparticles are considered an emerging class of nanofiller. Furthermore, cellulose nanofillers, metal sulfide nanofillers such as AgS, CuS, FeS nanoparticles, metal organic frameworks (MOFs) such as Zn-based MOF, Cu-based, Mn-based MOF, nanoclays, and silicate nanofillers have attracted a lot of attention due to their promising features and applications across several fields (Figure 4.2).

4.3.1 Cellulose Nanofillers

Cellulose is a naturally occurring polysaccharide with units connected by -1,4-glycosidic bonds (Seddiqi et al., 2021). Microfibrils in wood and plant cell walls, algal

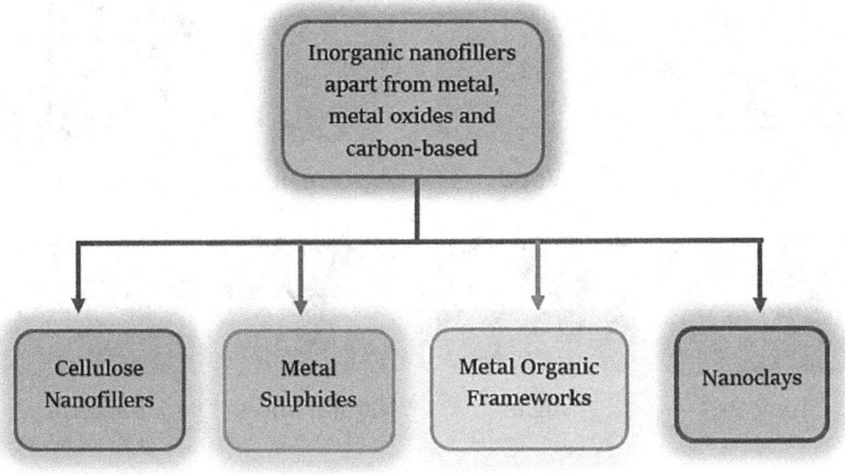

FIGURE 4.2 Various types of inorganic filler apart from metal, metal oxides, and carbon-based.

tissues, and tunicate epidermal cell membranes are sources of this unbranched biodegradable organic polymer most generally found in the kingdom plantae. It can also be produced by bacteria in the form of nanofiber networks (McCarthy, 2004). The presence of hydroxyl groups and strong hydrogen bonds distinguishes cellulose as a natural biopolymer with good physical and mechanical properties. Furthermore, cellulose is made comprised of amorphous and crystalline regions, the existence of which influences the material's modification and characteristics (Trache et al., 2020). Cellulose nanofillers are micron-sized crystal units having a whisker-like, rectangular cross-sectional area in the nanoscale, with varying dimensions depending on the cellulose source.

Cellulose is the most abundant biopolymer available in a wide variety of resources derived from sustainable raw materials such as plants and microorganisms to generate nanofillers for nanocomposites. Wood, cotton, sugarcane, ramie, sisal, flax, and wheat straw are examples of vegetable lignocellulosic materials, which are heterogeneous complexes of carbohydrate polymers (cellulose and hemicellulose) and lignin that produce cellulose. Acetobacters are microorganisms that produce cellulose by absorbing glucose, sugar, glycerol, or other organic substrates and convert them to pure cellulose in bacterial cellulose. Algae (autotrophs that live in both freshwater and marine habitats) form cellulose microfibrils within their polysaccharide cell walls made up of xylem and mannans and are mostly found in green algae (Cladophora and Valonia), brown algae (Phaeophyta), and red algae (Rhodophyta). Tunicates are sea animals with a mantle or tunic, which is integumentary tissue made up of cellulose microfibrils embedded in a protein matrix, and are known to perform cellulose production in their mature state, which are common cellulose sources (Santos et al., 2016) (Figure 4.3).

Numerous organic and inorganic polymers have used nanofillers extracted from cellulose as reinforcing materials. Because of the extremely large surface area, compatibility with natural polymers, good crystallinity of chains (depends on the source

FIGURE 4.3 Various sources of cellulose sources (Vigneshwaran et al., 2019).

and processing of cellulose), and strong mechanical attributes of cellulose, these nanofillers have reported to be efficient (Abbate dos Santos & Bruno Tavares, 2015; Babaee et al., 2015). Furthermore, the incorporation of cellulose-based nanofillers improves the nanocomposite's barrier, thermal, and mechanical properties, among other aspects. However, there have been some challenges to overcome in order to achieve excellent performance while using cellulose-based nanofillers in polymers to produce nanocomposites, such as efficient particle dispersion in the matrix, compatibility of nano-reinforcement in the matrix, and the development of improved manufacturing techniques for these materials (Santos et al., 2016).

Mireya Matos et al. produced waterborne coatings using cellulose whiskers (rigid rod microcrystalline cellulose nanoparticles) synthesized from hydrolyzed tunicate mantles (aqueous suspension of microcrystalline animal cellulose fillers in an epoxy matrix (Dufresnes & Gerard, 2001)). Cherian et al. developed cellulose nanocomposites with nanofibers isolated from pineapple leaf fibers and polyurethane (PU) to produce implants for medical application (Cherian et al., 2011). Freire et al described the preparation of edible films that can serve as protective coatings on food with modified surface of cellulose fibers by acylation using fatty acid to obtain nanocomposites with polyethylene (Freire et al., 2008).

4.3.2 Metal Sulfide Nanofillers

For the fabrication of polymer nanocomposites, several nanomaterials and nanofillers have been explored of which conjugated conducting polymers (CPs) have made steady development in applications such as chemical sensors, electrochemical supercapacitors, electrochromic and photovoltaic devices, light-emitting diodes, and batteries over the last decade (Shimpi et al., 2015). Researchers have been consistently working to improve the physical and structural features of CPs in these years. Important aspects like as polymerization method, oxidant type, pH, and temperature conditions during polymerization influence the physical and structural properties of CPs. However, in recent years, nanomaterials have been used to try to tune the properties of CPs (Sen et al., 2016). The eventual implementation of the resulting nanocomposites is often a result of the synergistic interaction between the CP and the nanomaterial, so identifying the appropriate nanomaterial for a CP is significant.

Due to the prevalence of definite properties at nanoscopic levels, metal-based chalcogenides are regarded the most promising semiconductor due to their excellent fluorescence; narrow optical band gap; and outstanding magnetic, thermal, mechanical, and structural stability. Quantum dots, nanocrystals, and metal-based chalcogenide nanoparticles with extremely complex structures, such as CdSe, PbS, CdSe-CdTe, CdSe-ZnTe, CdTe-CdSe, have also been proposed as biomaterials for use in biosensors, drug delivery, biolabeling, bioimaging, and diagnostics (Li & Wong, 2017). Patrick J. Gray fabricated polymer nanocomposites as effective food simulants from low-density polyethylene and CdSe quantum dots (QDs) (Gray et al., 2018).

Metallic sulfides containing chalcogenide sulfur such as AgS, CuS, FeS, and ZnS have recently attracted a lot of interest in the medical profession since they are significantly connected with a toxic-free metal (Dahoumane et al., 2016). The semiconducting nanofiller silver sulfide (Ag_2S) is widely used in the fabrication of photo-electrochemical devices and solar cells (Yeole et al., 2016). Ag2S nanoparticles were discovered to be an excellent material for demonstrating the prevention of microbiological development. Imaging, labeling, tissue imaging, diagnostics, and photodynamic treatment are a couple of applications utilizing Ag_2S functionalized QDs in the field of bioimaging and diagnostics. Also, the tracking and design of cells in vivo is another exciting use of Ag_2S QDs (Argueta-Figueroa et al., 2017). Kumari et al. (2014) discovered that Ag_2S nanoparticles have a reasonable antibacterial effect as well.

However, Goel et al investigated CuS nanoparticles, which are significantly developing as a viable phase for biosensing, photothermal healing, biomolecular imaging, and drug delivery. CuS nanoparticles played an important role in both in vitro and in vivo studies. CuS nanoparticles and derivatives have been widely used in molecular detection technology, including DNA detection, metabolites (glucose) detection, and food-borne pathogen detection, among others. CuS nanoparticles are gaining prominence in the field of biosensing due to their high conductivity and capacity, which facilitate electron transfer interactions with molecules (Goel et al., 2014). In addition, a CuS thin film-based immunosensor for the detection of anthropological immunoglobulin A (IgA) antibodies in serum was developed, using a goat antihuman IgA antimediator immobilized on the CuS thin film surface (Attarde & Pandit, 2020). Lokanatha Reddy synthesized zinc sulfide nanoparticles (ZnS NPs) (P. Lokanatha Reddy et al., 2019) and cadmium sulfide nanoparticles (CdS NPs) (Peram Lokanatha Reddy et al., 2020) from orange peel extract and fabricated a PVA-ZnS nanocomposite with improved dielectric characteristics that might be used in energy storage devices.

4.3.3 Metal Organic Frameworks

MOFs are novel crystalline nanoporous material defined as the self-assembly of metallic ions that serve as coordination centers and organic/inorganic-based ligands, which serve as linkers in metallic centers (Keskin & Kızılel, 2011). One of the most exciting major innovations in nanoporous science is porous coordination, often known as hybrid organic–inorganic coordination networks. MOFs have some unique features that make them more prolific, such as high porosity and surface area, which range from 1000 to 10,000 m^2/g, which is significantly greater than normal porous materials (Arenas-Vivo et al., 2019). Due to their outstanding drug-loading capacity, informal functionalization, optimal biodegradability, and strong biocompatibility, MOFs have been intensively investigated for drug delivery systems in modern development. Due to pore size adjustment and the possibility of adjusting functional groups, MOFs are considered to be the best material for drug delivery. Because of its remarkable properties, Mg(H4gal) MOF material (Hidalgo et al., 2017), Ag-based MOF material (Arenas-Vivo et al., 2019), and Zn-based MOFs (Briones et al., 2016) have attracted a lot of attention, especially as antibacterial agents and antioxidant carriers.

4.3.4 Nanoclays (Layered Silicate Nanofillers)

Layered silicates or clay minerals with residues of metal oxides and organic molecules that make up a class of materials are known as nanoclays. Clay is a natural, fine-grained ceramic material, with hydrous aluminum phyllosilicates comprising iron, magnesium, alkali metals, alkaline earths, and other cations (Kotal & Bhowmick, 2015). Research studies have revealed that the dispersion of exfoliated clays in polymer results in a significant increase in stiffness, fire retardancy, and barrier characteristics, even at low nanoparticle volume fractions (Zaïri et al., 2011). Because of their lamellar structure and high specific surface area (750 m^2/g), clays have been discovered to be efficient polymer reinforcing fillers. Due to their inexpensive cost, swelling qualities, and high cation exchange capabilities, smectite clays are the preferred choice for the manufacture of polymer nanocomposites. Montmorillonite, saponite, laponite, hectorite, sepiolite, and vermiculite are some of these clays (Schadler, 2003). Because of its extensive availability, well-known intercalation/exfoliation chemistry, high surface area, and reactivity, montmorillonite is the most extensively employed clay in polymer nanocomposites. The hydrated exchangeable cations inhabit the spaces between the lattices in montmorillonite (MMT), which is made up of two tetrahedral silica sheets with an alumina octahedral sheet in the middle (2:1 layered structure) (Chen, 2004).

The clay layers with thickness of 1 nm and a width of 100–500 nm, resulting in platelets with a high AR (Chen, 2004) are stacked to create a regular van der Waals gap between them, referred as an interlayer or gallery. Within the layers, isomorphic substitution occurs, resulting in generation of negative charges that are balanced by alkali and alkaline earth cations. Inside the galleries are sodium or calcium ions (Na^+ or Ca^{2+}). Cation exchange capacity (CEC) is a measure of cation exchange capacity. MMT is a hydrophilic mineral. Since most polymers are hydrophobic, the clay surface must be altered to produce the organophilic clay. This is commonly achieved by exchanging cations in the gallery using alkyl phosphonium or alkylammonium salts (Kotal & Bhowmick, 2015).

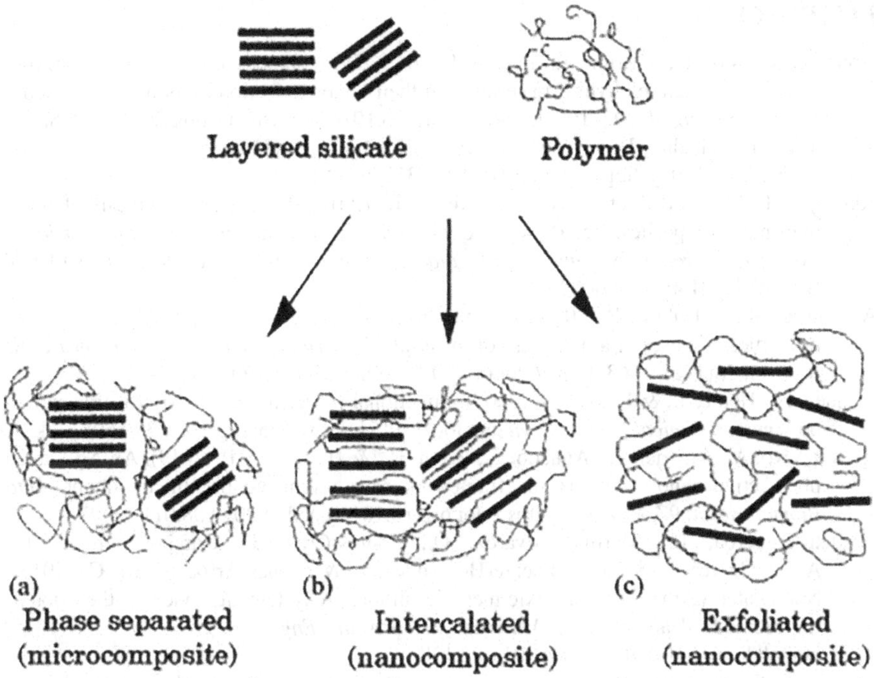

FIGURE 4.4 Different types of layered silicates (Alexandre & Dubois, 2000).

Based on the degree of separation of the clay layers, polymer–clay composites can be categorized into three types: conventional composites, intercalated nanocomposites, and exfoliated nanocomposites (Fu et al., 2019). The basal plane spacing d001 is the distance between a plane in one unit layer and its corresponding plane in the next unit layer; if the polymer does not enter the galleries, d001 of clay remains the same, and the composite is "conventional." The composite is "intercalated" when the polymer enters the galleries and causes an increase in d001 while the clay layers remain stacked. The composite is "exfoliated" when the clay layers are pulled entirely apart to generate a chaotic array. The basal planar spacing d001 is a reference to intercalation and exfoliation degrees, but it does not immediately reveal the amount of clay that is exfoliated. An intercalated clay's normal basal plane spacing is on the order of 1–4 nm (Tolle & Anderson, 2002) (Figure 4.4).

4.4 CONCLUSION

Surface area, AR, filler size, and homogenic dispersion of fillers are all significant aspects in developing new and better polymer-reinforced inorganic nanofillers. Cellulose is a naturally occurring biopolymer that can be developed from sustainable raw materials and used in edible coatings. Nanoclays, with their lamellar structure and high specific surface and stiffness, are another inorganic filler that has been intensively studied. Improvements in quality and the introduction of new fillers have the potential to broaden the variety of applications.

REFERENCES

Abbate dos Santos, F., & Bruno Tavares, M. I. (2015). Development of biopolymer/cellulose/silica nanostructured hybrid materials and their characterization by NMR relaxometry. *Polymer Testing*, *47*, 92–100. https://doi.org/10.1016/j.polymertesting.2015.08.008.

Ajayan, P. M., Schadler, L. S., & Braun, P. V. (Eds.). (2003). *Nanocomposite Science and Technology*. Wiley. https://doi.org/10.1002/3527602127

Akpan, E. I., Shen, X., Wetzel, B., & Friedrich, K. (2018). Design and synthesis of polymer nanocomposites. In *Polymer Composites with Functionalized Nanoparticles: Synthesis, Properties, and Applications*. Elsevier Inc. https://doi.org/10.1016/B978-0-12-814064-2.00002-0

Alexandre, M., & Dubois, P. (2000). Polymer-layered silicate nanocomposites: Preparation, properties and uses of a new class of materials. *Materials Science and Engineering: R: Reports*, *28*(1–2), 1–63. https://doi.org/10.1016/S0927-796X(00)00012-7

Althues, H., Henle, J., & Kaskel, S. (2007). Functional inorganic nanofillers for transparent polymers. *Chemical Society Reviews*, *36*(9), 1454. https://doi.org/10.1039/b608177k

Arenas-Vivo, A., Amariei, G., Aguado, S., Rosal, R., & Horcajada, P. (2019). An Ag-loaded photoactive nano-metal organic framework as a promising biofilm treatment. *Acta Biomaterialia*, *97*, 490–500. https://doi.org/10.1016/j.actbio.2019.08.011

Argueta-Figueroa, L., Martínez-Alvarez, O., Santos-Cruz, J., Garcia-Contreras, R., Acosta-Torres, L. S., de la Fuente-Hernández, J., & Arenas-Arrocena, M. C. (2017). Nanomaterials made of non-toxic metallic sulfides: A systematic review of their potential biomedical applications. *Materials Science and Engineering: C*, *76*, 1305–1315. https://doi.org/10.1016/j.msec.2017.02.120

Attarde, S. S., & Pandit, S. V. (2020). Anticancer potential of nanogold conjugated toxin GNP-NN-32 from Naja naja venom. *Journal of Venomous Animals and Toxins Including Tropical Diseases*, *26*. https://doi.org/10.1590/1678-9199-jvatitd–2019-0047

Babaee, M., Jonoobi, M., Hamzeh, Y., & Ashori, A. (2015). Biodegradability and mechanical properties of reinforced starch nanocomposites using cellulose nanofibers. *Carbohydrate Polymers*, *132*, 1–8. https://doi.org/10.1016/j.carbpol.2015.06.043

Briones, D., Fernández, B., Calahorro, A. J., Fairen-Jimenez, D., Sanz, R., Martínez, F., Orcajo, G., Sebastián, E. S., Seco, J. M., González, C. S., Llopis, J., & Rodríguez-Diéguez, A. (2016). Highly active anti-diabetic metal–organic framework. *Crystal Growth & Design*, *16*(2), 537–540. https://doi.org/10.1021/acs.cgd.5b01274

Chen, B. (2004). Polymer–clay nanocomposites: an overview with emphasis on interaction mechanisms. *British Ceramic Transactions*, *103*(6), 241–249. https://doi.org/10.1179/096797804X4592

Cherian, B. M., Leão, A. L., de Souza, S. F., Costa, L. M. M., de Olyveira, G. M., Kottaisamy, M., Nagarajan, E. R., & Thomas, S. (2011). Cellulose nanocomposites with nanofibres isolated from pineapple leaf fibers for medical applications. *Carbohydrate Polymers*, *86*(4), 1790–1798. https://doi.org/10.1016/j.carbpol.2011.07.009

Ciprari, D., Jacob, K., & Tannenbaum, R. (2006). Characterization of polymer nanocomposite interphase and its impact on mechanical properties. *Macromolecules*, *39*(19), 6565–6573. https://doi.org/10.1021/ma0602270

Dahoumane, S. A., Wujcik, E. K., & Jeffryes, C. (2016). Noble metal, oxide and chalcogenide-based nanomaterials from scalable phototrophic culture systems. *Enzyme and Microbial Technology*, *95*, 13–27. https://doi.org/10.1016/j.enzmictec.2016.06.008

DufresneS, M., & Gerard, J.-F. (2001). New waterborne epoxy coatings based on cellulose nanofillers. *Macromolecular Symposia*, *169*, 211–222.

Freire, C. S. R., Silvestre, A. J. D., Neto, C. P., Gandini, A., Martin, L., & Mondragon, I. (2008). Composites based on acylated cellulose fibers and low-density polyethylene: Effect of the fiber content, degree of substitution and fatty acid chain length on final properties. *Composites Science and Technology*, *68*(15–16), 3358–3364. https://doi.org/10.1016/j.compscitech.2008.09.008

Fu, S., Sun, Z., Huang, P., Li, Y., & Hu, N. (2019). Some basic aspects of polymer nanocomposites: A critical review. *Nano Materials Science*, *1*(1), 2–30. https://doi.org/10.1016/j.nanoms.2019.02.006

Goel, S., Chen, F., & Cai, W. (2014). Synthesis and biomedical applications of copper sulfide nanoparticles: from sensors to theranostics. *Small*, *10*(4), 631–645. https://doi.org/10.1002/smll.201301174

Gray, P. J., Hornick, J. E., Sharma, A., Weiner, R. G., Koontz, J. L., & Duncan, T. V. (2018). Influence of different acids on the transport of cdse quantum dots from polymer nanocomposites to food simulants. *Environmental Science & Technology*, *52*(16), 9468–9477. https://doi.org/10.1021/acs.est.8b02585

Gulotty, R., Castellino, M., Jagdale, P., Tagliaferro, A., & Balandin, A. A. (2013). Effects of functionalization on thermal properties of single-wall and multi-wall carbon nanotube–polymer nanocomposites. *ACS Nano*, *7*(6), 5114–5121. https://doi.org/10.1021/nn400726g

Gurun, B. (2010). Deformation studies of polymers and polymer/clay nanocomposites. *Materials Science*. https://doi.org/https://smartech.gatech.edu/handle/1853/37118

Haldorai, Y., Shim, J.-J., & Lim, K. T. (2012). Synthesis of polymer–inorganic filler nanocomposites in supercritical CO_2. *The Journal of Supercritical Fluids*, *71*, 45–63. https://doi.org/10.1016/j.supflu.2012.07.007

Hidalgo, T., Cooper, L., Gorman, M., Lozano-Fernández, T., Simón-Vázquez, R., Mouchaham, G., Marrot, J., Guillou, N., Serre, C., Fertey, P., González-Fernández, Á., Devic, T., & Horcajada, P. (2017). Crystal structure dependent in vitro antioxidant activity of biocompatible calcium gallate MOFs. *Journal of Materials Chemistry B*, *5*(15), 2813–2822. https://doi.org/10.1039/C6TB03101C

Imai, Y. (2020). Encyclopedia of Polymeric Nanomaterials. In S. Kobayashi & K. Müllen (Eds.), *Encyclopedia of Polymeric Nanomaterials*. Springer. https://doi.org/10.1007/978-3-642-36199-9

Keskin, S., & Kızılel, S. (2011). Biomedical applications of metal organic frameworks. *Industrial & Engineering Chemistry Research*, *50*(4), 1799–1812. https://doi.org/10.1021/ie101312k

Kotal, M., & Bhowmick, A. K. (2015). Polymer nanocomposites from modified clays: Recent advances and challenges. *Progress in Polymer Science*, *51*, 127–187. https://doi.org/10.1016/j.progpolymsci.2015.10.001

Kumari, P., Chandran, P., & Khan, S. S. (2014). Retracted: Synthesis and characterization of silver sulfide nanoparticles for photocatalytic and antimicrobial applications. *Journal of Photochemistry and Photobiology B: Biology*, *141*, 235–240. https://doi.org/10.1016/j.jphotobiol.2014.09.010

Lan, T., Kaviratna, P. D., & Pinnavaia, T. J. (1995). Mechanism of clay tactoid exfoliation in epoxy-clay nanocomposites. *Chemistry of Materials*, *7*(11), 2144–2150. https://doi.org/10.1021/cm00059a023

Lan, T., & Pinnavaia, T. J. (1994). Clay-reinforced epoxy nanocomposites. *Chemistry of Materials*, *6*(12), 2216–2219. https://doi.org/10.1021/cm00048a006

Li, S., Meng Lin, M., Toprak, M. S., Kim, D. K., & Muhammed, M. (2010). Nanocomposites of polymer and inorganic nanoparticles for optical and magnetic applications. *Nano Reviews*, *1*(1), 5214. https://doi.org/10.3402/nano.v1i0.5214

Li, Z., & Wong, S. L. (2017). Functionalization of 2D transition metal dichalcogenides for biomedical applications. *Materials Science and Engineering: C*, *70*, 1095–1106. https://doi.org/10.1016/j.msec.2016.03.039

McCarthy, S. P. (2004). Biodegradable polymers. In *Plastics and the Environment* (pp. 359–377). John Wiley & Sons, Inc. https://doi.org/10.1002/0471721557.ch9

Mohammed, D., Guillaume, E., & Chivas-Joly, C. (2011). Properties of Nanofillers in Polymer. In *Nanocomposites and Polymers with Analytical Methods*. InTech. https://doi.org/10.5772/21694

Reddy, P. Lokanatha, Deshmukh, K., Chidambaram, K., Ali, M. M. N., Sadasivuni, K. K., Kumar, Y. R., Lakshmipathy, R., & Pasha, S. K. K. (2019). Dielectric properties of polyvinyl alcohol (PVA) nanocomposites filled with green synthesized zinc sulphide (ZnS) nanoparticles. *Journal of Materials Science: Materials in Electronics*, *30*(5), 4676–4687. https://doi.org/10.1007/s10854-019-00761-y

Reddy, Peram Lokanatha, Deshmukh, K., Kovářík, T., Nambiraj, N. A., & Shaik, K. P. (2020). Green chemistry mediated synthesis of cadmium sulphide/polyvinyl alcohol nanocomposites: Assessment of microstructural, thermal, and dielectric properties. *Polymer Composites*, *41*(5), 2054–2067. https://doi.org/10.1002/pc.25520

Santos, F. A. dos, Iulianelli, G. C. V., & Tavares, M. I. B. (2016). The use of cellulose nanofillers in obtaining polymer nanocomposites: properties, processing, and applications. *Materials Sciences and Applications*, *07*(05), 257–294. https://doi.org/10.4236/msa.2016.75026

Schadler, L. S. (2003). Polymer-based and polymer-filled nanocomposites. In *Nanocomposite Science and Technology* (pp. 77–153). Wiley. https://doi.org/10.1002/3527602127.ch2

Seddiqi, H., Oliaei, E., Honarkar, H., Jin, J., Geonzon, L. C., Bacabac, R. G., & Klein-Nulend, J. (2021). Cellulose and its derivatives: Towards biomedical applications. *Cellulose*, *28*(4), 1893–1931. https://doi.org/10.1007/s10570-020-03674-w

Sen, T., Shimpi, N. G., & Mishra, S. (2016). Room temperature CO sensing by polyaniline/Co_3 O_4 nanocomposite. *Journal of Applied Polymer Science*, *133*(42). https://doi.org/10.1002/app.44115

Shimpi, N. G., Hansora, D. P., Yadav, R., & Mishra, S. (2015). Performance of hybrid nanostructured conductive cotton threads as LPG sensor at ambient temperature: preparation and analysis. *RSC Advances*, *5*(120), 99253–99269. https://doi.org/10.1039/C5RA16479F

Sundarram, S., Kim, Y.-H., & Li, W. (2015). Preparation and characterization of poly(ether imide) nanocomposites and nanocomposite foams. In *Manufacturing of Nanocomposites with Engineering Plastics* (pp. 61–85). Elsevier. https://doi.org/10.1016/B978-1-78242-308-9.00004-5

Tolle, T. B., & Anderson, D. P. (2002). Morphology development in layered silicate thermoset nanocomposites. *Composites Science and Technology*, *62*(7–8), 1033–1041. https://doi.org/10.1016/S0266-3538(02)00039-8

Trache, D., Tarchoun, A. F., Derradji, M., Hamidon, T. S., Masruchin, N., Brosse, N., & Hussin, M. H. (2020). Nanocellulose: From fundamentals to advanced applications. *Frontiers in Chemistry*, *8*. https://doi.org/10.3389/fchem.2020.00392

Vaisman, L., Wachtel, E., Wagner, H. D., & Marom, G. (2007). Polymer-nanoinclusion interactions in carbon nanotube based polyacrylonitrile extruded and electrospun fibers. *Polymer*, *48*(23), 6843–6854. https://doi.org/10.1016/j.polymer.2007.09.032

Vigneshwaran, N., Bharimalla, A. K., Arputharaj, A., & Patil, P. G. (2019). Nanocellulose from agro-residues and forest biomass for pulp and paper product. In *Nanoscience for Sustainable Agriculture* (pp. 355–372). Springer International Publishing. https://doi.org/10.1007/978-3-319-97852-916

Yeole, B., Sen, T., Hansora, D., & Mishra, S. (2016). Polypyrrole/metal sulphide hybrid nanocomposites: synthesis, characterization and room temperature gas sensing properties. *Materials Research*, *19*(5), 999–1007. https://doi.org/10.1590/1980-5373-MR-2015-0502

Zaïri, F., Gloaguen, J. M., Naït-Abdelaziz, M., Mesbah, A., & Lefebvre, J. M. (2011). Study of the effect of size and clay structural parameters on the yield and post-yield response of polymer/clay nanocomposites via a multiscale micromechanical modelling. *Acta Materialia*, *59*(10), 3851–3863. https://doi.org/10.1016/j.actamat.2011.03.009

5 To Comprehend the Applications of Inorganic Nanofiller-Derived Polymers in Packaging

Prakash Chander Thapliyal
CSIR-Central Building Research Institute

Neeraj Kumar
Indian Institute of Science

CONTENTS

5.1 INTRODUCTION

Polymer nanocomposites are extensively grown in the last few decades due to their remarkable improvement in electrical, structural, and mechanical properties. The use of nanofillers enhances the properties of host materials (polymer) as well as reduces the cost of polymeric products (Bhattacharya, 2016). However, the incorporation or addition of nanofillers into polymers leads to potential application in different fields including photocatalysts, energy storage devices, sensors/biosensors, drug delivery, and packaging (Kim et al., 2010; Sharma et al., 2018). A polymer composite consists of polymers as a matrix and stuff inorganic fillers that undergo a remarkable change in stress or modulus at strain over the virgin polymer. A broad

DOI: 10.1201/9781003279389-5

range of fillers are used for a reinforcement polymer matrix. Several traditional fillers are widely used as fillers in a polymer matrix such as glass fibers, calcium carbonate particles, talc, and carbon black in the micrometer range. Thus, the lack of interfacial interaction between the polymeric matrix and the filler directs to pathetic interfacial adhesion and results in failure (Bhattacharya, 2016). However, nanofillers lead the reinforcing of the polymer matrix predominantly due to the small particle size, and high reinforcement was observed in a particle dimension array of 100 nm and below. A good reinforcement can be achieved by using a small loading of nanofillers (Ain et al., 2021; Gheller et al., 2016; Edwards, 1990; Njuguna et al., 2008). Several methods of using nanofillers including clay, carbon nanotubes, silica, and nanoparticles (NPs) have been established to be effective ways to improve electrical, mechanical, and thermal properties of biopolymers. Nanofillers can be of three major kinds comprising zero-dimensional (silica NPs, metal NPs, metal oxide, quantum dots, etc.), one-dimensional (carbon nanotubes, nanorods (Shah et al., 2019), polymer nanowires, nanofibers, etc.), and two-dimensional (graphene and related materials, nano-clay, molybdenum disulfide, MXene, etc.) (Figure 5.1) (Thapliyal & Kumar, 2021; Fu et al., 2019; Shah et al., 2019; Imai, 2014). Nanofiller-incorporated polymer composites demonstrate outstanding mechanical and tribological properties. Several approaches proposed in the direction of improving mechanical along with barrier properties of biopolymers including by means of the crosslinkers, inorganic nanofillers, or blends with different polymers (Ain et al., 2021; Pires et al., 2021).

Nowadays, the most frequently used packing materials are created on or after the fossil fuels and are almost undegradable. Biodegradable packaging materials gained enormous attention in the packaging industry, which can help reduce environmental problems related to packaging waste (Sorrentino et al., 2007). Therefore, the novel synthesis and processing methods are required to achieve safe and green active packaging. In recent years, biopolymers have gained significant attention in industrial and

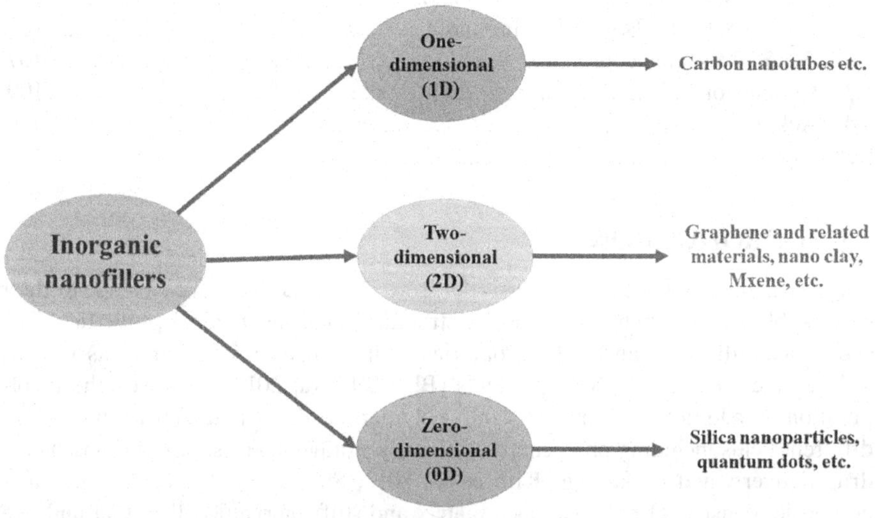

FIGURE 5.1 Schematic representation of inorganic nanofillers.

scientific research. Especially, polymers obtained from natural resources are gaining more attention to develop environmentally friendly materials. These materials can be very helpful in the areas of food coating, food packaging, and edible films for food plus encapsulation matrices for functional foods (Omerovic et al., 2021; Fabra et al., 2014). Packaging is an important aspect to control or to uphold high-quality and healthy foods. Thus, these materials provide a new platform for improving the food quality and shelf life while a minimized overall carbon footprint to food packaging (Basavegowda & Baek, 2021; Priyadarshi et al., 2021). Food packaging is highly important due to the raising demands for preservative-free and less-processed food. Polymers extracted from natural resources are polysaccharides (containing starch, chitin, chitosan, alginates, and cellulose), proteins (wheat, soy, corn, etc.), and lipids (bee wax and free fatty acids), while synthetic polymers are made of starting materials such as oil-based or biomass-derived monomers, natural fats, oils, sugar, and starch. However, some polymers are produced from the genetically modified microorganisms like polyhydroxyalcanoates or bacterial cellulose (Figure 5.2) (Gross & Kalra, 2002; Vroman & Tighzert, 2009; Fabra et al., 2014; Basavegowda & Baek, 2021; Priyadarshi et al., 2021).

Natural polymers are biodegradable and can be used as a substitute of synthetic polymers observed on or after the non-renewable petroleum resources. These biodegradable or biopolymers can be easily degraded into organic products, inorganic substances, methane, water, or biomass through the enzymatic reaction. Biodegradable materials boast affinity to degrade entirely by the accomplishment of microorganisms by secretion of enzymes in the natural environment. Degradation can also be executed through chemical, thermal, mechanical, biological, and radiation processes (Folino et al., 2020; Alim et al., 2022). Hence, the correct selection of materials is of prime importance for packaging technologies, and it is probable to maintain the quality of the product as well as freshness all through required time for its consumption along with commercialization. On applying the protective coating for packaging in the food industry, the freshness and shelf life of the product can be improved or increased (Sorrentino et al., 2007). In this chapter, we discussed the biopolymer or biodegradable materials and their composites with inorganic nanofillers and application in packaging.

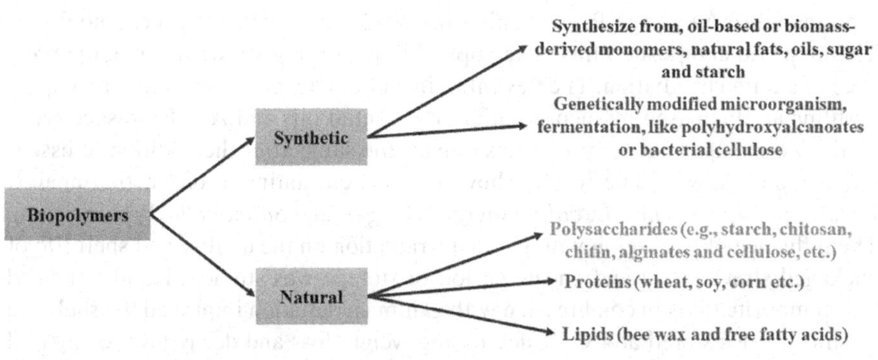

FIGURE 5.2 Representation of synthetic and natural polymers.

5.2 CHITOSAN

Chitosan is produced from chitin by a deacetylation process. Chitosan is a water-soluble cationic polymer containing a positive charge on its amino groups. This cationic polymer has a wide application due to its functional properties including ease of modification, low toxicity, antibacterial activity, high water permeability, biocompatibility, and biodegradability. Chitosan is soluble at a mild acidic condition at a pH less than 6.0–6.5, such as formic acid, acetic acid, lactic acid, and hydrochloric acid (Souza et al., 2020; Yadav et al., 2021; Garavand et al., 2022).

5.2.1 Nanofiller in Chitosan

Chitosan-related composites play a significant part in food packing and can be applied as films or edible coatings. Because of its high cost and lower performance (i.e., thermal and mechanical properties, and low resistance against water vapor), when compared to commercial petroleum-based polymers or plastic, the application of such biopolymer still limited. Several attempts have been made to improve the performance of biopolymers (Souza et al., 2020; Yadav et al., 2021; Garavand et al., 2022). In the food industry, chitosan has been widely used for coating fruit and meat products to minimize the water loss and food deterioration, as well as interruption in the ripening of fruits. The stability and antimicrobial activity of chitosan can be improved by using nanofillers such as metal NPs, metal oxide, and conducting polymers. Al-Naamani et al. (2016) developed chitosan–zinc oxide nanocomposites coating to improve antibacterial properties of the polyethylene. Wang et al. (2019) reported carboxymethyl chitosan–ZnO and sodium alginate-based biodegradable films for packaging. The carboxymethyl chitosan–ZnO NPs were synthesized by using the direct precipitation method, which demonstrated an average diameter of 100 nm. The incorporation of carboxymethyl chitosan–ZnO NPs improved the tensile strength in addition to water vapor resistance. Composite films demonstrate unique antibacterial activity against *E. coli* and *S. aureus*. Zhang et al. (2019) reported chitosan and TiO_2 and black plum peel extract-based multi-functional food packaging films. The improvement in the UV–Vis light, barrier properties against water, and mechanical strength of chitosan films was observed on incorporation of the black plum peel extract and TiO_2. Shankar et al. (2021) reported silver NPs (AgNPs) and chitosan-based films blended with essential oils. The prepared composite films were applied to packaging strawberries in the presence of gamma irradiation. The flexibility in addition to tensile strength of composite films are increased by incorporation of essential oils and AgNPs, respectively. While water vapor barrier properties remain the same after the addition of essential oils and AgNPs. The results show an excellent antimicrobial action against *Listeria monocytogenes*, *E. coli*, *Aspergillus niger*, and *Salmonella* Typhimurium. The influence of packaging with gamma irradiation on the quality and shelf life of packaged strawberries and on the period of storage was studied. Results showed that composite films in combination with gamma irradiation improved the shelf life of strawberries stored at 4°C by decreasing weight loss and decay level compared to control samples. Gamma irradiation helps enhance microbiological properties of the fruit by lowering the decay level. Essential oils, AgNPs, and gamma irradiation can stabilize the quality of strawberries by four days through preservation of the decay level. Kumar et al. (2018) developed chitosan, polyethylene glycol, gelatin, and AgNP-based hybrid nanocomposite films by using the solution casting method. The incorporation of AgNPs improves the mechanical properties and decreases the light transmittance in the visible region. Packaging of red grapes showed that the shelf life of fruits increased for extra two weeks in case of hybrid film. Wu et al. (2018) reported the chitosan-based packaging film with laponite-immobilized AgNPs to enhance the storage life of litchi, in which AgNPs were implanted in the interlayer of laponite due to the confinement effect. Moura et al. (2009) developed chitosan/tripolyphosphate NPs integrated in hydroxypropyl methylcellulose films. Mechanical and barrier properties of hydroxypropyl methylcellulose films improved significantly by addition of chitosan NPs. The tensile properties along with water vapor permeability of nanocomposite films also improved due to engagement of empty space in pores of the hydroxypropyl methylcellulose matrix by chitosan NPs. Chitosan-based composites demonstrate beneficial properties highly suitable for packaging materials and edible films for food packaging.

Zhang et al. (2017) and Xiao et al. (2004) prepared chitosan/TiO_2 nanocomposite films for food packaging and showed its effectiveness against pathogenic microbes in the packing of red grapes. Nano TiO_2 addition led to an increase in tensile strength (up to 89.64%) and elongation (up to 69.21%) as well as inhibition against four typical food-borne pathogenic microbes *E. coli* (Gram-negative bacteria), *S. aureus* (Gram-positive bacteria), *C. albicans* (fungi), and *A. niger* (molds).

Polyurethane–chitosan biodegradable food packaging composite films with 5% ZnO nanofiller showed improved tensile strength and stiffness along with enhanced antibacterial properties, barrier properties, and hydrophobicity of nanocomposite films. The shelf life of carrot pieces wrapped with a composite film extended up to nine days (Sarojini et al., 2019).

Ediyilyam et al. (2021) effectively formed packing films using naturally derived biodegradable polymer materials chitosan (CH) with gelatin (GE) to lessen environmental damage by synthetic packing materials. Antioxidant and antibacterial AgNPs were used as reinforcing nanofillers in the CH-GE matrix, resulting in an increase in the opacity of films, and CH-GE-AgNPs films emerge as capable antimicrobial packaging material so as to expand the shelf life of carrot.

5.2.2 Chitin

Chitin is produced through shellfish exoskeletons that are abundant in nature. Chitin and its derivatives are synthesis from the exoskeletons of crabs and prawns, crustaceans and shrimps, and squid pens and mushrooms by simple mechanical treatment. Both biopolymers have unique properties including high flexibility, tensile strength, biocompatibility, biodegradable, antifungal and antibacterial properties, and low toxicity. Chitin is insoluble in most solvents due to its high crystallization and the presence of hydrogen bonds between acetamide groups (Chang et al., 2010a; Sugimoto et al., 2010; Qin et al., 2016; Moustafa et al., 2022).

5.2.2.1 Chitin Composites

Chitin can be applied as a flame-retardant and a reinforcing agent into polymer matrices to find sustainable composite materials (Sugimoto et al., 2010). Chang et al. (2010b) reported chitin NPs and glycerol-plasticized potato starch-based composites prepared by casting and evaporation. The uniform dispersion and good interaction were observed between fillers and the matrix at low loading levels, which improves the tensile strength, glass transition temperature, storage modulus, and water vapor barrier properties of nanocomposite films. However, high loading (>5 wt.%) of chitin NPs leads to a negative effect on these properties. Moustafa et al. (2022) proposed the chitin–thermoplastic polyurethane to form the composite blend. The chitin–thermoplastic polyurethane blend mixed with ZnO-doped SiO_2 NPs to chelate chitin plus enhance properties of thermoplastic polyurethane composites. Thermal and mechanical properties, flammability, water vapor, hydrophobicity, oxygen barrier properties, and antimicrobial activity of thermoplastic polyurethane nanocomposites improved by addition of NPs (at 5 wt.%). The loading level of NPs (i.e., 5%–7%) showed the excellent antimicrobial activities against Gram-positive and Gram-negative bacteria, in addition to pathogenic fungi. Qin et al. (2016) reported the antibacterial and physiochemical properties of chitin nano-whiskers on maize starch-based films. The tensile strength of starch-based film improved from 1.64 to 3.69 MPa ($P<0.05$) by addition of 1% chitin. The water vapor permeability of the composite films decreased with increasing the chitin nano-whiskers loading from 0% to 2%. Peak temperature, onset temperature, and gelatinization enthalpy of nanocomposite films were higher than those of pure starch films. Nanocomposite films showed excellent antimicrobial activity. Shankar et al. (2015) prepared chitin nanofibril-reinforced carrageenan composite films by using the solution casting method. Chitin nanofibrils were synthesized via acid hydrolysis of chitin, followed by high-speed homogenization along with sonication. Fourier transmission infrared spectroscopy (FTIR) results showed that the chemical structure of chitin remains the same after acid hydrolysis. However, crystallinity of chitin nanofibrils was observed higher than that of chitin. The crystallite size of chitin nanofibrils and chitin were 6.27 and 4.73 nm, respectively. The modulus and tensile strength of carrageenan films were enhanced significantly ($P<0.05$) after the incorporation of chitin nanofibrils of up to 5 wt.%. However, a slight decrement in water vapor permeability and transparency of nanocomposite films was found. Carrageenan and chitin nanofibril composite films demonstrate exceptional antibacterial activity against the Gram-positive food-borne pathogen *Listeria monocytogenes*.

5.2.3 Cellulose

Cellulose is a polysaccharide that is found in abundant in nature. It is a natural biopolymer containing linear chains with many $\beta(1\rightarrow4)$ linked D-glucose units. Cellulose and cellulose NPs are obtained from natural resources in various sizes and shapes depending on the source of cellulose and synthesis methods. The exceptional mechanical properties, nanosize, availability, low cost, high aspect ratio, and renewability of cellulose make it a suitable candidate as a reinforcing material. However, fibers or NPs of cellulose can only improve a few properties of host polymers but cannot improve other properties such as UV barrier and antibacterial properties. UV light produces various undesirable chemical reactions in food products such as loss in food quality and reduced shelf life. Therefore, inorganic nanomaterials or nanofillers are required to further improve the properties of host polymers (Olsson et al., 2011; Saba et al., 2014; Shankar et al., 2017; Khosravi et al., 2020; Saedi & Rhim, 2020; Norrrahim et al., 2021; Saedi et al., 2021). Thus, to improve the properties of host materials, different nanofillers have been incorporated or added with cellulose and cellulose NPs.

5.2.3.1 Nanofillers in Cellulose

Cellulose nanofibers, carboxymethylcellulose, regenerated cellulose, cellulose nanocrystals, and other cellulose-based composite polymer films were developed for different applications. Antibacterial activities and tensile and thermal properties of the polymer matrix were improved by using nanofillers. Nanofillers in the polymer matrix are found effective to maintain the transparency of the packaging films (Fortunati et al., 2012; Moura et al., 2012; Volova et al., 2018). Another study suggested that the average size of nanofillers plays an important role in controlling various properties of nanocomposite films. Excellent antibacterial activities of thin films were observed with 41-nm AgNPs. Thus, the size of NPs also plays a significant role in maintaining the properties of the films. Poly(lactic acid)/cellulose nanocrystals/AgNPs, hydroxypropyl methylcellulose/AgNPs, and cellulose nanofibers/reduced graphene oxide (GO)-based composite films were prepared by Fortunati et al. (2012), Moura et al. (2012), and Jin et al. (2016), respectively. Thaigamani et al. (2019) reported the banana peel powder as fillers and cellulose as the matrix to prepare the composite film. Then, the film was dipped into the silver nitrate solution to grow AgNPs. The resulting hybrid nanocomposite film demonstrated higher thermal stability, tensile properties, and antimicrobial properties than the polymer matrix. Jin et al. (2016) proposed the biocompatible films of cellulose nanofibers/reduced GO by using the bacterial reduction method. The electrical properties and biochemical activity of the films improved by bacterial reduction of GO films and also showed excellent hydrophilicity. He et al. (2021) developed carboxymethyl cellulose (CMC)/cellulose nanocrystal-immobilized AgNPs paper coating for fruit packaging. Water vapor, air barrier, mechanical strength, and antibacterial activity of the nanocomposite-coated paper increased with the increasing level of cellulose nanocrystals/AgNPs. The nanocomposite-coated paper was applied to investigate the effect of package materials on strawberries under ambient conditions. Results demonstrated that the coated paper maintained the good quality of strawberries compared to unpackaged strawberries, and extended their shelf life up to seven days. Yadav et al. (2014) proposed GO/carboxymethylcellulose/alginate composite blends by using the solution mixing–evaporation method. Carboxymethylcellulose, GO, and alginate blends formed the homogeneous mixture. The tensile strength and Young's modulus of nanocomposites were improved by 40% and 1128%, respectively, by addition of 1 wt.% GO. Saedi et al. (2021) proposed a semi-transparent antimicrobial nanocomposite film by using regenerated cellulose and ZnO NPs for food packaging applications. The crystallinity, thermal stability, oxygen barrier, and UV properties of regenerated cellulose and ZnO NP-based films were appreciably ($P < 0.05$) amplified by using 7 wt.% of ZnO

NPs to regenerated cellulose. The composite films demonstrated excellent antibacterial activity against the bacteria mentioned. Antimicrobial tests demonstrated that loading of 3 wt.% of ZnO NPs reduced oxygen permeability and UV transmittance of the films by 37% and 32%, respectively, and efficiently prevented the growth of the pathogenic food-borne Gram-negative and Gram-positive bacteria.

In CMC nanocomposites containing sodium montmorillonite (MMT) and titanium dioxide (TiO_2), water vapor permeability reduced by 50% compared to that of neat CMC films in addition to a 34% reduction in moisture uptake of nanocomposite films (Achachlouei et al., 2018). Similar studies of MMT-reinforced materials reported recently with improved barrier along with mechanical properties, for example, mucilage/MMT films (Rohini et al., 2020) and thermoplastic starch/MMT/zinc oxide (ZnO) films (Vaezi et al., 2019).

5.2.4 Starch

Starch is a biodegradable natural material that has been widely used in different fields such as biomedical, pharmaceutical, food packaging, and agriculture industries due to their non-toxic nature, renewability, and easy availability. It is the second most abundant substance found in nature and is stored by plants. It can be extracted from various sources like some roots, cereal grains, tubes, fruit stones, and rhizomes (Yu & Moon, 2021; Guzmán et al., 2022; Ihsanullah et al., 2022). Starch has been considered a potential material in the packing industry due to its low cost, high yield, non-toxicity, renewable, and degradable. However, low water barrier properties and high brittleness of virgin starch films limited their application. Therefore, different types of nanocomposites have been developed with starch to design the suitable composite film, which has good water barrier, gas isolation, and mechanical strength and is effective for packaging. Starch-based NPs and nanocrystals were also used to improve mechanical and barriers properties of base polymers (Altaf et al., 2022; Su et al., 2022).

5.2.4.1 Nanofillers in Starch

Starch films can be prepared by using different starches including potato, rice, oat, maize, corn, wheat, barley, tapioca, and pearl millet starch by using 50% glycerin as a plasticizer. Various chemical methods and physical methods such as crosslinking (Subroto et al., 2020; Siroha et al., 2021; Ding et al., 2022; Fan et al., 2022), substitution (Xu et al., 2004; Mathew & Abraham, 2007), reduction (Xiong et al., 2014; Guzmán et al., 2022, Wang et al., 2022), UV irradiation (Pashkuleva et al., 2010; He et al., 2013), ultra-sonication, gamma irradiation (Kang et al., 1999; Bao et al., 2005; Kong at al., 2016), electron beam irradiation (Lepifre et al., 2004; Zain et al., 2019), high pressure (Shahbazi et al., 2018; Zhou et al., 2022), and plasma (Laricheh et al., 2022; Sun et al., 2022; Zhang et al., 2022) were developed to modify the starch or to prepare composites of starch. The casting method is a well-known method to obtain starch-based films from the starch solution in water (Tryznowska & Kaluza, 2021; Guzman et al., 2022). Starch and related composites can be used to make the coatings and edible films. Properties of starch-based films can be improved by using microfillers, nanofillers, blending with other biopolymers and plasticizers into a base polymer, and chemical and physical modifications of starch (Guzman et al., 2022). Sani et al. (2021) developed the potato starch/apple peel pectin/*Zataria multiflora* essential oil/zirconium oxide NP nanocomposite film for meat packaging. Chang et al. (2010a) reported the chitosan NP/glycerol-plasticized starch composite film for food packaging.

Ortega et al. (2017) reported the starch/AgNP-based composite film for food packaging. The composite film demonstrates excellent antimicrobial activity along with an increase in the shelf life of fresh cheese samples for 21 days. Narayanan et al. (2021) developed the starch-reduced GO–polyiodide nanocomposite for packaging. Outstanding antibacterial activity was found against the Gram-positive and Gram-negative bacteria.

Sharma et al. (2017) reported starch and poly(vinyl alcohol)-based bio-nanocomposites for packaging made by the combination of intercalation from solution and melt-mix methods. Laponite loading up to 10% enhanced the mechanical barrier properties of the starch/poly(vinyl alcohol) films due to good surface interaction between the matrix and layer of laponite RD. However, GO coating enhanced the water barrier property of starch/poly(vinyl alcohol)/laponite RD bio-nanocomposites. GP and laponite contents have significant effects on barrier properties of nanocomposite films. Water absorption was minimum with 10% laponite RD and a grapheme-to-lipid ratio of 25:75 proportion for enhanced barrier properties. Thus, the bio-nanocomposite film with improved mechanical, water vapor barrier, and controlled water resistance properties can be potential composite for food packaging field to enhance shelf life. In another study, Wu et al. (2019) developed the starch/poly(vinyl alcohol)/GO composite film by using the casting method by adding GO to the starch/poly(vinyl alcohol matrix. The tensile strength of the film increases with the increasing content of GO up to 2 mg/mL, but decreased afterward. The maximum tensile strength was 25.28 MPa (at 2 mg/mL, GO). The moisture resistance of the film increases with the increasing content of GO.

Mohanty et al. (2019) developed AgNP-decorated starch hybrid poly(ethyl methacrylate) composite films through the in situ polymerization method. The thermal stability, biodegradability, chemical resistance, and antimicrobial properties of composite films were improved due to appropriate dispersion of AgNPs and GO in the poly(ethyl methacrylate)/starch matrix. AgNPs were added in thermoplastic starch films for mechanical strength and moisture resistance for food packaging applications (Fahmy et al., 2020).

5.2.5 Miscellaneous

Owing to the only one of its kind luminescence of inorganic nanophosphors along with excellent processability of their dispersions in polymer solutions, inorganic/polymer nanocomposites show potential as materials for diverse lighting and display devices. Researchers synthesized CdSe nanodots and used in trioctylphosphine oxide, leading to a core–shell system covering the whole visible spectral range of emission colors with quantum efficiencies as high as 50% (Murray et al., 1993; Dabbousi et al., 1997). Colvin et al. (1994) prepared light-emitting diodes (LEDs) with tunable emission colors by incorporating CdSe nanocrystals with semiconducting polymers. Similarly, Tessler et al. (2002) developed InAs/polymer LEDs having near-infrared emission.

Zinc methacrylate was co-polymerized by Yang et al. (1997) with styrene which on treatment with H_2S gave a ZnS NP-incorporated polymer network and can be used in electroluminescent devices. GaN diodes were covered with lauryl methacrylate, having (CdSe)ZnS core–shell NPs to obtain a nanocomposite which on blue LED excitation showed photoluminescence (Lee et al., 2000). Similarly, Althues et al. (2006) dispersed ZnS:Mn NPs in acrylic acid to obtain transparent and luminescent nanocomposites with a quantum efficiency of 30% stable for years (Althues et al., 2006).

To enhance the durability of polymers, UV-absorbing pigments are added to polymers to arrange UV-protective coatings. Therefore, inorganic nanofillers owing to better photostability are of the material of choice. To boost UV resistance of polymers, up to 5 wt.% ZnO or TiO_2 NPs were used, and a major decrease in the carbonyl index was found for ZnO/PP and ZnO/PE nanocomposites (Ammala et al., 2002; Zhao & Li, 2006). To avoid polymeric degradation, inorganic coatings on the nanoparticle surface, made of Al_2O_3, ZrO_2, or SiO_2, can be applied (Allen et al., 2004). Ca doping in CeO2 nanofillers resulted in better UV absorption along with reduction in catalytic activity (Althues et al., 2007).

IR absorption along with reflection is central to the upcoming generation of energy-saving windows, heat-reflecting shields as well as security glasses. Indium tin oxide and antimony tin oxide are some well-known IR-absorbing pigments for transparent coatings (Song et al., 2006).

Pourhashem et al. (2017) developed SiO_2–GO nanohybrids using tetraethyl orthosilicate to modify GO nanosheets such that SiO_2 nanospheres covered the surface of GO sheets, and incorporating 0.1 wt.% nanohybrids on properties of epoxy coatings showed that an appreciable increase in pull-off adhesion strength of epoxy coatings to mild steel substrates and water contact angle on coatings. An innovative inorganic nanofiller sepiolite was reported recently, which, on modification, is then added into the epoxy resin matrix improving the corrosion resistance of coating significantly (Xiong et al., 2020).

Primarily polymer nanocomposites using carbon nanotubes (CNTs) as filler were reported by Ajayan et al. (1994). CNTs can be single-walled nanotubes, multi-walled nanotubes, and carbon nanofibers. CNTs have unique combination of mechanical, electrical, and thermal properties, making them brilliant candidates to substitute or else complement conventional nanofillers in fabrication of multifunctional polymer nanocomposites (Iijima, 1991; Iijima & Ichihashi, 1993; Gao et al., 1998; Cooper et al., 2001). CNTs are stronger than steel, lighter than aluminum, and extra conductive in comparison to copper (Heer, 2004; Uchida & Kumar, 2005).

Furthermore, in the uses of nanocomposites for food packaging, safety issues are constantly to fear. Consequently, toxicity of CNTs and migrations from polymer/CNT nanocomposites are studied by various researchers (Dong & Ma, 2015; Massoumi et al., 2015). Bearing in mind probable fitness risks, CNT nanocomposites have not yet been fully applied in food packaging to date. Nevertheless, understanding of CNT toxicity and surface functionalization will very much ease the applications of CNT nanocomposites in food packaging.

Layered silicate reserves are hydrous aluminum silicates and classified as either phyllosilicates or layered silicates. These consist of natural clays like hectorite, MMT, and saponite, as well as synthetic clays, for example, magadiite, laponite, mica, and fluorohectorite. Bentonite is a rock consisting of smectite or MMT group clays as a major component and auxiliary minerals such as kaolin, quartz, gypsum, and iron ore. MMT, beidellite, and nontronite are trioctahedral smectite clays, whereas hectorite and saponite are dioctahedral smectite clays (Giannelis et al., 1999).

Schmid et al. (2017) studied the effects of addition of MMT nanoplatelets on whey protein isolate-based nanocomposite films along with coatings for the development of nanocomposites with enhanced barrier and mechanical properties for packaging. On adding 15% (w/w protein) MMT into 10% (w/w dispersion), oxygen permeability decreased by 91% for glycerol-plasticized and 84% for sorbitol-plasticized coatings, and the water vapor transmission rate was reduced by 58% for sorbitol-plasticized cast films. On adding MMT nanofillers, Young's modulus and tensile strength increased 315% and 129%, respectively, suggesting that these developed nanocomposite films and coatings possess an immense prospective to substitute petro-based packing materials for at least oxygen barriers in multilayer-flexible packaging films.

Zorah et al. (2020) reinforced Poly lactic acid (PLA) with titania nanofillers and observed that the storage modulus increased from 3.13 to 3.26 GPa. Thermogravimetric analysis results showed that thermal stability of the PLA nanocomposites was also improved. Hence, the reinforced PLA nanocomposites can be potential substitute of conventional petrochemical based polymers widespread in food packing industries.

Calderaro et al. (2020) studied effects of a mixture of nanofillers and mixed organically modified MMT, sepiolite with nanotitanium dioxide (TiO_2) on water and oxygen vapor permeability coefficients, light transmission along with mechanical properties of poly(butylene adipate-co-terephthalate) targeting applications in packaging. Consequence of the grouping was more prominent on mechanical properties, that is, Young's modulus increased more than thrice when compared with clean polymer.

A crystalline polyamide Nylon-MXD6 was obtained from meta-xylenediamine and adipic acid for multilayer food packaging applications (Ammala, 2011). Mitsubishi Gas Chemical Company and Nanocor produced nanocomposites of multilayer PET with clay platelets in MXD6 nylon under the trade name "M9" (Sanchez et al., 2005) used in food as well as beverage packing to enhance the shelf life of carbonated beverages to reduce carbon dioxide loss from drinks and oxygen way in bottles.

5.3 CONCLUSION

The biopolymers are found very suitable for food packaging and edible coatings due to their biodegradable, renewable, and non-toxicity properties. Biodegradable materials are very effective to maintain the environmental waste. The biopolymers are facing some problems related to their weak barrier properties (water barrier and gas barrier), tensile strength, mechanical strength, and antimicrobial activity. These weak properties can limit their application in the food packaging and other industries. However, to overcome these limitations, various nanofillers, additives, and polymers are incorporated into the polymer matrix. In this chapter, we have discussed biopolymers and nanofillers and their related composites and nanocomposites. Nanocomposite-based packaging materials with improved thermal, gas, and water barrier properties, together with chemical resistance, biodegradable, and antimicrobial activity, make them suitable for food packaging and industrial packaging.

5.4 ACKNOWLEDGMENTS

One of the authors (PCT) wishes to show gratitude to the Director, CSIR-Central Building Research Institute, Roorkee, for his constant support. Another author (NK) acknowledges the grant from NPDF from DST and thanks Director, IISc, Bangalore.

REFERENCES

Achachlouei, B. F., & Zahedi, Y. (2018). Fabrication and characterization of CMC-based nanocomposites reinforced with sodium montmorillonite and TiO_2 nanomaterials. *Carbohydr. Polym.*, 199, 415–425.

Ain, Q. U., Sehgal, R., Wani, M. F., & Singh, M. K. (2021). An overview of polymer nanocomposites: Understanding of mechanical and tribological behavior. *IOP Conf. Ser. Mater. Sci. Eng.*, 1189(1), 012010. https://doi.org/10.1088/1757-899X/1189/1/012010.

Ajayan, P. M., Stephan, O., Colliex, C., & Trauth, D. (1994). Aligned carbon nanotube arrays formed by cutting a polymer resin nanotube composite. *Science,* 265, 1212–1214.

Alim, A. A. A., Shirajuddin, S. S. M., & Anuar, F. H. (2022). A review of nonbiodegradable and biodegradable composites for food packaging application. *J. Chem.*, 2022, 1–27. https://doi.org/10.1155/2022/7670819.

Allen, N. S., Edge, M., Ortega, A., Sandoval, G., Liauw, C. M., Verran, J., ... & McIntyre, R. B. (2004). Degradation and stabilisation of polymers and coatings: nano versus pigmentary titania particles. *Polymer Degradation and Stability*, 85(3), 927–946.

Altaf, A., Usmani, Z., Dar, A. H., & Dash, K. K. (2022). A comprehensive review of polysaccharide-based bionanocomposites for food packaging applications. *Discov. Food*, 2(1), 10. https://doi.org/10.1007/s44187-022-00011-x.

Althues, H., Henle J., & Kaskel, S. (2007). Functional inorganic nanofillers for transparent polymers. *Chem. Soc. Rev.*, 36, 1454–1465.

Althues, H., Palkovits, R., Rumplecker, A., Simon, P., Sigle, W., Bredol, M., Kynast, U., & Kaskel, S. (2006). Synthesis and characterization of transparent luminescent ZnS: Mn/PMMA nanocomposites. *Chem. Mater.*, 2006, 18(4), 1068–1072.

Ammala, A. (2011). Nylon-MXD6 resins for food packaging. In J. M. Lagaron (Ed.), *Multifunctional and Nanoreinforced Polymers for Food Packaging* (pp. 243–260), Woodhead Publishing Ltd., Cambridge, UK.

Ammala, A., Hill, A. J., Meakin, P., Pas, S. J., & Turney, T. W. (2002). Degradation studies of polyolefins incorporating transparent nanoparticulate zinc oxide UV stabilizers. *J. Nanopart. Res.*, 4, 167–174.

Bao, J., Ao, Z., & Jane, J. (2005). Characterization of physical properties of flour and starch obtained from gamma-irradiated white rice. *Starch - Starke*, 57(10), 480–487. https://doi.org/10.1002/star.200500422.

Basavegowda, N., & Baek, K. H. (2021). Advances in functional biopolymer-based nanocomposites for active food packaging applications. *Polymers*, 13(23), 4198. https://doi.org/10.3390/polym13234198.

Bhattacharya, M. (2016). Polymer nanocomposites: A comparison between carbon nanotubes, graphene, and clay as nanofillers. *Materials*, 9(4), 262. https://doi.org/10.3390/ma9040262.

Calderaro, M. P., Sarantopóulos, C. I. G. L., Sanchez, E. M. S., & Morales, A. R. (2020). PBAT/hybrid nanofillers composites-part 1: Oxygen and water vapor permeabilities, UV barrier and mechanical properties. *J Appl Polym Sci.*, 137, e49522.

Chang, P. R., Jian, R., Yu, J., & Ma, X. (2010a). Fabrication and characterisation of chitosan nanoparticles/plasticised-starch composites. *Food Chem.*, 120(3), 736–740. https://doi.org/10.1016/j.foodchem.2009.11.002.

Chang, P. R., Jian, R., Yu, J., & Ma, X. (2010b). Starch-based composites reinforced with novel chitin nanoparticles. *Carbohydr. Polym.*, 80(2), 420–425. https://doi.org/10.1016/j.carbpol.2009.11.041.

Colvin, V. L., Schlamp, M. C., & Alivisatos, A. P. (1994). Light-emitting diodes made from cadmium selenide nanocrystals and a semiconducting polymer. *Nature*, 370, 354–357.

Cooper, C. A., Young, R. J., & Halsall, M. (2001). Investigation into the deformation of carbon nanotubes and their composites through the use of Raman spectroscopy. *Compos. Part A- Applied Science and Manufacturing*, 32A, 401–411.

Dabbousi, B. O., Viejo, J. R., Mikulec, F. V., Heine, J. R., Mattoussi, H., Ober, R., Jensen, K. F., & Bawendi, M. G. (1997). Core-shell quantum dots: Synthesis and characterization of a size series of highly luminescent nanocrystallites. *J. Phys. Chem. B*, 101, 9463–9475.

Ding, L., Huang, Q., Xiang, W., Fu, X., Zhang, B., & Wu, J. Y. (2022). Chemical cross-linking reduces in vitro starch digestibility of cooked potato parenchyma cells. *Food Hydrocoll.*, 124, 107297. https://doi.org/10.1016/j.foodhyd.2021.107297.

Dong, J., & Ma, Q. (2015). Advances in mechanisms and signaling pathways of carbon nanotube toxicity. *Nanotoxicology*, 9(5), 1–19.

Ediyilyam, S., George, B., Shankar, S.S., Dennis, T.T., Waclawek, S., Cernik, M., & Padil, V.V.T. (2021). Chitosan/gelatin/silver nanoparticles composites films for biodegradable food packaging applications. *Polymers*, 13, 1680. https://doi.org/10.3390/ polym13111680.

Edwards, D. C. (1990). Polymer-filler interactions in rubber reinforcement. *J. Mater. Sci.*, 25(10), 4175–4185. https://doi.org/10.1007/BF00581070.

Fabra, M. J., Rubio, A. L., & Lagaron, J. M. (2014). Biopolymers for food packaging applications. In M. R. Aguilar & J. S. Roman (Eds.), *Smart Polymers and their Applications* (pp. 476–509). Spain, Elsevier.

Fahmy, H. M., Eldin, R. E. S., Serea, E. S. A., Gomaa, N. M., Elmagd, G. M. A., Salem, S. A., Elsayed, Z. A., Edrees, A., Eldin, E. S., & Shalan, A. E. (2020). Advances in nanotechnology and antibacterial properties of biodegradable food packaging materials. *RSC Adv.*, 10, 20467–20484.

Fan, J. X., Guo, X. N., & Zhu, K. X. (2022). Impact of laccase-induced protein cross-linking on the in vitro starch digestion of black highland barley noodles. *Food Hydrocoll.*, 124, 107298. https://doi.org/10.1016/j.foodhyd.2021.107298.

Folino, A., Karageorgiou, A., Calabro, P. S., & Komilis, D. (2020). Biodegradation of wasted bioplastics in natural and industrial environments: A review. *Sustainability*, 12(15), 6030. https://doi.org/10.3390/su12156030.

Fortunati, E., et al., (2012). Multifunctional bionanocomposite films of poly(lactic acid), cellulose nanocrystals and silver nanoparticles. *Carbohydr. Polym.*, 87(2), 1596–1605. https://doi.org/10.1016/j.carbpol.2011.09.066.

Fu, S., Sun, Z., Huang, P., Li, Y., & Hu, N. (2019). Some basic aspects of polymer nanocomposites: A critical review. *Nano Mater. Sci.*, 1(1), 2–30. https://doi.org/10.1016/j.nanoms.2019.02.006.

Gao, G., Cagin, T., & Goddard, W. A. (1998). Energetics, structure, mechanical and vibrational properties of single-walled carbon nanotubes. Nanotechnology, 9(3), 184–191.

Garavand, F., et al. (2022). A comprehensive review on the nanocomposites loaded with chitosan nanoparticles for food packaging. *Crit. Rev. Food Sci. Nutr.*, 62(5), 1383–1416.

Gheller, J., Ellwanger, M. V., & Oliveira, V. (2016). Polymer-filler interactions in a tire compound reinforced with silica. *J. Elastomers Plast.*, 48(3), 217–226. https://doi.org/10.1177/0095244314568470.

Giannelis, E. P., Krishnamoorti, R., & Manias, E. (1999). Polymer-silicate nanocomposites: Model systems for confined polymers and polymer brushes. *Adv. Polym. Sci.,* 138, 108–147.

Gross, R. A., & Kalra, B. (2002). Biodegradable polymers for the environment. *Science*, 297(5582), 803–807. https://doi.org/10.1126/science.297.5582.803.

Guzmán, L. G., Barjas, G. C., Hernández, C. G. S., Castaño, J., Lezama, A. Y. G., & Llamazares, S. R. (2022). Progress in starch-based materials for food packaging applications. *Polysaccharides*, 3(1), 136–177. https://doi.org/10.3390/polysaccharides3010007.

He, G. J., Liu, Q., & Thompson, M. R. (2013). Characterization of structure and properties of thermoplastic potato starch film surface cross-linked by UV irradiation. *Starch - Starke*, 65(3–4), 304–311. https://doi.org/10.1002/star.201200097.

He, Y., Li, H., Fei, X., & Peng, L. (2021). Carboxymethyl cellulose/cellulose nanocrystals immobilized silver nanoparticles as an effective coating to improve barrier and antibacterial properties of paper for food packaging applications. *Carbohydr. Polym.*, 252, 117156. https://doi.org/10.1016/j.carbpol.2020.117156.

Heer, W. A. D. (2004). Nanotubes and the pursuit of applications. *MRS Bull.,* 29(4), 281–285.

Ihsanullah, I., Bilal, M., & Jamal, A. (2022). Recent developments in the removal of dyes from water by starch-based adsorbents. *Chem. Rec.*, 2, e202100312. https://doi.org/10.1002/tcr.202100312.

Ijima, S. (1991). Helical microtubules of graphitic carbon. *Nature,* 354, 56–58.

Iijima, S., & Ichihashi, T. (1993). Single-shell carbon nanotubes of 1-nm diameter. *Nature,* 363, 603–605.

Imai, Y. (2014). Inorganic nano-fillers for polymers. In S. Kobayashi & K. Mullen (Eds.), *Encyclopedia of Polymeric Nanomaterials* (pp. 1–7). Springer, Berlin Heidelberg.

Jin, L., Zeng, Z., Kuddannaya, S., Wu, D., Zhang, Y., & Wang, Z. (2016). Biocompatible, free-standing film composed of bacterial cellulose nanofibers–graphene composite. *ACS Appl. Mater. Interfaces*, 8(1), 1011–1018. https://doi.org/10.1021/acsami.5b11241.

Kang, I. J. Byun, M. W., Yook, H. S., Bae, C. H., Lee, H. S., Kwon, J. H., & Chung, C. K (1999). Production of modified starches by gamma irradiation. *Radiat. Phys. Chem.*, 54(4), 425–430.

Khosravi, A., et al., (2020). Soft and hard sections from cellulose-reinforced poly(lactic acid)-based food packaging films: A critical review. *Food Packag. Shelf Life*, 23, 100429. https://doi.org/10.1016/j.fpsl.2019.100429.

Kim, H., Abdala, A. A., & Macosko, C. W. (2010). Graphene/polymer nanocomposites. *Macromolecules*, 43(16), 6515–6530. https://doi.org/10.1021/ma100572e.

Kong, X., Zhou, X., Sui, Z., & Bao, J. (2016). Effects of gamma irradiation on physicochemical properties of native and acetylated wheat starches. *Int. J. Biol. Macromol.*, 91, 1141–1150. https://doi.org/10.1016/j.ijbiomac.2016.06.072.

Kumar, S., Shukla, A., Baul, P. P., Mitra, A., & Halder, D. (2018). Biodegradable hybrid nanocomposites of chitosan/gelatin and silver nanoparticles for active food packaging applications. *Food Packag. Shelf Life*, 16, 178–184. https://doi.org/10.1016/j.fpsl.2018.03.008.

Laricheh, R., Fazel, M., & Goli, M. (2022). Corn starch structurally modified with atmospheric cold-plasma and its use in mayonnaise formulation. *J. Food Meas. Charact.* 16, 1859–1872. https://doi.org/10.1007/s11694-022-01296–3.

Lee, J., Sundar, V. C., Heine, J. R., Bawendi, M. G., & Jensen, K. F. (2000). Full colour emission from II–VI semiconductor quantum dot-polymer composites. *Adv. Mater.*, 12(15), 1102–1105.

Lepifre, S., Froment, M., Cazaux, F., Houot, S., Lourdin, D., Coqueret, X., Lapierre, C., & Baumberger, S. (2004). Lignin incorporation combined with electron-beam irradiation improves the surface water resistance of starch films. *Biomacromolecules*, 5(5), 1678–1686.

https://doi.org/10.1021/bm040005e.

Massoumi, B., Jafarpour, P., Jaymand, M., & Entezami, A. A. (2015). Functionalized multi-walled carbon nanotubes as reinforcing agents for poly(vinyl alcohol) and poly(vinyl alcohol)/starch nanocomposites: Synthesis, characterization, and properties. *Polymer International*, 64(5), 689–695.

Mathew, S., & Abraham, T. (2007). Physico-chemical characterization of starch ferulates of different degrees of substitution. *Food Chem.*, 105(2), 579–589. https://doi.org/10.1016/j.foodchem.2007.04.032.

Mohanty, F., & Swain, S. K. (2019). Nano silver embedded starch hybrid graphene oxide sandwiched poly(ethylmethacrylate) for packaging application. *Nano-Struct. Nano-Objects*, 18, 100300. https://doi.org/10.1016/j.nanoso.2019.100300.

Moura, M. R., Aouada, F. A., Bustillos, R. J. A., McHugh, T. H., Krochta, J. M., & Mattoso, L. H. C. (2009). Improved barrier and mechanical properties of novel hydroxypropyl methylcellulose edible films with chitosan/tripolyphosphate nanoparticles. *J. Food Eng.*, 92(4), 448–453.

Moura, M. R., Mattoso, L. H. C., & Zucolotto, V. (2012). Development of cellulose-based bactericidal nanocomposites containing silver nanoparticles and their use as active food packaging. *J. Food Eng.*, 109(3), 520–524. https://doi.org/10.1016/j.jfoodeng.2011.10.030.

Moustafa, H., Darwish, N. A., & Youssef, A. M. (2022). Rational formulations of sustainable polyurethane/chitin/rosin composites reinforced with ZnO-doped-SiO_2 nanoparticles for green packaging applications. *Food Chem.*, 371, 131193. https://doi.org/10.1016/j.foodchem.2021.131193.

Murray, C. B., Norris, D. J., & Bawendi, M. G. (1993). Synthesis and characterization of nearly monodisperse CdE (E = sulfur, selenium, tellurium) semiconductor nanocrystallites. *J. Am. Chem. Soc.*, 115, 8706–15.

Naamani, L. A., Dobretsov, S., & Dutta, J. (2016). Chitosan-zinc oxide nanoparticle composite coating for active food packaging applications. *Innov. Food Sci. Emerg. Technol.*, 38, 231–237. https://doi.org/10.1016/j.ifset.2016.10.010.

Narayanan, K. B., Park, G. T., & Han, S. S. (2021). Antibacterial properties of starch-reduced graphene oxide–polyiodide nanocomposite. *Food Chem.*, 342, 128385. https://doi.org/10.1016/j.foodchem.2020.128385.

Njuguna, J., Pielichowski, K., & Desai, S. (2008). Nanofiller-reinforced polymer nanocomposites. *Polym. Adv. Technol.*, 19(8), 947–959. https://doi.org/10.1002/pat.1074.

Norrahim, M. N. F., Ariffin, H., Anuar, T. A. T. Y., Hassan, M. A., Ibrahim, N. A., Yunus, W. M. Z. W., & Nishida, H. (2021). Performance evaluation of cellulose nanofiber with residual hemicellulose as a nanofiller in polypropylene-based nanocomposite. *Polymers*, 13(7), 1064. https://doi.org/10.3390/polym13071064.

Olsson, R. T., Fogelström, L., Sanz, M. M., & Henriksson, M. (2011). Cellulose nanofillers for food packaging. In J. M. Lagaron (Ed.), *Multifunctional and Nanoreinforced Polymers for Food Packaging* (pp. 86–107). USA, Elsevier. https://doi.org/10.1533/9780857092786.1.86.

Omerovic, N., et al., (2021). Antimicrobial nanoparticles and biodegradable polymer composites for active food packaging applications. *Compr. Rev. Food Sci. Food Saf.*, 20(3), 2428–2454. https://doi.org/ 10.1111/1541-4337.12727.

Ortega, F., Giannuzzi, L., Arce, V. B., & García, M. A. (2017). Active composite starch films containing green synthetized silver nanoparticles. *Food Hydrocoll.*, 70, 152–162.

Pashkuleva, I., Marques, A. P., Vaz, F., & Reis, R. L. (2010). Surface modification of starch based biomaterials by oxygen plasma or UV-irradiation. *J. Mater. Sci. Mater. Med.*, 21(1), 21–32.

Pires, J., Paula, C. D., Souza, V. G. L., Fernando, A. L., & Coelhoso, I. (2021). Understanding the barrier and mechanical behavior of different nanofillers in chitosan films for food packaging. *Polymers*, 13(5), 721.

Pourhashem, S., Vaezi, M. R., & Rashidi, A. (2017). Investigating the effect of SiO_2-graphene oxide hybrid as inorganic nanofiller on corrosion protection properties of epoxy coatings. *Surface and Coatings Technology*, 311, 282–294.

Priyadarshi, R., Roy, S., Ghosh, T., Biswas, D., & Rhim, J. W. (2021). Antimicrobial nanofillers reinforced biopolymer composite films for active food packaging applications: A review. *Sustain. Mater. Technol.*, e00353. https://doi.org/10.1016/j.susmat.2021.e00353.

Qin, Y., Zhang, S., Yu, J., Yang, J., Xiong, L., & Sun, Q. (2016). Effects of chitin nano-whiskers on the antibacterial and physicochemical properties of maize starch films. *Carbohydr. Polym.*, 147, 372–378.

Rohini, B., Ishwarya, S. P., Rajasekharan, R., & Kumar, A. K. V. (2020).°Cimum basilicum seed mucilage reinforced with montmorillonite for preparation of bionanocomposite film for food packaging applications. *Polym. Test.*, 87, 106465.

Saba, N., Tahir, P., & Jawaid, M. (2014). A review on potentiality of nano filler/natural fiber filled polymer hybrid composites. *Polymers*, 6(8), 2247–2273.

Saedi, S., & Rhim, J. W. (2020). Synthesis of hybrid nanoparticles for the preparation of carrageenan-based functional nanocomposite film. *Food Packag. Shelf Life*, 24, 100473.

Saedi, S., Shokri, M., Kim, J. T., & Shin, G. H. (2021). Semi-transparent regenerated cellulose/ZnONP nanocomposite film as a potential antimicrobial food packaging material. *J. Food Eng.*, 307, 110665.

Sanchez, C., Julian, B., Belleville, P., & Popall, M. (2005). Applications of hybrid organic-inorganic nanocomposites, *J. Mater. Chem.*, **15**, 3559–35912.

Sani, I. K., Geshlaghi, S. P., Pirsa, S., & Asdagh, A. (2021). Composite film based on potato starch/apple peel pectin/ZrO_2 nanoparticles/microencapsulated Zataria multiflora essential oil; investigation of physicochemical properties and use in quail meat packaging. *Food Hydrocoll.*, 117, 106719.

Sarojini, K. S., Indumathi, M. P., & Rajarajeswari, G. R. (2019). Mahua oil-based polyurethane/chitosan/nano ZnO composite films for biodegradable food packaging applications. *Int. J. Biol. Macromol.*, 124, 163–174.

Schmid, M., Merzbacher, S., Brzoska, N., Muller, K., & Jesdinszki, M. (2017). Improvement of food packaging related properties of whey protein isolate-based nanocomposite films and coatings by addition of montmorillonite nanoplatelets. *Front. Mater.*, 4, 35.

Shah, K., Wang, S. X., Soo, D., & Xu, J. (2019). One dimensional nanostructure engineering of conducting polymers for thermoelectric applications. *Appl. Sci.*, 9(7), 1422.

Shahbazi, M., Majzoobi, M., & Farahnaky, A. (2018). Physical modification of starch by high-pressure homogenization for improving functional properties of κ-carrageenan/starch blend film. *Food Hydrocoll.*, 85, 204–214.

Shankar, S., Khodaei, D., & Lacroix, M. (2021). Effect of chitosan/essential oils/silver nanoparticles composite films packaging and gamma irradiation on shelf life of strawberries. *Food Hydrocoll.*, 117, 106750.

Shankar, S., Oun, A. A., & Rhim, J. W. (2018). Preparation of antimicrobial hybrid nanomaterials using regenerated cellulose and metallic nanoparticles. *Int. J. Biol. Macromol.*, 107, 17–27.

Shankar, S., Reddy, J. P., Rhim, J. W., & Kim, H. Y. (2015). Preparation, characterization, and antimicrobial activity of chitin nanofibrils reinforced carrageenan nanocomposite films. *Carbohydr. Polym.*, 117, 468–475.

Sharma, B., Malik, P., & Jain, P. (2018). Biopolymer reinforced nanocomposites: A comprehensive review. *Mater. Today Commun.*, 16, 353–363.

Sharma, C., Manepalli, P. H., Thatte, A., Thomas, S., Kalarikkal, N., & Alavi, S. (2017). Biodegradable starch/PVOH/laponite RD-based bionanocomposite films coated with graphene oxide: Preparation and performance characterization for food packaging applications. *Colloid Polym. Sci.*, 295(9), 1695–1708.

Siroha, A. K., Bangar, S. P., Sandhu, K. S., Trif, M., Kumar, M., & Guleria, P. (2021). Effect of cross-linking modification on structural and film-forming characteristics of pearl millet (Pennisetum glaucum L.) starch. *Coatings*, 11(10), 1163.

Song, J. E., Kim, Y. H., & Kang, Y. S. (2006). Preparation of indium tin oxide nanoparticles and their application to near IR-reflective film. *Curr. Appl. Phys.*, 6, 791–795.

Sorrentino, A., Gorrasi, G., & Vittoria, V. (2007). Potential perspectives of bio-nanocomposites for food packaging applications. *Trends Food Sci. Technol.*, 18(2), 84–95.

Souza, V. G. L., Pires, J. R. A., Rodrigues, C., Coelhoso, I. M., & Fernando, A. L. (2020). Chitosan composites in packaging industry—Current trends and future challenges. *Polymers*, 12(2), 417.

Su, C., Li, D., Wang, L., & Wang, Y. (2022). Biodegradation behavior and digestive properties of starch-based film for food packaging: A review. *Crit. Rev. Food Sci. Nutr.*, 1–23.

Subroto, E., Indiarto, R., Djali, M., & Rosyida, H. D. (2020). Production and application of crosslinking- modified starch as fat replacer: A review. *Int. J. Eng. Trends Technol.*, 68(12), 26–30.

Sugimoto, M., Kawahara, M., Teramoto, Y., & Nishio, Y. (2010). Synthesis of acyl chitin derivatives and miscibility characterization of their blends with poly(ε-caprolactone). *Carbohydr. Polym.*, 79(4), 948–954.

Sun, X., et al., (2022). Modification of multi-scale structure, physicochemical properties, and digestibility of rice starch via microwave and cold plasma treatments. *LWT*, 153, 112483.

Tessler, N., Medvedev, V., Kazes, M., Kan, S., & Banin, U. (2002). Efficient near-infrared polymer nanocrystal light-emitting diodes. *Science*, 295, 1506–1508.

Thapliyal, P. C., & Kumar, N. (2021). Functionalization of graphene based biopolymer nanocomposites for packaging and building applications. In B. Sharma & P. Jain (Eds.), *Graphene based Biopolymer Nanocomposites* (pp. 251–271). Springer, Singapore.

Thiagamani, S. M. K., Rajini, N., Siengchin, S., Rajulu, A. V., Hariram, N., & Ayrilmis, N. (2019). Influence of silver nanoparticles on the mechanical, thermal and antimicrobial properties of cellulose-based hybrid nanocomposites. *Compos. Part B Eng.*, 165, 516–525.

Tryznowska, Z. Ż., & Kałuża, A. (2021). The Influence of starch origin on the properties of starch films: Packaging performance. *Materials*, 14(5), 1146.

Uchida, T., & Kumar, S. (2005). Single wall carbon nanotube dispersion and exfoliation in polymers. *J. Appl. Polym. Sci.*, 98(3), 985–989.

Vaezi, K., Asadpour, G., & Sharifi, H. (2019). Effect of ZnO nanoparticles on the mechanical, barrier and optical properties of thermoplastic cationic starch/montmorillonite biodegradable films. *Int. J. Biol. Macromol.*, 124, 519–529.

Volova, T. G., et al., (2018). Antibacterial properties of films of cellulose composites with silver nanoparticles and antibiotics. *Polym. Test.*, 65, 54–68.

Vroman, I., & Tighzert, L. (2009). Biodegradable polymers. *Materials*, 2(2), 307–344.

Wang, H., et al., (2019). Preparation and characterization of multilayer films composed of chitosan, sodium alginate and carboxymethyl chitosan-ZnO nanoparticles. *Food Chem.*, 283, 397–403.

Wang, Z., Mhaske, P., Farahnaky, A., Kasapis, S., & Majzoobi, M. (2022). Cassava starch: Chemical modification and its impact on functional properties and digestibility, a review. *Food Hydrocoll.*, 107542.

Wu, Z., et al. (2019). Physical properties and structural characterization of starch/polyvinyl alcohol/graphene oxide composite films. *Int. J. Biol. Macromol.*, 123, 569–575.

Wu, Z., Huang, X., Li, Y. C., Xiao, H., & Wang, X. (2018). Novel chitosan films with laponite immobilized Ag nanoparticles for active food packaging. *Carbohydr. Polym.*, 199, 210–218.

Xiao, L., Green, A. N. M., Haque, S. A., Mills, A., & J. R. Durrant. (2004). Light-driven oxygen scavenging by titania/polymer nanocomposite films. *J. Photochem. Photobiol. A*, 162, 253–259.

Xiong et al., Z., (2014). Surface hydrophobic modification of starch with bio-based epoxy resins to fabricate high-performance polylactide composite materials. *Compos. Sci. Technol.*, 94, 16–22.

Xiong, H., Qia, F., Zhao, N., Yuan, H., Wan, P., Liao, B., & Ouyang, X. (2020). Effect of organically modified sepiolite as inorganic nanofiller on the anti-corrosion resistance of epoxy coating. *Materials Letters*, 260, 126941.

Xu, Y., Miladinov, V., & Hanna, M. A. (2004). Synthesis and characterization of starch acetates with high substitution. *Cereal Chem. J.*, 81(6), 735–740.

Yadav, M., Rhee, K. Y., & Park, S. J. (2014). Synthesis and characterization of graphene oxide/carboxymethylcellulose/alginate composite blend films. *Carbohydr. Polym.*, 110, 18–25.

Yadav, S., Mehrotra, G. K., & Dutta, P. K. (2021). Chitosan based ZnO nanoparticles loaded gallic-acid films for active food packaging. *Food Chem.*, 334, 127605.

Yang, Y., Huang, J., Liu, S., & Shen, J. (1997). Preparation, characterization and electroluminescence of ZnS nanocrystals in a polymer matrix. *J. Mater. Chem.*, 7, 131–133.

Yu, J. K., & Moon, Y. S. (2021). Corn starch: quality and quantity improvement for industrial uses. *Plants*, 11(1), 92.

Zain, A. H. M., Wahab, M. K. A., & Ismail, H. (2019). Influence of electron beam radiation on structural and mechanical performance of thermoplastic cassava starch. *Mater. Res. Express*, 6(10), 105335.

Zhang, K., et al., (2022). Low-pressure plasma modification of the rheological properties of tapioca starch. *Food Hydrocoll.*, 125, 107380.

Zhang, X., Liu, Y., Yong, H., Qin, Y., Liu, J., & Liu, J. (2019). Development of multifunctional food packaging films based on chitosan, TiO_2 nanoparticles and anthocyanin-rich black plum peel extract. *Food Hydrocoll.*, 94, 80–92.

Zhao, H., & Li, R. K. Y. (2006). A study on the photo-degradation of zinc oxide (ZnO) filled polypropylene nanocomposites. *Polymer*, 47, 3207–3217.

Zhang, X., Xiao, G., Wang, Y., Zhao, Y., Su, H., & Tan, T. (2017). Preparation of chitosan-TiO_2 composite film with efficient antimicrobial activities under visible light for food packaging applications. *Carbohydr. Polym.*, 169, 101–107.

Zhou, X., Chen, J., Wang, S., & Zhou, Y. (2022). Effect of high hydrostatic pressure treatment on the formation and in vitro digestion of Tartary buckwheat starch/flavonoid complexes. *Food Chem.*, 382, 132324.

Zieba, T., et al., (2021). The Annealing of acetylated potato starch with various substitution degrees. *Molecules*, 26(7), 2096.

Zorah, M., Mustapa, I. R., Daud, N., Nahida, J. H., Sudin, N. A. S., Majhool, A. A., & Mahmoudi, E. (2020). Improvement thermomechanical properties of polylactic acid via titania nanofillers reinforcement. *J. Adv. Res. Fluid Mech. Ther. Sci.*, 70(1), 97–111.

6 Inorganic Nanofiller-Incorporated Polymeric Nanocomposites for Biomedical Applications

Mohammad Fahim Tazwar,
Syed Talha Muhtasim, and Tousif Reza
Military Institute of Science and Technology (MIST)

Yashdi Saif Autul
Worcester Polytechnic Institute (WPI)

Mohammed Enamul Hoque
Military Institute of Science and Technology (MIST)

CONTENTS

DOI: 10.1201/9781003279389-6

6.1 INTRODUCTION

Polymer nanocomposites (PNC) are 3D structures composed of a polymer or copolymer having nanofillers scattered within the polymer matrix. Nanofillers may take on a variety of morphologies such as fibers, spheroids, or platelets but should have one dimension less than 100 nm. In recent years, low concentrations of these nanofillers in polymers have been shown to increase mechanical, thermal, corrosion resistance, and flammability characteristics without impairing the processability of the nanocomposites (NCs).[1,2] The material characteristics of these composites are determined by the type of nanofillers, concentration of filler and polymer, size, dispersion, and the orientation of nanofillers within the polymer matrices.[3] Nanofillers may be categorized into two main categories: organic and inorganic. Globally, substantial research into polymeric composites and their applications has been carried out over the last few decades. Today, both organic and inorganic nanofiller-based PNC are being employed in a variety of sectors, including electronics, structural, biomedicine, energy storage, food packaging, aerospace, automotive, etc.

Inorganic nanofillers such as graphene, nanoclays, carbon nanotubes (CNTs), metallic nanoparticles (NPs) such as platinum (Pt), titanium (Ti), palladium (Pd), gold (Au), iron (Fe), silver (Ag), and metal oxide NPs such as iron (III) oxide (Fe_2O_3), cerium (IV) oxide (CeO_2), cobalt (III) oxide (Co_2O_3), iron (II) oxide (FeO), nickel oxide (NiO), and titanium dioxide (TiO_2) have been employed as fillers in various PNC. PNC have been broadly employed in biomedical applications due to their biodegradation rate, biocompatibility, and decomposition into harmless components. Natural polymers comprising gelatin, chitosan, proteins, collagen, and synthetic polymers include polyacrylic acid (PAA), polyaniline (PANI), polyethylene glycol (PEG), polylactic acid (PLA), polyurethane (PU), poly(D, L-lactide) (PDLA), polylactic-co-glycolic acid (PLGA), polypyrrole (PPy), polygalacturonic acid (PgA), branched polyethylenimine (BPEI), polymethyl methacrylate (PMMA), poly(ε-caprolactone) (PCL), and polyvinyl alcohol (PVA) are vastly used in NCs for biomedical applications.

Biomedical applications need stringent safety measures due to their direct interaction with live organisms within the body[4]. The materials utilized in this scenario should be biocompatible, biodegradable, and nontoxic to the live cells to which they will be administered. Due to the suitable physiomechanical attributes of these composites, these nanomaterials have been implemented in diverse biomedical applications namely drug delivery systems,[5–8] cancer therapy,[9–12] tissue engineering,[13–15] implants,[16,17] and biosensors[18–20] as shown in Figure 6.1. The aim of this chapter is to review the biomedical applications of inorganic nanofiller-based PNC and to yield an understanding of how these NCs might be used in different aspects of biomedical engineering.

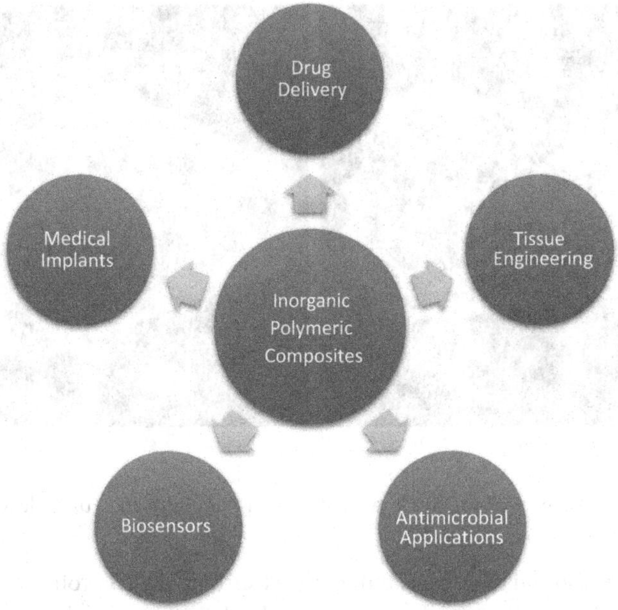

FIGURE 6.1 Schematic diagram showing the various biomedical applications of inorganic polymeric composites.

6.2 BIOMEDICAL APPLICATIONS OF DIFFERENT INORGANIC NANOFILLER POLYMERIC NANOCOMPOSITES

6.2.1 Carbon Nanotubes (CNTs)

Carbon is used in a broad variety of domains, including material science, biology, chemistry, and physics. Coatings and fillers made of carbon biomaterials have been utilized in implants because of their inherent features, such as resistance to corrosiveness, low coefficient of friction, hardness, and chemical inertness which make them suitable for implants.[14,21] The biocompatibility of these materials makes them suitable for use in medicine.[22] CNTs, due to their unique characteristics, have been researched for use in the biomedical field. CNTs have diverse characteristics depending on the configuration of carbon atoms. Because of their unique structural, mechanical, and electrical qualities, they provide more alternatives and functions.[23] CNTs are created by winding graphene nanosheets (GNS) into nanoscale cylinders. CNTs are classified into two categories: single-walled carbon nanotubes (SWCNTs) and multiwalled carbon nanotubes (MWCNTs).[23] SWCNTs and MWCNTs have distinct graphene cylinder arrangements.[24] Structures of SWCNTs and MWCNTs are shown in Figure 6.2. On top of being extremely strong, CNTs possess excellent qualities such as high strength, higher surface area, and improved chemical and thermal resilience. CNTs have a one-dimensional structure that makes them very efficient for biomedical applications.[6,25] The strong electrical characteristics of these

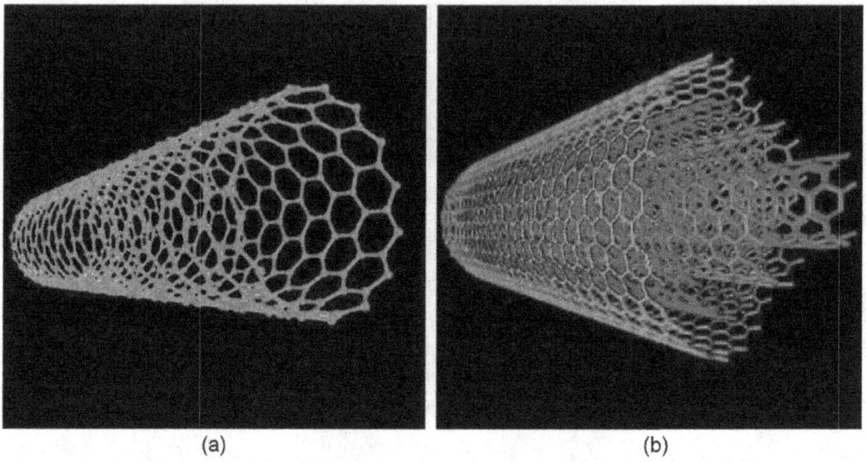

FIGURE 6.2 Structure of (a) single-walled carbon nanotube and (b) multiwalled carbon nanotube.[41]

materials have allowed them to be utilized as electronically controlled regulators and as optical and electronic nano-biosensors for the detection of biological metabolites such as RNA, DNA, proteins as well as cells and microbes. CNTs possess compelling ocular properties in the infrared spectrum,[26] Raman scattering,[27] photoluminescence,[28] and optoacoustic attributes which help in the surveillance of CNTs in biological environments. Because of their superior ability to carry drugs into cells and to display enhanced pharmacological action, CNTs are the preferred drug delivery system for many medicines.[12] Owing to these characteristics, CNTs are highly desirable for usage as nanofillers in PNC materials to improve such materials' structural and mechanical properties. To date, composite materials combining SWCNTs or MWCNTs into polymer matrices, metals, or ceramic matrices have been developed.[29–33] Multiple chitosan/CNT polymeric composites have been synthesized for biomedical applications because of the unique characteristics of both chitosan and CNTs.[34–37] CNT-based polymers have unique properties that allow their usage in a wide variety of biomedical applications. Biosensors, neurological electrodes, antimicrobial coatings, and tissue engineering prosthetics are some of the medical applications for CNT composites that have been looked into.[38–40] Some applications of CNT PNC in the biomedical field are shown in Table 6.1.

6.2.1.1 Bone Tissue Engineering

Regeneration and replacement of damaged tissues is the focus of tissue engineering, which develops materials and bone grafts that can repair, maintain, and increase tissue functions in the body.[61] Owing to CNTs' unique physiomechanical capabilities, several research works have been conducted to analyze their prospects in reinforcing matrix materials, utilized in hard tissue implants and scaffolds. The goal of scaffolds is to offer a suitable foundation for cellular development and to assist tissue and organ regeneration.[15] PVA-based NCs (PVA/SWCNTs) with varying proportions of single-wall carbon nanotubes (0, 0.5, 1, 2, and 3 wt. %) were synthesized by Anis et al and

TABLE 6.1
Biomedical Applications of CNT Polymeric Composites

Matrix	Polymer Nanocomposite	Applications	References
Chitosan/folic acid	CNT-(Fe)/HA/FA nanocomposite	Magnetic-targeted drug delivery carrier	42
Polyethylenimine-betaine	Polyethylenimine-betaine/ MWCNT	Gene therapy and chemotherapy	43
Gelatin chitosan	MWCNTs/gelatin chitosan nanocomposites	Drug release	44
Starch	Starch/MWCNT glucose nanocomposites	Drug delivery	45
Polydimethylsiloxane	nNiO/MWCNT nanocomposite	Microfluidic biosensor platform	46
Polycaprolactone (PCL)	MWCNT/PCL composite	3D Jaw and bone scaffolds	47,48
Polylactic acid (PLA)	PLA/CNTs-COOH nanocomposites	Nanocrystalline materials and jaw tissue scaffolds	49,50
Poly(lactide-co-glycolide) (PLGA)	CNT/PLGA composite	Jaw tissue and repair scaffolds	51–53
Chitosan	MWCNT/CS	Nanocomposite films and jaw skin scaffolds	13,54
Silk fibroin	Silk fibroin/MWCNT nanocomposite	Nanocomposite films	55
Collagen	MWCNT/coated collagen sponge	3D CNT-covered jaw and jaw repair materials	56
Microbial cellulose	BC/MWCNT nanocomposites	Bone tissue scaffolds	57
Collagen–hydroxyapatite and gelatin–chitosan	MWCNT/Col/HA	Jaw scaffolds	58
Calcium phosphate	RGO/CNT/CPC	Injectable jaw substitute materials	59
β-tricalcium phosphate	β-TCP/CNT nanocomposite	Jaw repair materials	60

tested the cytotoxicity and pro-inflammatory response of these composites.[62] NCs containing SWCNTs up to concentrations of 1 wt. % might be useful for regenerating cartilage, bone or neural tissue as well as enhancing the physical characteristics of PVA. According to Mikael et al., PLGA membranes with just 3% MWCNTs have significantly better compressive characteristics and a modulus greater than those of pure PLGA scaffolds.[53] In addition, it demonstrated excellent cell uptake, mineralization, and propagation properties. Gupta et al. established that integrating SWCNTs with PLGA resulted in a unique SWCNT/PLGA composite with superior cell proliferation and mineralization.[63] When it comes to bone tissue engineering, hydroxyapatite (HA) is a popular choice because of its excellent biocompatibility and osteoconductivity.[64] Numerous investigations have been conducted on the strengthening of HA coatings using CNTs to increase their mechanical characteristics. Singh et al. synthesized a NC of HA modified with PMMA and MWCNTs.[65] They synthesized this composite using the freeze granulation procedure to ensure uniform dispersion of MWCNTs. The uniform distribution is critical for the efficient transmission of load to MWCNTs. Their investigation demonstrated that 0.1% MWCNTs provided the composite with the optimal mechanical characteristics. This substance might be used in bone cement and implant coatings. CNTs and silk fibronectin were employed by Yao et al. in their work to modify HA scaffolds by cross-linking and cryogenic processing in order to administer the drug dexamethasone (DEX). Nanotubes have increased the biocompatibility and physical capacity of nHA or PA66 scaffolds, making them highly applicable for bone restoration. Using stem cells and DEX at its greatest dosage of 1 mg/mL, this DEX-loaded support exhibited osteogenesis-stimulating results. When drugs are immobilized in small pores or bonded to the layers, CNTs, like bone marrow mesenchymal stem cells, have a larger area that allows for increased drug loading.[66] Carbon nanotube reinforced HAP-medicated gold NPs studied by Zhao et al. for their tissue regeneration ability and biocompatibility.[67] The resultant composite was compatible with biological environments and exhibited high cell proliferation. Composite scaffolds made of CNTs and HA that resemble superparamagnetic materials were developed by Lu et al.[68] An optimal pore diameter of 20–300 nm was found in the fluid-like structure, with small pores of 20–300 nm and big particles of 1–2 mm. CNT–HA scaffolds with a saturation magnetization of emu/gram were found to exhibit the superparamagnetic behavior, allowing them to attract and retain stem cells and other physiologically active biomolecules in vivo as a development factor. Using a thermally induced phase separation technique, Jell et al. were able to develop a porous CNT–PU foam NCs that mimicked the porosity nature of bones.[69] Scientists were able to evenly disperse CNTs throughout the polyurethane matrix by using this method. Their research found that adding 5 wt. % CNTs to PU enhanced its compressive strength by 200%. In addition to being biocompatible, this NC enhanced osteoblast production of the potent angiogenesis factor, the vascular endothelial growth factor by increasing the weight percent of CNTs within the polymer matrix. This suggests that by altering the CNT composition in the polymer matrix, it is possible to manipulate cellular activity (Figure 6.3).

6.2.1.2 Antimicrobial Applications

Several CNT composites have been explored for use in various antimicrobial applications. Aslan et al. synthesized SWCNT and PLGA composites for antimicrobial purposes. On SWCNT/PLGA composites, it was discovered that the survival and metabolic activity of *Staphylococcus epidermidis* and *Escherichia coli*, which are two pathogens frequently found in medicinal implants, were dramatically decreased. It took SWCNT/PLGA NCs only 1 hour to kill up to 98% of the pathogens, compared to just 15%–20% on pure PLGA.[71] An electroactive antimicrobial bandage was developed using SWCNTs in another study.[72] In this composite, the resistance to electrical stimulation of cells was low enough to promote cell proliferation and speed up the healing process. Chitosan-CNT-based NCs have been developed to give moderate isoniazid dosages to treat bacterial ulcers. Analysis in a wet lab found that the release of isoniazid was significantly extended to a week and that the isoniazid's mechanism of discharge remained stable while retaining its physiological functions. Mycobacterium TB was destroyed by isoniazid NPs in a tubercular ulcer model in a

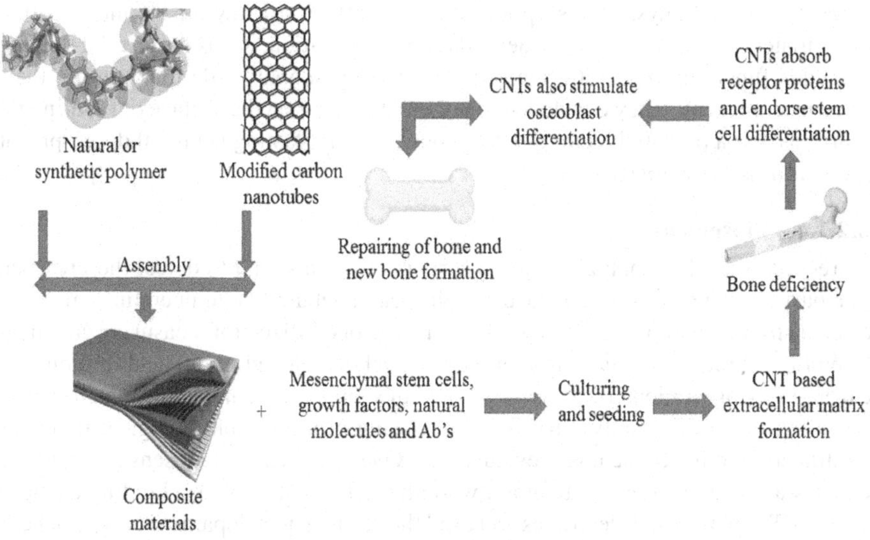

FIGURE 6.3 Scaffolds for bone tissue engineering using carbon nanotube-based nanocomposites.[70]

mouse. These NPs can be used to treat skeletal deficiency and secondary wounds by loading them with isoniazid (INH–CS–CNT).[73]

6.2.1.3 Neural Applications

For neural implant applications, several researchers have examined the compatibility and viability of CNTs and CNT-based NCs in biological environments. A technique for cultivating embryonic rat brain neurons on MWCNTs was established by Mattson et al. in one of the earliest studies in this approach.[74] Neuronal development can also be influenced by the conductivity of CNT-based composites. An increase in neuronal development and neurite outgrowth was seen for SWCNT-PEG composite films with a limited range of conductivity. Jan et al. compared CNT composite films with two additional cutting-edge materials for covering brain electrodes, namely poly (3,4-ethylene dioxythiophene) (PEDOT) and iridium oxide (IrOx).[75] The findings of this work demonstrated that the layer-by-layer-assembled CNT polymeric coatings outperformed IrOx and PEDOT coatings in terms of impedance reduction, charge transfer facilitation, and cathodic charge capability increase on typical iridium–platinum wire electrodes. Other studies looked at how CNT-composites could be used to improve electrode neural interfaces. In order to enhance the effectiveness of their Pt microelectrodes, Lu et al. coated PPy/SWCNT films on them. Besides in vitro testing, these electrodes were transplanted into the brain of rodents for testing in vivo.[76] Electrodes coated with PPy/SWCNT films demonstrated great biocompatibility in these experiments. Having these properties makes electrodes coated with PPy/SWCNT films a highly desired option for implantable neural electrodes. Laminin, a protein known to promote neuronal growth, has been utilized by Nazeri et al. to improve PLGA/CNT electrospun nanofibrous scaffolds via either poly(dopamine)

(PD) or by direct physical adsorption as a straightforward way for the modification of biomaterials.[77] After a 4-week degradation investigation, it was discovered that the scaffolds functionalized with laminin were biodegradable while maintaining their structural integrity. They concluded that the PD coating on the surface of biomimetic scaffolds was a potential technique for producing long-lasting neuronal development for neural tissue engineering.

6.2.1.4 Biosensors

To record or further analyze a signal, sensors create a signal to detect the presence of a particular material or to measure a physical attribute like temperature, mass, or electrical/optical properties.[78] Biosensors are sensors utilized for measuring chemical or biological reactions. Biosensors are extensively employed in the field of biomedical engineering. A biomedical sensor should meet many fundamental characteristics, including being minimally intrusive and causing minimum physiological injury or modification to the tissue under evaluation.[79] One of the earliest biosensors to utilize CNTs was developed by Britto et al., who utilized bromoform as a binding agent to create MWCNT-pasted electrodes and used the NC to detect dopamine (a type of neurotransmitter).[80] In a recent study, the surface of carbon paste electrodes was modified using MWCNTs, PANI nanofibers, and chitosan to create a DNA biosensor.[81] The synergistic action of MWCNT and PANI led to the formation of a biocompatible and highly conductive PANI/MWCNT/chitosan NC, which significantly improved probe DNA immobilization on the electrode surface. Glassy carbon electrodes were coated with PPy/SWCNT NC films by Li et al. using a solution comprising of PPy and SWCNTs.[82] At a potential of +1.8 V, this composite film was then oxidized. Uric acid, dopamine, and ascorbic acid were all readily oxidized by the resultant overoxidized PPy-/SWCNT-modified glassy carbon electrodes. Such electrodes are ideal for biosensor applications because of their desirable features. Pushpanjali et al. used an electrochemical sensor composed of a carbon nanotube paste electrode modified with a fine coating of electrochemically polymerized L-Leucine to assess the prevalence of cetirizine (CTZ) in the proximity of paracetamol.[83] The electrooxidation of CTZ on the functionalized electrode surface, in comparison to the bare carbon nanotube paste electrode, resulted in a significant transition in the maximum potential, as well as an increase in the peak current which can be accredited to the greater surface area of the functionalized electrode.

6.2.1.5 COVID-19 Detection

Pinals and colleagues developed a nanosensor based on SWCNTs that had been noncovalently functionalized with ACE2, a host protein that has a strong affinity for the spike proteins of the SARS-CoV-2 virus.[84] These ACE2-SWCNT nanosensors were also shown to maintain their ability to detect SARS-CoV-2 virus-like particles (35 mg/L) in a surface-immobilized configuration, with a fluorescence turn-on responsiveness of 73% within 5 seconds. Rapid testing of SARS-Cov-2 virus can be employed by developing these nanosensors as optical detection tools. Clinical nasopharyngeal samples were tested for the presence of SARS-CoV-2 antigens by Shao et al. using a field-effect transistor based on highly purified semiconductive SWCNTs.[85] As a proof of concept for a quick COVID-19 antigen detection tool with high precision and specificity at a cheap cost, the sensors showed excellent performance in distinguishing negative and positive clinical samples.

6.2.1.6 Drug Administration

CNTs have been employed in the development of profuse medicine administration systems for the treatment of several ailments. CNTs possess several unique properties (high mechanical strength, high drug loading capacity, excellent chemical stability, and high surface area) that allow them to be good carriers for medication administration. MWCNTs are particularly well suited for application in the creation of high-performance gene and drug methodologies in brain tissue cells because of their capacity to interact with these cells. Using PEG and MWCNTs, researchers have developed cocoon-like NPs that might be employed as nano-biomaterials and loaded with curcumin, a natural anticancer medicine.[86] H. Li et al. synthesized a magnetically targeted medicine delivery system utilizing a HA/CNT-Fe NC and chemically modified the NC with chitosan and folic acid for guided delivery of the anticancer medicine doxorubicin (DOX).[42] Sukhodub et al. developed a bioactive composite material employing MWCNTs filled with iron (Fe) and HA.[87] The findings of the study enabled the possible application of the composite in drug delivery systems. Another study fabricated a starch-based NC film containing MWCNTs and surface modified the NC using glucose (Gl).[45] The starch-/MWCNT-Gl NC was then converted into NPs for use in drug delivery. Carbon nanotube porin (CNTP)-coated liposomes are effective carriers for direct cytoplasmic drug administration because they facilitate lipid membrane fusing and full merging of membrane phase with the contents of the interior vesicle.[88] The anticancer medicine, DOX, was effectively delivered to cancerous cells using liposomes containing membrane-bound CNTPs and loaded with the drug, eliminating close to 90% of them.

6.2.2 Graphene

Graphene is a two-dimensional honeycomb lattice nanostructure comprised of a single layer of carbon atoms, making it an allotrope of carbon.[89] Graphene possesses exceptional material properties such as a high tensile strength of 130 ± 10 GPa and a high elastic modulus (1 TPa).[90] The biocompatibility and toxicity of graphene and graphene-based materials are influenced by their varied physicochemical features.[91] Its hydrophobicity and limited solubility in most solvents necessitate the functionalization of graphene for its broad range of applications due to its zero-band gap, sharp edges, and necessity for doping.[92,93] Graphene oxide (GO), reduced graphene oxide (RGO), graphene quantum dots, and graphene NCs with inorganic, polymer, and organic nanomaterials are all members of the graphene-based nanoparticle (GBN) family.[94,95] The integration of graphene into NCs has been described in various literature, as well as the considerable improvement in the composites' specific conductance and mechanical stability.[96,97] Nanoscale carbon sheets can considerably enhance the physiomechanical attributes of polymers at very low loadings when integrated properly. The incorporation of graphene into polymer composites enhances the electrical and mechanical aspects of biomaterials and promotes cell adhesion and cellular proliferation.[98] Biopolymers formed by incorporating graphene into carbohydrates like chitin, cyclodextrins, and cellulose provides thermostability, oxygen-repellent capabilities, flame resistance, and gelation qualities with improved biocompatibility and stability.[99] The polymeric matrix's coatings protect cells from harm caused by the sharp edges of graphene when its NPs are introduced into cells.[100] Because of these characteristics, graphene and graphene-based nanoparticles are great candidates for usage as nanofillers in polymeric composites to improve their structural and physiomechanical attributes. In PNC, mostly GO is used as nanofillers instead of pristine graphene due to the higher compatibility of GO with both organic and synthetic polymers.[101,102] Figure 6.4 illustrates the structures of different forms of graphene.

6.2.2.1 Drug/Gene Delivery and Cancer Treatment

Graphene-based NCs are favorable materials to be utilized in biomedicine and targeted drug administration systems due to the ample sp^2 hybridized carbon area of graphene. An anti-inflammatory medicine, ibuprofen, was transported by a chitosan-grafted GO in a work carried out by Rana et al.[104] Nanographene oxide (NGO) and PEG-based biocompatible NCs were employed by Liu et al. to noncovalently bind hydrophobic aromatic molecules of SN-38 (an antineoplastic drug), through π–π stacking.[11] The resultant NGO/PEG/SN-38 composite was highly water-soluble while retaining the same cancer cell cytotoxicity as pure SN-38 molecules in other organic solvents. In another study, Sun et al. synthesized an NGO-PEG NC and integrated a potent antineoplastic drug for in vitro killing of cancerous cells, DOX within it.[8] A PEG-NGO NC with a redox-responsive removable PEG casing was developed by

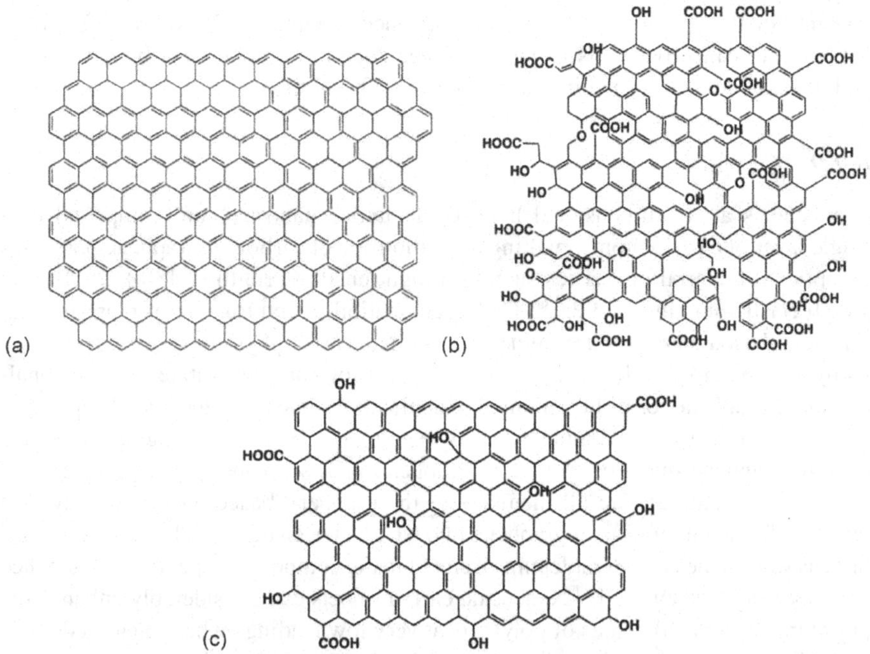

FIGURE 6.4 Molecular structure of (a) graphene (Gr), (b) graphene oxide (GO), and (c) reduced graphene oxide (RGO).[103]

Wen et al. for the swift discharge of doxorubicin hydrochloride.[105] In another study, DOX was loaded into a composite made out of gelatin as the polymer matrix and GNS as the reinforcement material.[106] The Ge-GNS/DOX composite was highly toxic to MCF-7 cancer cells and displayed prolonged release through gelatin in vitro, which may have the benefit of improving therapeutic effectiveness. To facilitate photothermally stimulated cytosolic delivery of drugs and successive delivery of drugs by endosomal rupture, researchers analyzed RGO, doped with PEG and BPEI as a nano-template.[107] The resulting RGO-BPEI-PEG NC had a higher load capacity for DOX than unreduced GO-BPEI-PEG. Paclitaxel (PTX), an anticancer medication, was successfully delivered using an improved PEG-functionalized GO created by Xu et al.[108] Comparatively, A549 and MCF-7 cells were much more sensitive to the GO–PEG–PTX nano combination than pure PTX. Nanocarriers based on PEGylated graphene oxide (PEG-GO) have been successfully employed to carry proteins into cells, as demonstrated by Shen et al.[109] Their work also demonstrated that PEG-GO NCs might be an effective protein nanocarrier because of their payload capacity, biocompatibility, and preservation of the proteins' biological functions. Kazempour et al. recently disclosed surface functionalization of GO using PEG-4000.[110] Dual delivery of two medicines, cisplatin (Pt) and DOX, was achieved by Pei et al. employing a four-armed PEG-functionalized nanographene oxide.[111] Superparamagnetic, PEG-functionalized GO/IO-NP NCs for cancer treatment were developed by Ma et al.[112] The GO–IONP–PEG–DOX composite was loaded with DOX and utilized in vitro to deliver DOX to 4T1 breast cancer cells. For the transport of DOX to human alveolar-basal epithelial cancer cell line A549, mimic drugs, O-benzylhydroxylamine (BHA) and pyridine-2-thione (PT) were conjoined into a new tri-block copolymer using the reversible addition–fragmentation chain transfer polymerization process by Song et al., and integration of graphene into the polymer was done through π–π stacking.[113] Dual drug release mechanisms were studied in the graphene/tri-block copolymer NC for the release of BHA and PT from the composite in vitro. In theory, this new composite might be used as an intracellular medicine administration system that utilizes mechanisms of dual release of drugs. Co-delivery of two anticancer medicines camptothecin (CPT) and 3,3′diindolylmethane for breast cancer treatment was created by Deb et al. using a chitosan-functionalized NC.[114] The resulting medicine-loaded CS/FA/GO NC was found to possess high anticancer activity due to the synergistic effect of the dual release of medicines. Abbasian et al. designed and developed a novel medicine administration system using chitosan-*graft*-poly(methacrylic acid) (CS-*g*-PMAA) polymers and GO NPs.[115] DOX was integrated into the CS-*g*-PMAA/GO NC and pH-dependent controlled release behaviors were examined. Results indicated that the synthesized CS-*g*-PMAA/GO composite can be used as medicine administration system for chemotherapy. A range of drug/gene delivery uses for graphene-based PNC are shown in Table 6.2.

6.2.2.2 Tissue Engineering and Regenerative Medicine

As a branch of biomedical science, tissue engineering utilizes cells, engineering, and material processes as well as appropriate biochemical or physicochemical parameters to repair, sustain, enhance, or replace various kinds of biological tissues. Blood vessels, stem cells, and extracellular matrices may be cultured and specialized on scaffolds to

TABLE 6.2
Drug/Gene Applications of Graphene Polymeric Nanocomposites

Nanocomposites	Drug/Gene	Applications	References
PEG-GO	DOX	Faster release of DOX	105
Gelatin-GNS	DOX	High toxicity against MCF-7 cancer cells	106
Folic acid-GO	DOX and CPT	High toxicity against MCF-7 cancer cells	116
PEG-GO	SN38	High cytotoxicity toward cancer cells	11
Chitosan-GO	Ibuprofen	Controlled delivery of ibuprofen	104
PEG-GO	Chlorin e6	Photodynamic destruction of cancer cells	117
PEI-GO	DOX, Si-RNA	Enhanced chemotherapy potency	118
PEI-GO	pGL-3	Targeted gene delivery	119
RGO-BPEI-PEG	Plasmid DNA	Controlled gene delivery	120

be used in regenerative medicine.[121] When a scaffold is implanted, tissue and cells in the surrounding region connect to it based on the surface structure of the scaffold. Tissue regeneration is aided by early interaction between the regenerating tissue and cells and the nanostructures. Due to their beneficial properties such as high elasticity and adaptation to uneven surfaces, graphene-based NCs may play a critical role in extended proliferation, appropriate adhesion, and higher differentiation of cells.[122] A GO/$CaCO_3$ NC was synthesized by S. Kim et al. which showed improved in vitro bone bioactivity and enhanced osteoblast cell survival.[123] Hybrid films containing PLGA and collagen (Col) and decorated with GO have recently been proposed as dermal tissue engineering scaffolds made from dual electrospun GO-coated GO-PLGA/Col films.[124] Using PAG NPs, Baniasadi et al. have developed conductive porous scaffolds that might be used to regenerate neural tissues in vitro and in vivo.[125] Biomedical applications for the conductive NC include scaffolds for tissue regeneration and nervous tissue healing, as well as other devices requiring electric conductivity. Kumar and Chatterjee synthesized a hybrid NC by integrating strontium (Sr) and RGO NPs into PCL matrix.[126] They found that integrating strontium (Sr) into the NC scaffold increased osteoblast differentiation and proliferation compared to pure PCL/RGO scaffolds.

6.2.2.3 Biosensors

In recent years, graphene-based NCs have shown tremendous potential for use in a range of biosensors that allow for the precise, responsive, and selective detection of biomarkers, among other applications. Biosensors are capable of sensing various biological matters such as glucose,[127] uric acid,[128] dopamine,[129] hemoglobin, DNA, ascorbic acid, and others. A biosensor based on RGO was developed by Cai et al. using peptide nucleic acid (PNA)-DNA hybridization. The generated RGO/PNA-DNA biosensor exhibited ultrasensitivity and excellent specificity, suggesting that it might be used as a diagnostic tool in disease diagnostics. J. W. Park et al. developed a high-sensitivity H_2O_2 sensor using graphene- PPy NCs.[130] Rapid response to H_2O_2 was exhibited by the sensor and showed high sensitivity toward H_2O_2 in a solution comprising chemicals prevalent in biofluids. Feng et al. synthesized an electrochemical biosensor based on GO for the identification of cancerous cells.[131]

6.2.3 Metal and Metal Oxide Polymeric Nanocomposites

Metallic NPs are often utilized as nanofillers in polymer/metal NC due to their optical, electrical, and mechanical properties. Metallic NPs embedded in polymer matrices are an effective approach to make use of their benefits. Polymer/metal NC may be made by in situ reduction of metal salts or compounds in a polymer matrix or via in situ polymerization of a monomer in the presence of metal NPs.[132] There is a broad range of polymeric substrate physiochemical characteristics, shape, or texture that may be formed in metal and metal oxide NC supported by polymers.[4] When it comes to PNC properties, the kind of polymer matrix, nanoparticle type, size/concentration/dispersion ratio, and dispersion level of nanofillers within the polymer matrix all go hand in hand.[133] Palladium (Pd), platinum (Pt), Gold (Au), titanium (Ti), and silver (Ag) are the most common metals to be used as nanofillers due to their remarkable physiochemical properties and compatibility with biological environments.[134]

6.2.3.1 Antimicrobial Applications

Metal-based composites have been researched for use in practical antimicrobial applications in recent years. Polymer/metal NC has also been shown in investigations on cell culture and bacterial adhesion to limit antibacterial activity and increase biocompatibility when compared to pure polymers.[135] Ag NPs are excellent antimicrobial agents but due to their cytotoxicity cannot directly be used in biological environments.[136] Embedding Ag NPs in polymer matrices can help mitigate its cytotoxic effects.[137] NPs of Ag in a polymer facilitated proliferation and adhesion of L929 cells.[138] Drugs comprising metal NPs used to delay or even impede the biogenesis of fungus and other harmful microbes were demonstrated in a nanocomposite film based on Cu/PNC.[139] When it comes to the effectiveness of Ag NPs, Morones et al. found that the smaller the nanoparticle, the more effective it was in killing *Escherichia coli* germs.[140] It was also shown that the combination of Ag/chitosan NC showed more antimicrobial effect over two strains of *Staphylococcal aureus* than any of the constituents acting alone.[141] The chitosan matrix surrounded the NPs with a positively charged coating, which increased the proportion of finer particles entering the cell wall, reduced particle accumulation, and facilitated reciprocity with negative charges on the exterior of the cell. Nuge et al. synthesized various metal/gelatin NC to be used as antibacterial nanomedicine to prevent infections in wounds and manage the inflammatory reaction.[142] They found that Ag/Ge, Fe/Ge, Zn/Ge, and Cu/Ge composites showed excellent antimicrobial activity. Surface plasma therapy of Cu/PP and Ag/PP NC increases their antimicrobial activities against infectious bacteria, according to España-Sánchez et al., who also found a surge in exposed metal NPs and an amplification in film roughness due to the treatment.[143] Other studies have demonstrated the antibacterial activities of Ag/polyethylene and Cu/polyethylene NC.[144–146] Some possible uses for polymer/Ag NC' antibacterial properties include wound dressing, food preservation, disinfection, and antifouling paint. They might also be used in water treatment applications.

6.2.3.2 Nanobiosensors

Metal/polymer NC have also been used as biosensors. It is the chemical and structural changes caused by the metal NPs within the polymer matrix, along with the various electrical configurations, that determine the sensing characteristics of metal/polymer NC.[147] The electrical conductivity of PANI/Ag composites changed in response to ethanol vapor. Higher ethanol sensitivity was found in the prepared PANI/Ag samples compared to pure PANI. Ag concentration also affected the sensor response, reaction time, and repeatability. The 2.5 mol% Ag sample demonstrated the greatest long-term stability and the maximum sensing capabilities.[148] Mazeiko et al. synthesized a gold NC (Au-NPs/PANI) using PANI to encase gold NPs Au-NPs and applied the NC on carbon electrode surfaces.[19] The findings demonstrated that Au-NPs/PANI NC preserve glucose oxidase catalytic activity and appear to be ideal for designing amperometric biosensors. Njagi et al. developed and improved a universal technique for inactivating enzymes in chemically generated Au-PPy NC, and they demonstrated their usefulness in amperometric biosensors.[20] They experimented with three enzyme systems namely, polyphenol oxidase, cytochrome *c,* and glucose oxidase. Sensors developed by biosensors allowed for fast, easy, and precise detection of phenol and glucose at a cheap cost and with great sensitivity. Other forms of biosensors, such as SPEEK/Pd, Co, Pt, Cu, and Ni NC as glucose biosensors,[147] Au/PEDOT NC for uric acid and dopamine detection in ascorbic acid[149], and PANI/Ag NC for sensing of urea have also been studied.[150]

6.2.3.3 Drug Delivery

Metallic nanomaterials have been investigated in the realm of targeted medicine administration because they have the potential to significantly enhance diagnostics and therapy. Magnetic NPs are unusual as drug carriers in that they may be guided and confined using a magnetic field.[5,151] Superparamagnetic PNC have been identified as a possible approach for achieving targeted medication administration using external magnetic fields in the treatment of complicated disorders. PNC comprised the enclosed bioactive payload and surface modifiers that are used to deliver a therapeutic substance to the desired place and at the desired rate. D. Chen et al. created an iron oxide/polymethacrylic acid/silica (FeO/PMMA/SiO_2) biocompatible magnetic NC that was created using anticancer medication DOX as a model drug.[9] A NC composed of kappa carrageenan-g-poly (acrylic acid) and superparamagnetic iron oxide (C-g-PAA/SPION) was produced by Bardajee et al., and it was proven to be efficacious in the administration of the medication deferasirox.[152] The composite was also tested against *Staphylococcus aureus* and *Escherichia coli* bacteria and was found to be an effective antimicrobial agent. The delivery of the prevalent antibiotic ciprofloxacin was achieved using a SPION/PVA PNC modified with PMMA.[153] The photocatalytic and magnetic capabilities of TiO_2 NPs are enhanced by their high surface areas. Hydrophilicity, durability, and antibacterial capabilities were also shown to be excellent characteristics of these materials.[154] Chen et al. fabricated PLA-based NC using PLA nanofibers combined with TiO_2 NPs. The resulting nTiO_2/PLA NC could be utilized to facilitate permeation and pileup of drugs on leukemia K562 cells.[155]

6.2.3.4 Cancer Therapy

There are several methods in which biocompatible metallic NCs might improve cancer therapy by serving as medication or gene delivery systems. Delivering anticancer medications to the tumor's exact location, while sparing healthy cells in the area, is an essential biomedical application known as targeted drug delivery. Drugs are encapsulated in the NPs and then external magnetic fields are utilized to navigate them to the specified area.[4] Iron-based NCs can be used to cure cancer,[156,157] mend damaged tissue,[158] and sort cells[159] using magnetic fields. SPIONs have the potential to be utilized as a cancer diagnostic MRI contrast agent. Early identification of cancer or tumor markers in tissue or blood enables rapid illness identification, therapy, and treatment for cancer patients with a good chance of long-term survival. In magnetic resonance imaging, SPIONs coated with PVA,[160] dextran,[161] and PEG[162] have been utilized to detect cancer cells. Several other magnetic NCs such as SiO_2-coated magnetic polystyrene comprising Fe_3O_4 and CdS/CdTe NPs, Fe3O4/chitosan[163], and magnetite/chitosan-L-glutamic acid[164] have also been utilized in targeted medicine administration systems.

Table 6.3 shows a variety of biomedical applications for metal and metal oxide polymeric nanocomposites.

6.2.4 Nanoclays

Clays are categorized into different groups depending on their particle form and chemical composition, including halloysite, smectite, illite, chlorite, and kaolinite.[186] Nanoclays have been investigated and developed for use in various sectors because of their vast accessibility, cheap price, and less effect on the environment.[187] Clay minerals are gradually being utilized as natural nanomaterials as nanotechnology advances.[188] Nanoclays are layered mineral silicate NPs comprising layered structures which can be stacked to build intricate clay crystalline structures.[189] Octahedral and/or tetrahedral sheets make up an individual layer unit.[190] The arrangement of these sheets influences a variety of nanoclays' defining and distinguishing characteristics. Nanoclays are classified into around 30 distinct varieties based on their mineralogical composition, and their distinct features enable them to be used in a variety of applications.[191] Table 6.4 illustrates the three fundamental sheet configurations seen in common nanoclay materials: 1:1, 2:1, and 2:1:1. The crystalline structures of nanoclays are shown in Figure 6.5. Halloysite is a natural aluminosilicate nanotube with typical measurements of 15 nm × 1000 nm.[192] Because of its structure like a hollow tube, halloysite nanoclay is utilized in numerous fields, including food packaging and medical applications, and even for rheology modification.[193] Because of its high specific surface area and capability to be exfoliated into layers of nanoscale thickness, montmorillonite (MMT) is another extensively used filler for composite materials and the development of other functional nanomaterials.[194–197] Due to these features, MMT is an excellent nanofiller for the fabrication of polymer composites with configurable physiomechanical properties.[197]

TABLE 6.3
Biomedical Applications of Metal-Based Polymeric Composites

Nanofiller	Matrix	Polymer Nanocomposite	Applications	References
Fe_3O_4	Cross-linked chitosan	Fe_3O_4/CHIT	Reduction of tumor cell proliferation	165
Fe_3O_4	Hyperbranched polyurethane	Hyperbranched polyurethane/Fe_3O_4	Shape memory materials	166
Fe_3O_4	Chitosan/polyvinyl alcohol (PVA)	Fe_3O_4/CS/ PVA	Bone regeneration	167
Fe_3O_4	Polyethylene oxide (PEO)	Fe_3O_4/PEO	Drug carrier	168
Fe_3O_4	Poly(L-lactic acid) (PLLA)	Fe_3O_4/PLLA	Amplified cell proliferation	169
Nickel oxide NPs and MWCNTs	Poly dimethylsiloxane	nNiO-MWCNT/PDMS	Microfluidic biosensor platform	46
Nickel ferrite NPs	Chitosan	n-$NiFe_2O_4$-CH	Antimicrobial agents	170
E-CS-Fe_3O_4	Chitosan	E-CHS-Fe_3O_4/CS	Hyperthermia treatment of tumor cells	171
CeO_2	PVA	CeO_2/PVA	Catalyst	172
Co_2O_3	PVA	CO_3O_4/PVA	Biomarker	173
Superparamagnetic Fe_2O_3 NPs	HA and PLA	Fe_2O_3/HA-PLA	White rabbit model of lumbar transverse defects	174
Superparamagnetic iron oxide nanoparticles (SPIONs)	Poly(D, L-lactic acid) (PDLLA)	SPION/PDLA	Bone regeneration	175
Tantalum isopropyl oxide	PVA	Ta_2O_5/PVA	Implants	176
TiO_2	Polyhydroxyethyl methacrylate	TiO_2-polyhydroxyethyl methacrylate	Ophthalmological	177
Graphene-silver hybrid particles	Poly(ε-caprolactone)	Poly(ε-caprolactone)/ graphene oxide and poly(ε-caprolactone)/ Ag	Cryogenic applications	178
Graphene oxide-iron oxide hybrid NPs	Polyethylene glycol	Graphene oxide-iron oxide/PEG	Magnetically targeted drug delivery	112
Calcium phosphate NPs	PLA	Dicalcium phosphate anhydrate/polylactic acid	Scaffolds for bone tissue restoration	179
Macrocyclic Gd (III)	Amphiphilic peptide (PA)	Gd/PA	MRI	180

(*Continued*)

TABLE 6.3 (*Continued*)
Biomedical Applications of Metal-Based Polymeric Composites

Nanofiller	Matrix	Polymer Nanocomposite	Applications	References
Gadolinium (Gd)	Hydroxyapatite nanocrystals within PCL	Gd/HA/PCL	Bone tissue regeneration monitoring by MRI	181
Au NPs	Polyoxoborate Matrix	Gold–oxoborate Nanocomposites	Antibacterial activity	182
Au NPs	Amino acid	Gold/amino acid	Biosensors	18
MNPs	Poly(ε-caprolactone) (PCL)	PCL/MNP	Bone regeneration	183
MNPs	Hydroxyapatite (HA)	HA/MNP	Bone repair	184
Cu NPs	Cellulose	Copper-based cellulose nanocomposites	Antimicrobial inhibition of a drug-resistant wound pathogen	185

TABLE 6.4
Classification of Clay Minerals[198]

Structure of Clay	Clay Category	Clay Members
1:1	Chrysotile, halloysite, rectorite, kaolinite	Brindleyite, dickite, fraipontite, kellyite, lizardite, cronstedtite, nacrite, berthierine
2:1	Smectite, mica, brittle mica, vermiculite, pyrophyllite talc	Bityite, montmorillonite, sepiolite, hectorite, bentonite, laponite, vermiculite, pyrophyllite, talc, muscovite, paragonite, clintonite
2:1:1	Chlorite	Amesite, cookeite

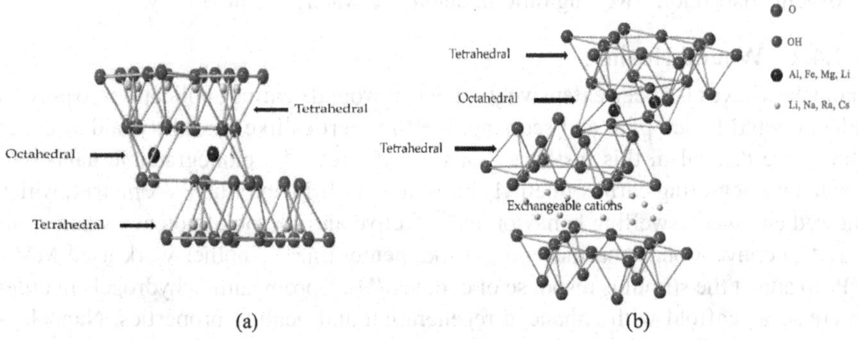

FIGURE 6.5 Crystalline structures of nanoclays: (a) Type 1:1; (b) Type 2:1.[198]

In situ polymerization, solution blending, and melt blending are three synthesis methods for the synthesis of nanoclay/PNC, and each has its unique set of procedures. The homogeneous dissipation of nanoclay in the polymer matrix is a pivotal step in the amalgamation of nanoclay/PNC.[198] In recent years, nanoclays have been intensively researched for a broad spectrum of biomedical purposes, including medicine administration, wound healing, scaffold fabrication, enzyme immobilization, bone cement, cancer treatment, and tissue engineering.[199–201] The unique features of nanoclay/PNC such as the large surface area/volume, compatibility with biopolymers, and high retention capacities are driving this expanding interest.[202]

6.2.4.1 Drug Delivery

Nanoclay/PNC have established themselves as nanocarriers for drug delivery applications.[203] They possess effective multifunctional capabilities that are required for both targeting and modulating the distribution of encapsulated anticancer and antibacterial medicines.[204] Anticancer drugs loaded in biodegradable polymer/nanoclay composites may be easily customized to supply the necessary dosages to specific tissues or organs, hence improving the antitumor activity of the drugs.[205] Saha et al. performed an in vitro discharge investigation of drugs using methylcellulose/pectin/MMT (MC/PEC/MMT) NC films for controlled transdermal drug delivery.[206] According to the results, decreasing the amount of nanoclay and pectin in the methylcellulose matrix slowed down the drug release. Gorrasi et al. investigated the mechanism of sodium benzoate release from a nanoclay/ PCL NC by incorporating halloysite nanoclays into a biodegradable PCL matrix.[207] With the inclusion of the halloysite nanoclay, the elastic moduli of the composites enhanced while the rate of thermal degradation was slowed down. It was found that a cleaning approach may successfully separate the sodium benzoate molecules, allowing for more exact modulation of the release of drugs from the NCs. PDL/MMT nanoclay composites were developed by Othman et al. for use as drug carriers.[208] The PDLA matrix's medicine encapsulation efficiency and medicine-loading capacity were both improved by the incorporation of MMT into the polymer. It also sped up the pace at which drugs were released into a simulated intestinal fluid. Pacelli et al. created a double network hydrogel using gellan gum (GG) and synthetic Laponite nanoclays. The created tough and stretchy hydrogel is a unique platform for delivering drugs with enhanced ofloxacin adsorption, swelling ratio in deionized water, and flexibility.[209]

6.2.4.2 Wound Healing

Nanoclays have also been extensively studied in wound healing applications to prevent infection and lessen pain and scarring.[210–214] Properties like flexibility and swelling ability are critical in this instance. Sabaa et al. created a biodegradable nanocomposite by integrating carboxymethyl chitosan and MMT into the PVA matrix, which showed enhanced swelling behavior and effective antimicrobial potency when compared to conventional pharmaceuticals like penicillin.[215] Another work used MMT NPs to adjust the stimulus response of collagen/N-isopropylamide hydrogels in order to create a scaffold with enhanced regeneration and healing properties. Nanoclays increased the mechanical and chemical stability of the material, making it acceptable for wound healing.[216] Halloysite nanotube (HNT)/chitosan oligosaccharide nanocomposites were investigated by Sandri et al. for their promising potential in wound healing.[213] These nanocomposites were harmless to normal human fibroblasts and promoted cell proliferation. Substantial hair follicle regeneration and revascularization were seen during in vivo wound healing. As a result, this technique might be utilized to treat skin lesions and burns that are difficult to cure.

6.2.4.3 Tissue Engineering

As the world's population grows older and suffers from more chronic diseases, the field of tissue engineering is becoming more essential. Tissue engineering encompasses not only the creation of artificial tissues and organs but also the production and synthesis of biologically active compounds that aid in tissue healing and the preservation of vital functions in living organisms.[217] Composites of polymer and nanoclay may aid in the healing and/or replacement of damaged tissues and organs.[218] Nanoclay fillers provide adjustable physical and mechanical qualities depending on the intended application.[219–223] Bonifacio et al. developed a hydrogel consisting of GG, HNTs, and glycerol for soft tissue regeneration such as skin, liver, and pancreatic healing.[224] Glycerol added to GG increased the viscosity of the substance, but HNTs reduced water absorption by 30%–35%.[224] Using the solution casting process, De Silva et al. demonstrated that the inclusion of 5% HNTs to chitosan/HNT membranes enhanced both the thermal stability and the physiomechanical attributes of the NC.[225] Nitya et al. synthesized a fibrous PCL/HNT NC by electrospinning to be utilized as scaffolds for bone regeneration.[221] The scaffolds exhibited improved protein adsorption, cell proliferation, and mineralization. HNTs were also used by Zhou et al. in order to change the surface morphology and nano-topography of PVA NC films. As a result, cell adhesion was significantly increased, and the physiomechanical attributes of the films were also significantly improved.[226]

An MMT-CNT hybrid nanostructure was developed by Shuai et al. by integrating ammonium salt-grafted CNT with MMT NPs.[227] While the lamellar MMT worked as a steric impediment to prevent CNT from entangling itself, the tubular CNT decreased the stacking of MMT. This resulted in a synergistic action that improved the dispersion of the NPs within the hybrid nanostructure. Due to these improvements in mechanics, it was found that integrating the MMT-CNT hybrid NPs to a scaffold made of poly(L-lactic) acid (PLLA) boosted both the compressive strength (risen by 58.20%) as well as the modulus (risen by 63.27%) and the tensile strength (risen by 113.04%) and modulus (risen by 111.46%) compared to scaffolds made with pure PLLA.[227] Katti et al. developed a biopolymer using chitosan combined with HA and MMT, which had an intercalated structure that improved nanomechanical characteristics and heat stability of the NC.[220] Ambre et al. developed scaffolds for bone tissue engineering applications based on chitosan/polygalacturonic acid (CSPgA) composites including customized MMT NPs. Human osteoblasts grew and proliferated, according to the findings. The results of the customized MMT were equivalent to the osteoconductive HA findings. The porosity of these composites rose by 90%, allowing nutrients to flow freely throughout the scaffold.[228] Other new approaches to using clay nanomaterials for tissue engineering have been looked into, as well. For example, Payne et al. combined an amino acid with sodium MMT (Na MMT) for mineralizing synthetic HA, which replicated the properties of biogenic HA found in the bone of a human.[223]

6.2.4.4 Bioimaging and Biosensors

Because of their influence on pharmacokinetics and biodistribution, a plethora of nanoclay/PNC have recently been introduced as mediums of contrast for several bioimaging methodologies, such as fluorescence imaging, computed tomography (CT) scan, magnetic resonance imaging (MRI), X-ray, and photoacoustic (PA) imaging.[229] Lin et al. used polydopamine/Fe_3O_4 composites to develop a new theragnostic agent with promising features including near-infrared absorption and great effectiveness in fluorescence quenching.[230] This can also be employed in MRI, PA, and mRNA detection imaging, among other bioimaging systems. The introduction of nanoclay/polymer composite biosensors has sparked curiosity in a variety of areas, such as the identification of diseases, environmental monitoring, and food analysis due to their synergistic effects and hybrid features.[231–234] However, there are still some compatibility concerns with the use of nanoclay/PNC as biosensors.[235–237] Turkmen et al. recently integrated platinum NPs into nanoclay/polymer composites for detecting glucose directly, which has been utilized to monitor the level of glucose in a human blood serum sample.[238]

6.2.4.5 Enzyme Immobilization

In recent years, researchers have looked into nanoclays' ability to immobilize enzymes. Bugatti et al. effectively enclosed lysozyme which is an antimicrobial protein inside nanoclays (HNTs) and integrated it into PLA NC.[239] Due to the influence of the curvy route, the inclusion of HNTs improved mechanical and barrier characteristics. Oliveira et al. created an enzyme-inhibiting biosensor for the detection of glyphosate levels using atemoya peroxidase incapacitated on functionalized nanoclay.[240] Lipase B from Candida antarctica was immobilized by Tzialla et al. on laponite and two types of MMTs.[241] In low-water medium, the results showed an increase in activity and stability. MMT was also used to immobilize microbial phytases from *Escherichia coli* and *Aspergillus niger*.[242]

Table 6.5 illustrates the versatility of nanoclay polymeric nanocomposites in biomedical applications.

6.3 CONCLUSION

Enhanced research in the domain of nanotechnology may lead to the development of novel nanomaterials that may revolutionize material science. PNC provide a great platform for investigating novel capabilities not available in traditional materials. The interfacial contact between the matrix and nanofiller determines the majority of the PNC's chemical, mechanical, and physical characteristics. Recent years have seen an increase in the number of studies into the potential uses of inorganic nanofiller polymeric composites due to their various physical and chemical properties. This study covers current achievements in biomedical applications of various PNC based on inorganic nanofillers. Recent years have seen an increase in interest in numerous areas of inorganic nanofiller-based composites, including their production

TABLE 6.5
Biomedical Applications of Nanoclay Polymeric Composites

Nanofiller	Polymer	Applications	References
Laponite	Pluronic-based multi-block copolymer	Low-molecular-weight protein delivery	243
Laponite	Poly(N-isopropylacrylamide)	Swelling/de-swelling applications	244,245
Laponite	PEO	Controlling cell adhesion	246
Laponite	PEG	Craniofacial, dental, and orthopedic applications	247
Hectorite/laponite	N,N-Dimethylacrylamide /Monoethanolamine (MEA)	Stimuli-responsive nanocomposites	248
Montmorillonite	Gelatin/Cellulose	Enhanced cell proliferation and biodegradation	249
Montmorillonite	Chitosan-g-lactic	Controlled drug release and cell proliferation	250
Montmorillonite	Chitosan/gelatin	Matrix for tissue engineering	251
Montmorillonite	PLLA	Scaffolds for bone tissue engineering	252
Montmorillonite	PLGA/PEG	Drug release of aspirin	253
Montmorillonite	PLGA	Controlled drug delivery of venlafaxine hydrochloride	254
Montmorillonite	Chitosan/PGA	Bone tissue engineering	255
Montmorillonite	PU	Drug delivery system for triamcinolone acetonide	256
Montmorillonite	PLA	Functionalized nanocomposites for docetaxel delivery	257
Montmorillonite	PLGA	Delivery of anticancer medicine paclitaxel	10
Montmorillonite	PU	Ocular medicine delivery of dexamethasone	258
Halloysite	Chitosan/poloxamer	Periodontitis medication	259
Halloysite	Chitosan	Nanocarriers for medicine delivery in cancer treatment	260
Na-Montmorillonite	PVA	Antimicrobial applications	261
Modified layered silicates	PU	Stomach acid -reducing medication	262

and mechanical properties, as well as cell-based research and biocompatibility testing. Inorganic nanomaterials such as graphene, HA, CNTs, and nanoclays are generally biocompatible and nontoxic. As a result, these materials are suitable for utilization in biological environments. Today, PNC of these materials are utilized in various biomedical applications such as medical implants, tissue engineering and regenerative medicine, antimicrobial applications, biosensors, drug delivery systems, etc. Utilizing inorganic nanofillers such as HA, CNTs, metal NPs, graphene and their intercalated NC hybrids, new PNC with changeable mechanical properties for applications in the biomedical field might be produced. As long as further research is done, inorganic nanofiller-based polymer composite materials may find usage in additional biomedical applications and provide new chances for the synthesis of engineered biomaterials in the near future. Innovative diagnostic and therapeutic approaches for a wide range of ailments may one day be made possible with the help of nanotechnology in medicine.

REFERENCES

1. Bitinis N, Hernandez M, Verdejo R, Kenny JM, Lopez-Manchado MA. Recent advances in clay/polymer nanocomposites. *Adv Mater.* 2011 Nov 23;23(44):5229–36.
2. Müller K, Bugnicourt E, Latorre M, Jorda M, Echegoyen Sanz Y, Lagaron J. Review on the processing and properties of polymer nanocomposites and nanocoatings and their applications in the packaging, automotive and solar energy fields. *Nanomaterials.* 2017 Mar 31;7(4):74.
3. Zaïri F, Gloaguen JM, Naït-Abdelaziz M, Mesbah A, Lefebvre JM. Study of the effect of size and clay structural parameters on the yield and post-yield response of polymer/clay nanocomposites via a multiscale micromechanical modelling. *Acta Materialia.* 2011 Jun;59(10):3851–63.
4. Sagadevan S, Fareen A, Hoque ME, Chowdhury ZZ, Johan MohdRB, Rafique RF. Nanostructured polymer biocomposites: pharmaceutical applications. In: *Nanostructured Polymer Composites for Biomedical Applications* [Internet]. Elsevier; 2019 [cited 2022 Feb 2]. p. 227–59. Available from: https://linkinghub.elsevier.com/retrieve/pii/B9780128167717000120
5. Dobson J. Magnetic nanoparticles for drug delivery. *Drug Dev Res.* 2006 Jan;67(1):55–60.
6. Elhissi AMA, Ahmed W, Hassan IU, Dhanak Vinod R, D'Emanuele A. Carbon nanotubes in cancer therapy and drug delivery. *J Drug Del.* 2012 Oct 18;2012:1–10.
7. Lvov YM, DeVilliers MM, Fakhrullin RF. The application of halloysite tubule nanoclay in drug delivery. *Exp Opin Drug Del.* 2016 Jul 2;13(7):977–86.
8. Sun X, Liu Z, Welsher K, Robinson JT, Goodwin A, Zaric S. Nano-graphene oxide for cellular imaging and drug delivery. *Nano Res.* 2008 Sep;1(3):203–12.
9. Chen D, Jiang M, Li N, Gu H, Xu Q, Ge J. Modification of magnetic silica/iron oxide nanocomposites with fluorescent polymethacrylic acid for cancer targeting and drug delivery. *J Mater Chem.* 2010;20(31):6422.
10. Dong Y, Feng S-S. Poly(d,l-lactide-co-glycolide)/montmorillonite nanoparticles for oral delivery of anticancer drugs. *Biomaterials.* 2005 Oct;26(30):6068–76.
11. Liu Z, Robinson JT, Sun X, Dai H. PEGylated nanographene oxide for delivery of water-insoluble cancer drugs. *J Am Chem Soc.* 2008 Aug 1;130(33):10876–7.
12. Zhang W, Zhang Z, Zhang Y. The application of carbon nanotubes in target drug delivery systems for cancer therapies. *Nanoscale Res Lett.* 2011 Dec;6(1):555.
13. Abarrategi A, Gutiérrez MC, Moreno-Vicente C, Hortigüela MJ, Ramos V, López-Lacomba JL, et al. Multiwall carbon nanotube scaffolds for tissue engineering purposes. *Biomaterials.* 2008 Jan;29(1):94–102.
14. Laurencin CT, Nair LS, editors. *Nanotechnology and Tissue Engineering: The Scaffold* [Internet]. CRC Press; 2008 [cited 2022 Jan 10]. Available from: https://www.taylorfrancis.com/books/9781420051834
15. O'Brien FJ. Biomaterials & scaffolds for tissue engineering. *Mat Today.* 2011 Mar;14(3):88–95.

16. Mortier J, Engelhardt M. Fremdkörperreaktion bei Karbonfaserstiftimplantation im Kniegelenk - Kasuistik und Literaturübersicht. *Z Orthop Ihre Grenzgeb.* 2000 Dec 31;138(05):390–4.
17. Zhang Q, Su K, Chan-Park MB, Wu H, Wang D, Xu R. Development of high refractive ZnS/PVP/PDMAA hydrogel nanocomposites for artificial cornea implants. *Acta Biomaterialia.* 2014 Mar;10(3):1167–76.
18. Liu Y, Yuan M, Liu L, Guo R. A facile electrochemical uricase biosensor designed from gold/amino acid nanocomposites. *Sen Actu B: Chem.* 2013 Jan;176:592–7.
19. Mazeiko V, Kausaite-Minkstimiene A, Ramanaviciene A, Balevicius Z, Ramanavicius A. Gold nanoparticle and conducting polymer-polyaniline-based nanocomposites for glucose biosensor design. *Sen Actu B: Chem.* 2013 Dec;189:187–93.
20. Njagi J, Andreescu S. Stable enzyme biosensors based on chemically synthesized Au–polypyrrole nanocomposites. *Biosens Bioelect.* 2007 Sep;23(2):168–75.
21. Lu L, Garcia CA, Mikos AG. In vitro degradation of thin poly(DL-lactic-co-glycolic acid) films. *J Biomed Mater Res.* 1999 Aug;46(2):236–44.
22. Cui FZ, Li DJ. A review of investigations on biocompatibility of diamond-like carbon and carbon nitride films. *Surf Coatings Technol.* 2000 Sep;131(1–3):481–7.
23. Iijima S. Helical microtubules of graphitic carbon. *Nature.* 1991 Nov;354(6348):56–8.
24. Sinha N, Yeow JT-W. Carbon Nanotubes for Biomedical Applications. *IEEE Trans Nanobiosci.* 2005 Jun;4(2):180–95.
25. Vashist SK, Zheng D, Al-Rubeaan K, Luong JHT, Sheu F-S. Advances in carbon nanotube based electrochemical sensors for bioanalytical applications. *Biotechnol Adv.* 2011 Mar;29(2):169–88.
26. Murakami Y, Einarsson E, Edamura T, Maruyama S. Polarization dependence of the optical absorption of single-walled carbon nanotubes. *Phys Rev Lett.* 2005 Mar 2;94(8):087402.
27. Dresselhaus MS, Dresselhaus G, Saito R, Jorio A. Raman spectroscopy of carbon nanotubes. *Phys Rep.* 2005 Mar;409(2):47–99.
28. Lefebvre J, Austing DG, Bond J, Finnie P. Photoluminescence imaging of suspended single-walled carbon nanotubes. *Nano Lett.* 2006 Aug 1;6(8):1603–8.
29. Chen WX, Tu JP, Wang LY, Gan HY, Xu ZD, Zhang XB. Tribological application of carbon nanotubes in a metal-based composite coating and composites. *Carbon.* 2003 Feb;41(2):215–22.
30. Dalton AB, Collins S, Muñoz E, Razal JM, Ebron VH, Ferraris JP. Super-tough carbon-nanotube fibres. *Nature.* 2003 Jun;423(6941):703–703.
31. Kumar S, Doshi H, Srinivasarao M, Park JO, Schiraldi DA. Fibers from polypropylene/nano carbon fiber composites. *Polymer.* 2002 Mar;43(5):1701–3.
32. Li X, Liu X, Dong W, Feng Q, Cui F, Uo M, et al. In vitro evaluation of porous poly(L-lactic acid) scaffold reinforced by chitin fibers. *J Biomed Mater Res.* 2009 Aug;90B(2):503–9.
33. Xia Z, Riester L, Curtin WA, Li H, Sheldon BW, Liang J. Direct observation of toughening mechanisms in carbon nanotube ceramic matrix composites. *Acta Materialia.* 2004 Feb;52(4):931–44.
34. Kittana N, Assali M, Abu-Rass H, Lutz S, Hindawi R, Ghannam L. Enhancement of wound healing by single-wall/multi-wall carbon nanotubes complexed with chitosan. *IJN.* 2018 Nov;13:7195–206.
35. Mirmusavi MH, Zadehnajar P, Semnani D, Karbasi S, Fekrat F, Heidari F. Evaluation of physical, mechanical and biological properties of poly 3-hydroxybutyrate-chitosan-multiwalled carbon nanotube/silk nano-micro composite scaffold for cartilage tissue engineering applications. *Int J Biol Macromol.* 2019 Jul;132:822–35.

36. Mombini S, Mohammadnejad J, Bakhshandeh B, Narmani A, Nourmohammadi J, Vahdat S, et al. Chitosan-PVA-CNT nanofibers as electrically conductive scaffolds for cardiovascular tissue engineering. *Int J Biol Macromol.* 2019 Nov;140:278–87.
37. Wang S, Li Y, Zhao R, Jin T, Zhang L, Li X. Chitosan surface modified electrospun poly(ε-caprolactone)/carbon nanotube composite fibers with enhanced mechanical, cell proliferation and antibacterial properties. *Int J Biol Macromol.* 2017 Nov;104:708–15.
38. Armentano I, Dottori M, Fortunati E, Mattioli S, Kenny JM. Biodegradable polymer matrix nanocomposites for tissue engineering: A review. *Poly Degrad Stab.* 2010 Nov;95(11):2126–46.
39. Su H, Yuan R, Chai Y, Zhuo Y, Hong C, Liu Z. Multilayer structured amperometric immunosensor built by self-assembly of a redox multi-wall carbon nanotube composite. *Electrochimica Acta.* 2009 Jul;54(17):4149–54.
40. Zhang D, Kandadai MA, Cech J, Roth S, Curran SA. Poly(L-lactide) (PLLA)/multiwalled carbon nanotube (MWCNT) composite: Characterization and biocompatibility evaluation. *J Phys Chem B.* 2006 Jul 1;110(26):12910–5.
41. Tang R, Shi Y, Hou Z, Wei L. Carbon nanotube-based chemiresistive sensors. *Sensors.* 2017 Apr 18;17(4):882.
42. Li H, Sun X, Li Y, Li B, Liang C, Wang H. Preparation and properties of carbon nanotube (Fe)/hydroxyapatite composite as magnetic targeted drug delivery carrier. *Mat Sci Engg.* 2019 Apr;97:222–9.
43. Cao Y, Huang H-Y, Chen L-Q, Du H-H, Cui J-H, Zhang LW. Enhanced lysosomal escape of ph-responsive polyethylenimine–betaine functionalized carbon nanotube for the codelivery of survivin small interfering RNA and doxorubicin. *ACS Appl Mater Interfaces.* 2019 Mar 13;11(10):9763–76.
44. Sharmeen S, Rahman AFMM, Lubna MM, Salem KS, Islam R, Khan MA. Polyethylene glycol functionalized carbon nanotubes/gelatin-chitosan nanocomposite: An approach for significant drug release. *Bioactive Mater.* 2018 Sep;3(3):236–44.
45. Mallakpour S, khodadadzadeh L. Ultrasonic-assisted fabrication of starch/MWCNT-glucose nanocomposites for drug delivery. *Ultrason Sonochem.* 2018 Jan;40:402–9.
46. Ali MdA, Srivastava S, Solanki PR, Reddy V, Agrawal VV, Kim C. Highly efficient bienzyme functionalized nanocomposite-based microfluidics biosensor platform for biomedical application. *Sci Rep.* 2013 Dec;3(1):2661.
47. Huang B, Vyas C, Roberts I, Poutrel Q-A, Chiang W-H, Blaker JJ. Fabrication and characterisation of 3D printed MWCNT composite porous scaffolds for bone regeneration. *Mat Sci Engg.* 2019 May;98:266–78.
48. Pan L, Pei X, He R, Wan Q, Wang J. Multiwall carbon nanotubes/polycaprolactone composites for bone tissue engineering application. *Coll Surf B: Biointer.* 2012 May;93:226–34.
49. Shao S, Zhou S, Li L, Li J, Luo C, Wang J, et al. Osteoblast function on electrically conductive electrospun PLA/MWCNTs nanofibers. *Biomaterials.* 2011 Apr;32(11):2821–33.
50. Zhou Y, Lei L, Yang B, Li J, Ren J. Preparation and characterization of polylactic acid (PLA) carbon nanotube nanocomposites. *Polymer Testing.* 2018 Jul;68:34–8.
51. Cheng Q, Rutledge K, Jabbarzadeh E. Carbon nanotube–poly(lactide-co-glycolide) composite scaffolds for bone tissue engineering applications. *Ann Biomed Eng.* 2013 May;41(5):904–16.
52. Lin C, Wang Y, Lai Y, Yang W, Jiao F, Zhang H. Incorporation of carboxylation multiwalled carbon nanotubes into biodegradable poly(lactic-co-glycolic acid) for bone tissue engineering. *Coll Surf B: Biointer.* 2011 Apr;83(2):367–75.
53. Mikael PE, Amini AR, Basu J, Josefina Arellano-Jimenez M, Laurencin CT, Sanders MM, et al. Functionalized carbon nanotube reinforced scaffolds for bone regenerative engineering: fabrication, *in vitro* and *in vivo* evaluation. *Biomed Mater.* 2014 Mar 31;9(3):035001.

54. Venkatesan J, Ryu B, Sudha PN, Kim S-K. Preparation and characterization of chitosan–carbon nanotube scaffolds for bone tissue engineering. *Int J Biol Macromol.* 2012 Mar;50(2):393–402.
55. Cho SY, Yun YS, Kim ES, Kim MS, Jin H-J. Stem cell response to multiwalled carbon nanotube-incorporated regenerated silk fibroin films. *Nanosci Nanotechnol.* 2011 Jan 1;11(1):801–5.
56. Hirata E, Uo M, Takita H, Akasaka T, Watari F, Yokoyama A. Multiwalled carbon nanotube-coating of 3D collagen scaffolds for bone tissue engineering. *Carbon.* 2011 Aug;49(10):3284–91.
57. Gutiérrez-Hernández JM, Escobar-García DM, Escalante A, Flores H, González FJ, Gatenholm P, et al. In vitro evaluation of osteoblastic cells on bacterial cellulose modified with multi-walled carbon nanotubes as scaffold for bone regeneration. *Mat Sci Engg.* 2017 Jun;75:445–53.
58. Jing Z, Wu Y, Su W, Tian M, Jiang W, Cao L. Carbon nanotube reinforced collagen/hydroxyapatite scaffolds improve bone tissue formation in vitro and in vivo. *Ann Biomed Eng.* 2017 Sep;45(9):2075–87.
59. Wang S, Sun X, Wang Y, Sun K, Bi J. Properties of reduced graphene/carbon nanotubes reinforced calcium phosphate bone cement in a microwave environment. *J Mater Sci: Mater Med.* 2019 Mar;30(3):37.
60. Mirjalili F, Mohammadi H, Azimi M, Hafezi M, Abu Osman NA. Synthesis and characterization of β-TCP/CNT nanocomposite: Morphology, microstructure and in vitro bioactivity. *Ceramics Int.* 2017 Jul;43(10):7573–80.
61. Anita Lett J, Sagadevan S, Fatimah I, Hoque ME, Lokanathan Y, Léonard E. Recent advances in natural polymer-based hydroxyapatite scaffolds: Properties and applications. *Eur Poly J.* 2021 Apr;148: 110360.
62. Anis B, Khalil WKB, Kamel NA, Abd El-Messieh SL. Preparation, characterization, and genotoxicity of Polyvinyl alcohol-single-wall carbon nanotubes (PVA/SWCNTs) nanocomposites for tissue engineering applications. *Adv Nat Sci: Nanosci Nanotechnol.* 2021 Dec 1;12(4): 045017.
63. Gupta A, Woods MD, Illingworth KD, Niemeier R, Schafer I, Cady C. Single walled carbon nanotube composites for bone tissue engineering: SWCNT composites for bone regeneration. *J Orthop Res.* 2013 Sep;31(9):1374–81.
64. Hench LL, Polak JM. Third-generation biomedical materials. *Science.* 2002 Feb 8;295(5557):1014–7.
65. Singh MK, Shokuhfar T, Gracio JJ de A, de Sousa ACM, Fereira JMDF, Garmestani H. Hydroxyapatite modified with carbon-nanotube-reinforced poly(methyl methacrylate): A nanocomposite material for biomedical applications. *Adv Funct Mater.* 2008 Mar 11;18(5):694–700.
66. Yao M-Z, Huang-Fu M-Y, Liu H-N, Wang X-R, Sheng X, Gao J-Q. Fabrication and characterization of drug-loaded nano-hydroxyapatite/polyamide 66 scaffolds modified with carbon nanotubes and silk fibroin. *IJN.* 2016 Nov;11:6181–94.
67. Zhao S, Cui W, Rajendran NK, Su F, Rajan M. Investigations of gold nanoparticles-mediated carbon nanotube reinforced hydroxyapatite composite for bone regenerations. *J Saudi Chem Soc.* 2021 Jul;25(7): 101261.
68. Lu XY, Qiu T, Wang XF, Zhang M, Gao XL, Li RX. Preparation of foam-like carbon nanotubes/hydroxyapatite composite scaffolds with superparamagnetic properties. *Appl Surf Sci.* 2012 Dec;262:227–30.
69. Jell G, Verdejo R, Safinia L, Shaffer MSP, Stevens MM, Bismarck A. Carbon nanotube-enhanced polyurethane scaffolds fabricated by thermally induced phase separation. *J Mater Chem.* 2008;18(16):1865.

70. Lekshmi G, Sana SS, Nguyen V-H, Nguyen THC, Nguyen CC, Le QV. Recent progress in carbon nanotube polymer composites in tissue engineering and regeneration. *IJMS.* 2020 Sep 3;21(17):6440.
71. Aslan S, Loebick CZ, Kang S, Elimelech M, Pfefferle LD, Van Tassel PR. Antimicrobial biomaterials based on carbon nanotubes dispersed in poly(lactic-co-glycolic acid). *Nanoscale.* 2010;2(9):1789.
72. Simmons TJ, Lee S-H, Park T-J, Hashim DP, Ajayan PM, Linhardt RJ. Antiseptic single wall carbon nanotube bandages. *Carbon.* 2009 May;47(6):1561–4.
73. Chen G, Wu Y, Yu D, Li R, Luo W, Ma G. Isoniazid-loaded chitosan/carbon nanotubes microspheres promote secondary wound healing of bone tuberculosis. *J Biomater Appl.* 2019 Feb;33(7):989–96.
74. Mattson MP, Haddon RC, Rao AM. Molecular functionalization of carbon nanotubes and use as substrates for neuronal growth. *JMN.* 2000;14(3):175–82.
75. Jan E, Hendricks JL, Husaini V, Richardson-Burns SM, Sereno A, Martin DC. Layered carbon nanotube-polyelectrolyte electrodes outperform traditional neural interface materials. *Nano Lett.* 2009 Dec 9;9(12):4012–8.
76. Lu Y, Li T, Zhao X, Li M, Cao Y, Yang H. Electrodeposited polypyrrole/carbon nanotubes composite films electrodes for neural interfaces. *Biomaterials.* 2010 Jul;31(19):5169–81.
77. Nazeri N, Karimi R, Ghanbari H. The effect of surface modification of poly-lactide-co -glycolide/carbon nanotube nanofibrous scaffolds by laminin protein on nerve tissue engineering. *J Biomed Mater Res.* 2021 Feb;109(2): 159–69.
78. Mahbub T, Hoque ME. Introduction to nanomaterials and nanomanufacturing for nano-sensors. In: *Nanofabrication for Smart Nanosensor Applications* [Internet]. Elsevier; 2020 [cited 2022 Feb 8]. p. 1–20. Available from: https://linkinghub.elsevier.com/retrieve/pii/B9780128207024000015
79. Rabbani M, Hoque ME, Mahbub ZB. Nanosensors in biomedical and environmental applications: Perspectives and prospects. In: *Nanofabrication for Smart Nanosensor Applications* [Internet]. Elsevier; 2020 [cited 2022 Feb 2]. p. 163–86. Available from: https://linkinghub.elsevier.com/retrieve/pii/B9780128207024000076
80. Britto PJ, Santhanam KSV, Ajayan PM. Carbon nanotube electrode for oxidation of dopamine. *Bioelectrochem Bioener.* 1996 Oct;41(1):121–5.
81. Yang T, Zhou N, Zhang Y, Zhang W, Jiao K, Li G. Synergistically improved sensitivity for the detection of specific DNA sequences using polyaniline nanofibers and multi-walled carbon nanotubes composites. *Biosens Bioelect.* 2009 Mar;24(7):2165–70.
82. Li Y, Wang P, Wang L, Lin X. Overoxidized polypyrrole film directed single-walled carbon nanotubes immobilization on glassy carbon electrode and its sensing applications. *Biosens Bioelect.* 2007 Jun;22(12):3120–5.
83. Pushpanjali PA, Manjunatha JG, Hareesha N, Souza ESD, Charithra MM, Prinith NS. Voltammetric analysis of antihistamine drug cetirizine and paracetamol at poly(L-Leucine) layered carbon nanotube paste electrode. *Surf Interfaces.* 2021 Jun;24: 101154.
84. Pinals RL, Ledesma F, Yang D, Navarro N, Jeong S, Pak JE. Rapid SARS-CoV-2 spike protein detection by carbon nanotube-based near-infrared nanosensors. *Nano Lett.* 2021 Mar 10;21(5): 2272–80.
85. Shao W, Shurin MR, Wheeler SE, He X, Star A. Rapid detection of SARS-CoV-2 antigens using high-purity semiconducting single-walled carbon nanotube-based field-effect transistors. *ACS Appl Mater Interfaces.* 2021 Mar 3;13(8): 10321–7.
86. R. H, M. J, Haridoss P, Sharma CP. Novel nano-cocoon like structures of polyethylene glycol–multiwalled carbon nanotubes for biomedical applications. *Nano-Struct Nano-Objects.* 2018 Feb;13:30–5.

87. Sukhodub LB, Sukhodub LF, Prylutskyy YuI, Strutynska NYu, Vovchenko LL, Soroca VM. Composite material based on hydroxyapatite and multi-walled carbon nanotubes filled by iron: Preparation, properties and drug release ability. *Mat Sci Engg.* 2018 Dec;93:606–14.
88. Ho NT, Siggel M, Camacho KV, Bhaskara RM, Hicks JM, Yao Y-C. Membrane fusion and drug delivery with carbon nanotube porins. *Proc Natl Acad Sci USA.* 2021 May 11;118(19):e2016974118.
89. Randviir EP, Brownson DAC, Banks CE. A decade of graphene research: Production, applications and outlook. *Mat Today.* 2014 Nov;17(9):426–32.
90. Galpaya D, Wang M, Liu M, Motta N, Waclawik E, Yan C. Recent advances in fabrication and characterization of graphene-polymer nanocomposites. *Graphene.* 2012;01(02):30–49.
91. Magne TM, de Oliveira Vieira T, Alencar LMR, Junior FFM, Gemini-Piperni S, Carneiro SV. Graphene and its derivatives: Understanding the main chemical and medicinal chemistry roles for biomedical applications. *J Nanostruct Chem.* 2021 Sep 6 [cited 2022 Feb 7]; Available from: https://link.springer.com/10.1007/s40097-021-00444-3
92. Liu C-C, Zhao J-J, Zhang R, Li H, Chen B, Zhang L-L. Multifunctionalization of graphene and graphene oxide for controlled release and targeted delivery of anticancer drugs. *Am J Transl Res.* 2017;9(12):5197–219.
93. Loh KP, Bao Q, Ang PK, Yang J. The chemistry of graphene. *J Mater Chem.* 2010;20(12):2277.
94. Dasari Shareena TP, McShan D, Dasmahapatra AK, Tchounwou PB. A review on graphene-based nanomaterials in biomedical applications and risks in environment and health. *Nano-Micro Lett.* 2018 Jul;10(3):53.
95. Gonçalves G, Vila M, Portolés M-T, Vallet-Regi M, Gracio J, Marques PAAP. Nano-graphene oxide: A potential multifunctional platform for cancer therapy. *Adv Healthcare Mater.* 2013 Aug;2(8):1072–90.
96. Compton OC, Nguyen ST. Graphene oxide, highly reduced graphene oxide, and graphene: Versatile building blocks for carbon-based materials. *Small.* 2010 Mar 22;6(6):711–23.
97. Kuilla T, Bhadra S, Yao D, Kim NH, Bose S, Lee JH. Recent advances in graphene based polymer composites. *Prog Polymer Sci.* 2010 Nov;35(11):1350–75.
98. Pinto AM, Gonçalves IC, Magalhães FD. Graphene-based materials biocompatibility: A review. *Coll Surf B: Biointer.* 2013 Nov;111:188–202.
99. Shende P, Pathan N. Potential of carbohydrate-conjugated graphene assemblies in biomedical applications. *Carbohydrate Polymers.* 2021 Mar;255: 117385.
100. Sharma H, Mondal S. Functionalized graphene oxide for chemotherapeutic drug delivery and cancer treatment: A promising material in nanomedicine. *IJMS.* 2020 Aug 30;21(17): 6280.
101. Dikin DA, Stankovich S, Zimney EJ, Piner RD, Dommett GHB, Evmenenko G. Preparation and characterization of graphene oxide paper. *Nature.* 2007 Jul;448(7152):457–60.
102. Vickery JL, Patil AJ, Mann S. Fabrication of graphene-polymer nanocomposites with higher-order three-dimensional architectures. *Adv Mater.* 2009 Jun 5;21(21):2180–4.
103. Kumar CV, Pattammattel A. Discovery of graphene and beyond. In: *Introduction to Graphene* [Internet]. Elsevier; 2017 [cited 2022 Feb 3]. p. 1–15. Available from: https://linkinghub.elsevier.com/retrieve/pii/B9780128131824000015
104. Rana VK, Choi M-C, Kong J-Y, Kim GY, Kim MJ, Kim S-H. Synthesis and drug-delivery behavior of chitosan-functionalized graphene oxide hybrid nanosheets: Synthesis and drug-delivery behavior of chitosan-functionalized. *Macromol Mater Eng.* 2011 Feb 14;296(2):131–40.

105. Wen H, Dong C, Dong H, Shen A, Xia W, Cai X. Engineered redox-responsive PEG detachment mechanism in pegylated nano-graphene oxide for intracellular drug delivery. *Small.* 2012 Mar 12;8(5):760–9.
106. Liu K, Zhang J-J, Cheng F-F, Zheng T-T, Wang C, Zhu J-J. Green and facile synthesis of highly biocompatible graphene nanosheets and its application for cellular imaging and drug delivery. *J Mater Chem.* 2011;21(32):12034.
107. Kim H, Lee D, Kim J, Kim T, Kim WJ. Photothermally triggered cytosolic drug delivery via endosome disruption using a functionalized reduced graphene oxide. *ACS Nano.* 2013 Aug 27;7(8):6735–46.
108. Xu Z, Wang S, Li Y, Wang M, Shi P, Huang X. Covalent functionalization of graphene oxide with biocompatible poly(ethylene glycol) for delivery of paclitaxel. *ACS Appl Mater Interfaces.* 2014 Oct 8;6(19):17268–76.
109. Shen H, Liu M, He H, Zhang L, Huang J, Chong Y. PEGylated graphene oxide-mediated protein delivery for cell function regulation. *ACS Appl Mater Interfaces.* 2012 Nov 28;4(11):6317–23.
110. Kazempour M, Namazi H, Akbarzadeh A, Kabiri R. Synthesis and characterization of PEG-functionalized graphene oxide as an effective pH-sensitive drug carrier. *Artif Cells Nanomed Biotechnol.* 2019 Dec 4;47(1):90–4.
111. Pei X, Zhu Z, Gan Z, Chen J, Zhang X, Cheng X. PEGylated nano-graphene oxide as a nanocarrier for delivering mixed anticancer drugs to improve anticancer activity. *Sci Rep.* 2020 Dec;10(1): 2717.
112. Ma X, Tao H, Yang K, Feng L, Cheng L, Shi X. A functionalized graphene oxide-iron oxide nanocomposite for magnetically targeted drug delivery, photothermal therapy, and magnetic resonance imaging. *Nano Res.* 2012 Mar;5(3):199–212.
113. Song Z, Xu Y, Yang W, Cui L, Zhang J, Liu J. Graphene/tri-block copolymer composites prepared via RAFT polymerizations for dual controlled drug delivery via pH stimulation and biodegradation. *Eur Polymer J.* 2015 Aug;69:559–72.
114. Deb A, Andrews NG, Raghavan V. Natural polymer functionalized graphene oxide for co-delivery of anticancer drugs: In-vitro and in-vivo. *Int J Biol Macromol.* 2018 Jul;113:515–25.
115. Abbasian M, Roudi M-M, Mahmoodzadeh F, Eskandani M, Jaymand M. Chitosan-grafted-poly(methacrylic acid)/graphene oxide nanocomposite as a pH-responsive de novo cancer chemotherapy nanosystem. *Int J Biol Macromol.* 2018 Oct;118:1871–9.
116. Zhang L, Xia J, Zhao Q, Liu L, Zhang Z. Functional graphene oxide as a nanocarrier for controlled loading and targeted delivery of mixed anticancer drugs. *Small.* 2010 Feb 22;6(4):537–44.
117. Tian B, Wang C, Zhang S, Feng L, Liu Z. Photothermally enhanced photodynamic therapy delivered by nano-graphene oxide. *ACS Nano.* 2011 Sep 27;5(9):7000–9.
118. Zhang L, Lu Z, Zhao Q, Huang J, Shen H, Zhang Z. Enhanced chemotherapy efficacy by sequential delivery of siRNA and anticancer drugs using PEI-grafted graphene oxide. *Small.* 2011 Feb 18;7(4):460–4.
119. Chen B, Liu M, Zhang L, Huang J, Yao J, Zhang Z. Polyethylenimine-functionalized graphene oxide as an efficient gene delivery vector. *J Mater Chem.* 2011;21(21):7736.
120. Kim H, Kim WJ. Photothermally controlled gene delivery by reduced graphene oxide-polyethylenimine nanocomposite. *Small.* 2014 Jan 15;10(1):117–26.
121. Chen F-M, Jin Y. Periodontal tissue engineering and regeneration: Current approaches and expanding opportunities. *Tissue Engg Part B: Rev.* 2010 Apr;16(2):219–55.
122. Zhang Y, Nayak TR, Hong H, Cai W. Graphene: A versatile nanoplatform for biomedical applications. *Nanoscale.* 2012;4(13):3833.
123. Kim S, Ku SH, Lim SY, Kim JH, Park CB. Graphene-biomineral hybrid materials. *Adv Mater.* 2011 May 3;23(17):2009–14.

124. Lee EJ, Lee JH, Shin YC, Hwang D-G, Kim JS, Jin OS. Graphene oxide-decorated PLGA/collagen hybrid fiber sheets for application to tissue engineering scaffolds. *Biomater Res.* 2014;18(1):18–24.
125. Baniasadi H, Ramazani S.A. A, Mashayekhan S. Fabrication and characterization of conductive chitosan/gelatin-based scaffolds for nerve tissue engineering. *Int J Biol Macromol.* 2015 Mar;74:360–6.
126. Kumar S, Chatterjee K. Strontium eluting graphene hybrid nanoparticles augment osteogenesis in a 3D tissue scaffold. *Nanoscale.* 2015;7(5):2023–33.
127. Shan C, Yang H, Song J, Han D, Ivaska A, Niu L. Direct electrochemistry of glucose oxidase and biosensing for glucose based on graphene. *Anal Chem.* 2009 Mar 15;81(6):2378–82.
128. Lian H, Sun Z, Sun X, Liu B. Graphene doped molecularly imprinted electrochemical sensor for uric acid. *Anal Lett.* 2012 Nov 30;45(18):2717–27.
129. Hou S, Kasner ML, Su S, Patel K, Cuellari R. Highly sensitive and selective dopamine biosensor fabricated with silanized graphene. *J Phys Chem C.* 2010 Sep 9;114(35):14915–21.
130. Park JW, Park SJ, Kwon OS, Lee C, Jang J. Polypyrrole nanotube embedded reduced graphene oxide transducer for field-effect transistor-type H_2O_2 biosensor. *Anal Chem.* 2014 Feb 4;86(3):1822–8.
131. Feng L, Chen Y, Ren J, Qu X. A graphene functionalized electrochemical aptasensor for selective label-free detection of cancer cells. *Biomaterials.* 2011 Apr;32(11):2930–7.
132. Son WK, Youk JH, Park WH. Antimicrobial cellulose acetate nanofibers containing silver nanoparticles. *Carbohyd Polym.* 2006 Sep;65(4):430–4.
133. Winey KI, Vaia RA. Polymer nanocomposites. *MRS Bull.* 2007 Apr;32(4):314–22.
134. Dzhardimalieva GI, Uflyand IE. Preparation of metal-polymer nanocomposites by chemical reduction of metal ions: functions of polymer matrices. *J Polym Res.* 2018 Dec;25(12):255.
135. Prabhakar PK, Raj S, Anuradha PR, Sawant SN, Doble M. Biocompatibility studies on polyaniline and polyaniline–silver nanoparticle coated polyurethane composite. *Coll Surf B: Biointer.* 2011 Aug;86(1):146–53.
136. Vimala K, Samba Sivudu K, Murali Mohan Y, Sreedhar B, Mohana Raju K. Controlled silver nanoparticles synthesis in semi-hydrogel networks of poly(acrylamide) and carbohydrates: A rational methodology for antibacterial application. *Carbohyd Polym.* 2009 Feb;75(3):463–71.
137. Hsu S, Liu, Dai S, Fu. Antibacterial properties of silver nanoparticles in three different sizes and their nanocomposites with a new waterborne polyurethane. *IJN.* 2010 Nov;Volume 5:1017–28.
138. Liu B-S, Huang T-B. Nanocomposites of genipin-crosslinked chitosan/silver nanoparticles: Structural reinforcement and antimicrobial properties. *Macromol Biosci.* 2008 Oct 8;8(10):932–41.
139. Cioffi N, Torsi L, Ditaranto N, Tantillo G, Ghibelli L, Sabbatini L. Copper nanoparticle/polymer composites with antifungal and bacteriostatic properties. *Chem Mater.* 2005 Oct 1;17(21):5255–62.
140. Morones JR, Elechiguerra JL, Camacho A, Holt K, Kouri JB, Ramírez JT. The bactericidal effect of silver nanoparticles. *Nanotechnology.* 2005 Oct 1;16(10):2346–53.
141. Potara M, Jakab E, Damert A, Popescu O, Canpean V, Astilean S. Synergistic antibacterial activity of chitosan–silver nanocomposites on *Staphylococcus aureus.* *Nanotechnology.* 2011 Apr 1;22(13):135101.
142. Nuge T, Tshai K, Lim S, Nordin N, Hoque M. Preparation and characterization of CU-, FE-, AG-, ZN-and NI-doped gelatin nanofibers for possible applications in antibacterial nanomedicine. *J Eng Sci Technol.* 2017;12(3):68–81.

143. España-Sánchez BL, Ávila-Orta CA, Padilla-Vaca F, Neira-Velázquez MG, González-Morones P, Rodríguez-González JA. Enhanced antibacterial activity of melt processed poly(propylene) Ag and Cu nanocomposites by argon plasma treatment: Antibacterial activity of PP/Ag and PP/Cu nanocomposites by APT. *Plasma Process Polym.* 2014 Apr;11(4):353–65.
144. Morley KS, Webb PB, Tokareva NV, Krasnov AP, Popov VK, Zhang J. Synthesis and characterisation of advanced UHMWPE/silver nanocomposites for biomedical applications. *Euro Polymer J.* 2007 Feb;43(2):307–14.
145. Tamayo LA, Zapata PA, Vejar ND, Azócar MI, Gulppi MA, Zhou X. Release of silver and copper nanoparticles from polyethylene nanocomposites and their penetration into Listeria monocytogenes. *Mat Sci Engg.* 2014 Jul;40:24–31.
146. Zhang W, Zhang Y-H, Ji J-H, Zhao J, Yan Q, Chu PK. Antimicrobial properties of copper plasma-modified polyethylene. *Polymer.* 2006 Oct;47(21):7441–5.
147. Muraviev DN, Ruiz P, Muñoz M, Macanás J. Novel strategies for preparation and characterization of functional polymer-metal nanocomposites for electrochemical applications. *Pure Appl Chem.* 2008 Jan 1;80(11):2425–37.
148. Barkade SS, Naik JB, Sonawane SH. Ultrasound assisted miniemulsion synthesis of polyaniline/Ag nanocomposite and its application for ethanol vapor sensing. *Coll Surf A: Physicochem Engg Asp.* 2011 Mar;378(1–3):94–8.
149. Crespilho FN, Iost RM, Travain SA, Oliveira ON, Zucolotto V. Enzyme immobilization on Ag nanoparticles/polyaniline nanocomposites. *Biosens Bioelect.* 2009 Jun;24(10):3073–7.
150. Mathiyarasu J, Senthilkumar S, Phani KLN, Yegnaraman V. PEDOT-Au nanocomposite film for electrochemical sensing. *Mat Lett.* 2008 Feb;62(4–5):571–3.
151. Pankhurst QA, Connolly J, Jones SK, Dobson J. Applications of magnetic nanoparticles in biomedicine. *J Phys D: Appl Phys.* 2003 Jul 7;36(13):R167–81.
152. Bardajee GR, Hooshyar Z, Rastgo F. Kappa carrageenan-g-poly (acrylic acid)/SPION nanocomposite as a novel stimuli-sensitive drug delivery system. *Colloid Polym Sci.* 2013 Dec;291(12):2791–803.
153. Bajpai AK, Gupta R. Magnetically mediated release of ciprofloxacin from polyvinyl alcohol based superparamagnetic nanocomposites. *J Mater Sci: Mater Med.* 2011 Feb;22(2):357–69.
154. Sharif A, Mondal S, Hoque ME. Polylactic acid (PLA)-based nanocomposites: processing and properties. In: Sanyang ML, Jawaid M, editors. *Bio-based Polymers and Nanocomposites* [Internet]. Cham: Springer International Publishing; 2019 [cited 2022 Feb 8]. p. 233–54. Available from: http://link.springer.com/10.1007/978-3-030-05825-8_11
155. Chen C, Lv G, Pan C, Song M, Wu C, Guo D. Poly(lactic acid) (PLA) based nanocomposites—a novel way of drug-releasing. *Biomed Mater.* 2007 Dec;2(4):L1–4.
156. Sonvico F, Mornet S, Vasseur S, Dubernet C, Jaillard D, Degrouard J. Folate-conjugated iron oxide nanoparticles for solid tumor targeting as potential specific magnetic hyperthermia mediators: Synthesis, physicochemical characterization, and in vitro experiments. *Bioconjugate Chem.* 2005 Sep 1;16(5):1181–8.
157. Wu P-C, Wang W-S, Huang Y-T, Sheu H-S, Lo Y-W, Tsai T-L. Porous iron oxide based nanorods developed as delivery nanocapsules. *Chem Eur J.* 2007 May 7;13(14):3878–85.
158. Gupta AK, Gupta M. Synthesis and surface engineering of iron oxide nanoparticles for biomedical applications. *Biomaterials.* 2005 Jun;26(18):3995–4021.
159. Groman EV, Bouchard JC, Reinhardt CP, Vaccaro DE. Ultrasmall mixed ferrite colloids as multidimensional magnetic resonance imaging, cell labeling, and cell sorting agents. *Bioconjugate Chem.* 2007 Nov 1;18(6):1763–71.
160. Kayal S, Ramanujan RV. Doxorubicin loaded PVA coated iron oxide nanoparticles for targeted drug delivery. *Mat Sci Engg.* 2010 Apr;30(3):484–90.

161. Tassa C, Shaw SY, Weissleder R. Dextran-coated iron oxide nanoparticles: A versatile platform for targeted molecular imaging, molecular diagnostics, and therapy. *Acc Chem Res*. 2011 Oct 18;44(10):842–52.
162. Allard-Vannier E, Cohen-Jonathan S, Gautier J, Hervé-Aubert K, Munnier E, Soucé M. Pegylated magnetic nanocarriers for doxorubicin delivery: A quantitative determination of stealthiness in vitro and in vivo. *Eur J Pharm Biopharm*. 2012 Aug;81(3):498–505.
163. Arias JL, Reddy LH, Couvreur P. Fe3O4/chitosan nanocomposite for magnetic drug targeting to cancer. *J Mater Chem*. 2012;22(15):7622.
164. Santos DP, Ruiz MA, Gallardo V, Zanoni MVB, Arias JL. Multifunctional antitumor magnetite/chitosan-l-glutamic acid (core/shell) nanocomposites. *J Nanopart Res*. 2011 Sep;13(9):4311–23.
165. Lin T-C, Lin F-H, Lin J-C. *In vitro* characterization of magnetic electrospun IDA-grafted chitosan nanofiber composite for hyperthermic tumor cell treatment. *J Biomat Sci, Polymer Edn*. 2013 Jun;24(9):1152–63.
166. Kalita H, Karak N. Bio-based hyperbranched polyurethane/Fe_3O_4 nanocomposites as shape memory materials: Fe_3O_4-based polyurethane nanocomposites as shape memory materials. *Polym Adv Technol*. 2013 Sep;24(9):819–23.
167. Wei Y, Zhang X, Song Y, Han B, Hu X, Wang X. Magnetic biodegradable Fe_3O_4/CS/PVA nanofibrous membranes for bone regeneration. *Biomed Mater*. 2011 Oct 1;6(5):055008.
168. Gaysinsky S, Davidson PM, McClements DJ, Weiss J. Formulation and characterization of phytophenol-carrying antimicrobial microemulsions. *Food Biophysics*. 2008 Mar;3(1):54–65.
169. Shan D, Shi Y, Duan S, Wei Y, Cai Q, Yang X. Electrospun magnetic poly(l-lactide) (PLLA) nanofibers by incorporating PLLA-stabilized Fe3O4 nanoparticles. *Mat Sci Engg*. 2013 Aug;33(6):3498–505.
170. Singh J, Roychoudhury A, Srivastava M, Chaudhary V, Prasanna R, Lee DW. Highly efficient bienzyme functionalized biocompatible nanostructured nickel ferrite–chitosan nanocomposite platform for biomedical application. *J Phys Chem C*. 2013 Apr 25;117(16):8491–502.
171. Lin T-C, Lin F-H, Lin J-C. In vitro feasibility study of the use of a magnetic electrospun chitosan nanofiber composite for hyperthermia treatment of tumor cells. *Acta Biomaterialia*. 2012 Jul;8(7):2704–11.
172. Yang X, Shao C, Liu Y, Mu R, Guan H. Nanofibers of CeO_2 via an electrospinning technique. *Thin Solid Films*. 2005 May;478(1–2):228–31.
173. Sahoo S, Dilnawaz F, Krishnakumar S. Nanotechnology in ocular drug delivery. *Drug Discovery Today*. 2008 Feb;13(3–4):144–51.
174. Meng J, Xiao B, Zhang Y, Liu J, Xue H, Lei J. Super-paramagnetic responsive nanofibrous scaffolds under static magnetic field enhance osteogenesis for bone repair in vivo. *Sci Rep*. 2013 Dec;3(1):2655.
175. Li L, Yang G, Li J, Ding S, Zhou S. Cell behaviors on magnetic electrospun poly-d, l-lactide nanofibers. *Mat Sci Engg*. 2014 Jan;34:252–61.
176. Shukla A, Patra MK, Mathew M, Songara S, Singh VK, Gowd GS. Preparation and characterization of biocompatible and water-dispersible superparamagnetic iron oxide nanoparticles (SPIONs). *Adv Sci Lett*. 2010 Jun 1;3(2):161–7.
177. Reinhardt HM, Recktenwald D, Kim H-C, Hampp NA. High refractive index TiO2-PHEMA hydrogel for ophthalmological applications. *J Mater Sci*. 2016 Nov;51(22):9971–8.
178. Kumar S, Raj S, Jain S, Chatterjee K. Multifunctional biodegradable polymer nanocomposite incorporating graphene-silver hybrid for biomedical applications. *Mater Design*. 2016 Oct;108:319–32.

179. Chae T, Yang H, Ko F, Troczynski T. Bio-inspired dicalcium phosphate anhydrate/poly(lactic acid) nanocomposite fibrous scaffolds for hard tissue regeneration: *In situ* synthesis and electrospinning: *In situ* synthesis and electrospinning. *J Biomed Mater Res.* 2014 Feb;102(2):514–22.
180. Preslar AT, Parigi G, McClendon MT, Sefick SS, Moyer TJ, Haney CR. Gd(III)-labeled peptide nanofibers for reporting on biomaterial localization *in vivo*. *ACS Nano.* 2014 Jul 22;8(7):7325–32.
181. Ganesh N, Ashokan A, Rajeshkannan R, Chennazhi K, Koyakutty M, Nair SV. Magnetic resonance functional nano-hydroxyapatite incorporated poly(caprolactone) composite scaffolds for *in situ* monitoring of bone tissue regeneration by MRI. *Tissue Engg Part A.* 2014 Oct;20(19–20):2783–94.
182. Wybrańska K, Paczesny J, Serejko K, Sura K, Włodyga K, Dzięcielewski I. Gold–oxoborate nanocomposites and their biomedical applications. *ACS Appl Mater Interfaces.* 2015 Feb 25;7(7):3931–9.
183. Singh RK, Patel KD, Lee JH, Lee E-J, Kim J-H, Kim T-H. Potential of magnetic nanofiber scaffolds with mechanical and biological properties applicable for bone regeneration. *PLoS One.* 2014 Apr 4;9(4):e91584.
184. Zeng X, Zeng X, Hu, Xie, Lan, Wu. Magnetic responsive hydroxyapatite composite scaffolds construction for bone defect reparation. *IJN.* 2012 Jul;3365.
185. Cady NC, Behnke JL, Strickland AD. Copper-based nanostructured coatings on natural cellulose: Nanocomposites exhibiting rapid and efficient inhibition of a multi-drug resistant wound pathogen, baumannii, and mammalian cell biocompatibility in vitro. *Adv Funct Mater.* 2011 Jul 8;21(13):2506–14.
186. Nazir MS, Kassim MHM, Mohapatra L, Gilani MA, Raza MR, Majeed K. Characteristic properties of nanoclays and characterization of nanoparticulates and nanocomposites. In: *Nanoclay Reinforced Polymer Composites.* Springer; 2016. p. 35–55.
187. Müller K, Bugnicourt E, Latorre M, Jorda M, Echegoyen Sanz Y, Lagaron JM. Review on the processing and properties of polymer nanocomposites and nanocoatings and their applications in the packaging, automotive and solar energy fields. *Nanomaterials.* 2017;7(4):74.
188. Rytwo G. Clay minerals as an ancient nanotechnology: Historical uses of clay organic interactions, and future possible perspectives. *Macla.* 2008;9:15–7.
189. Lee SM, Tiwari D. Organo and inorgano-organo-modified clays in the remediation of aqueous solutions: An overview. *Appl Clay Sci.* 2012;59:84–102.
190. Uddin MK. A review on the adsorption of heavy metals by clay minerals, with special focus on the past decade. *Chem Engg J.* 2017;308:438–62.
191. Savic Gajic I, Stojiljkovic S, Savic I, Gajic D. Industrial application of clays and clay minerals. *ACS Appl Mater Interfaces.* 2014;379–402.
192. Yu F, Deng H, Bai H, Zhang Q, Wang K, Chen F. Confine clay in an alternating multilayered structure through injection molding: a simple and efficient route to improve barrier performance of polymeric materials. *ACS Appl Mater Interfaces.* 2015;7(19):10178–89.
193. Morgan AB, Gilman J. Polymer-clay nanocomposites: Design and application of multifunctional materials. *Mater Matters.* 2007;2:20–5.
194. Brostow W, Dutta M, Ricardo de Souza J, Rusek P, Marcos de Medeiros A, Ito EN. Nanocomposites of poly (methyl methacrylate)(PMMA) and montmorillonite (MMT) Brazilian clay: A tribological study. *Express Polymer Lett.* 2010;4(9).
195. Jin Y-H, Park H-J, Im S-S, Kwak S-Y, Kwak S. Polyethylene/clay nanocomposite by in-situ exfoliation of montmorillonite during ziegler-natta polymerization of ethylene. *Macromolecular Rapid Commun.* 2002;23(2):135–40.

196. Raji M, Mekhzoum MEM, Bouhfid R. Nanoclay modification and functionalization for nanocomposites development: Effect on the structural, morphological, mechanical and rheological properties. In: *Nanoclay Reinforced Polymer Composites*. Springer; 2016. p. 1–34.
197. Tyan H-L, Wu C-Y, Wei K-H. Effect of montmorillonite on thermal and moisture absorption properties of polyimide of different chemical structures. *J Appl Polymer Sci*. 2001;81(7):1742–7.
198. Guo F, Aryana S, Han Y, Jiao Y. A Review of the synthesis and applications of polymer–nanoclay composites. *Appl Sci*. 2018 Sep 19;8(9):1696.
199. Mortimer GM, Jack KS, Musumeci AW, Martin DJ, Minchin RF. Stable non-covalent labeling of layered silicate nanoparticles for biological imaging. *Mat Sci Engg*. 2016 Apr;61:674–80.
200. Rawtani D, Agrawal YK. Multifarious applications of halloysite nanotubes: A review. *Rev Adv Mater Sci*, 2012; 30: 282–95.
201. Wu K, Feng R, Jiao Y, Zhou C. Effect of halloysite nanotubes on the structure and function of important multiple blood components. *Mat Sci Engg*. 2017 Jun;75:72–8.
202. Kerativitayanan P, Tatullo M, Khariton M, Joshi P, Perniconi B, Gaharwar AK. Nanoengineered osteoinductive and elastomeric scaffolds for bone tissue engineering. *ACS Biomater Sci Eng*. 2017 Apr 10;3(4):590–600.
203. Gul S, Kausar A, Muhammad B, Jabeen S. Research progress on properties and applications of polymer/clay nanocomposite. *Polymer-Plastics Technol Engg*. 2016 May 2;55(7):684–703.
204. Mishra DK, Yadav KS, Prabhakar B, Gaud RS. Nanocomposite for cancer targeted drug delivery. In: *Applications of Nanocomposite Materials in Drug Delivery* [Internet]. Elsevier; 2018 [cited 2022 Jan 18]. p. 323–37. Available from: https://linkinghub.elsevier.com/retrieve/pii/B9780128137413000145
205. Rajan M, Murugan M, Ponnamma D, Sadasivuni KK, Munusamy MA. Poly-carboxylic acids functionalized chitosan nanocarriers for controlled and targeted anti-cancer drug delivery. *Biomed Pharmacother*. 2016 Oct;83:201–11.
206. Saha NR, Sarkar G, Roy I, Rana D, Bhattacharyya A, Adhikari A. Studies on methylcellulose/pectin/montmorillonite nanocomposite films and their application possibilities. *Carbohydrate Polymers*. 2016 Jan;136:1218–27.
207. Gorrasi G, Attanasio G, Izzo L, Sorrentino A. Controlled release mechanisms of sodium benzoate from a biodegradable polymer and halloysite nanotube composite: Controlled release mechanisms of sodium benzoate. *Polym Int*. 2017 May;66(5):690–8.
208. Othman R, Vladisavljević GT, Thomas NL, Nagy ZK. Fabrication of composite poly(d,l-lactide)/montmorillonite nanoparticles for controlled delivery of acetaminophen by solvent-displacement method using glass capillary microfluidics. *Coll Surf B: Biointer*. 2016 May;141:187–95.
209. Pacelli S, Paolicelli P, Avitabile M, Varani G, Di Muzio L, Cesa S. Design of a tunable nanocomposite double network hydrogel based on gellan gum for drug delivery applications. *Euro Polymer J*. 2018 Jul;104:184–93.
210. Heydary HA, Karamian E, Poorazizi E, Khandan A, Heydaripour J. A Novel nano-fiber of iranian gum tragacanth-polyvinyl alcohol/nanoclay composite for wound healing applications. *Procedia Mat Sci*. 2015;11:176–82.
211. Liu M, Shen Y, Ao P, Dai L, Liu Z, Zhou C. The improvement of hemostatic and wound healing property of chitosan by halloysite nanotubes. *RSC Adv*. 2014;4(45):23540–53.
212. Noori S, Kokabi M, Hassan ZM. Nanoclay enhanced the mechanical properties of poly(vinyl alcohol)/chitosan /montmorillonite nanocomposite hydrogel as wound dressing. *Procedia Mat Sci*. 2015;11:152–6.

213. Sandri G, Aguzzi C, Rossi S, Bonferoni MC, Bruni G, Boselli C. Halloysite and chitosan oligosaccharide nanocomposite for wound healing. *Acta Biomaterialia.* 2017 Jul;57:216–24.
214. Yang C, Xue R, Zhang Q, Yang S, Liu P, Chen L. Nanoclay cross-linked semi-IPN silk sericin/poly(NIPAm/LMSH) nanocomposite hydrogel: An outstanding antibacterial wound dressing. *Mat Sci Engg.* 2017 Dec;81:303–13.
215. Sabaa MW, Abdallah HM, Mohamed NA, Mohamed RR. Synthesis, characterization and application of biodegradable crosslinked carboxymethyl chitosan/poly(vinyl alcohol) clay nanocomposites. *Mat Sci Engg.* 2015 Nov;56:363–73.
216. Nistor MT, Vasile C, Chiriac AP. Hybrid collagen-based hydrogels with embedded montmorillonite nanoparticles. *Mat Sci Engg.* 2015 Aug;53:212–21.
217. Hoque ME, Shehryar M, Islam KMN. Processing and characterization of cockle shell calcium carbonate (caco3) bioceramic for potential application in bone tissue engineering. *J Material Sci Eng [Internet].* 2013 [cited 2022 Feb 8];02(04). Available from: http://www.omicsgroup.org/journals/processing-and-characterization-of-cockle-shell-calcium-carbonate-bioceramic-for-potential-application-in-bone-tissue-engineering-2169-0022.1000132.php?aid=21866
218. Wang X, Jiang M, Zhou Z, Gou J, Hui D. 3D printing of polymer matrix composites: A review and prospective. *Comp Part B: Engg.* 2017 Feb;110:442–58.
219. Aliabadi M, Dastjerdi R, Kabiri K. HTCC-modified nanoclay for tissue engineering applications: A synergistic cell growth and antibacterial efficiency. *BioMed Res Int.* 2013;2013:1–7.
220. Katti KS, Katti DR, Dash R. Synthesis and characterization of a novel chitosan/montmorillonite/hydroxyapatite nanocomposite for bone tissue engineering. *Biomed Mater.* 2008 Sep;3(3):034122.
221. Nitya G, Nair GT, Mony U, Chennazhi KP, Nair SV. In vitro evaluation of electrospun PCL/nanoclay composite scaffold for bone tissue engineering. *J Mater Sci: Mater Med.* 2012 Jul;23(7):1749–61.
222. Olad A, Farshi Azhar F. The synergetic effect of bioactive ceramic and nanoclay on the properties of chitosan–gelatin/nanohydroxyapatite–montmorillonite scaffold for bone tissue engineering. *Ceramics Int.* 2014 Aug;40(7):10061–72.
223. Payne SA, Katti DR, Katti KS. Probing electronic structure of biomineralized hydroxyapatite inside nanoclay galleries. *Micron.* 2016 Nov;90:78–86.
224. Bonifacio MA, Gentile P, Ferreira AM, Cometa S, De Giglio E. Insight into halloysite nanotubes-loaded gellan gum hydrogels for soft tissue engineering applications. *Carbohydrate Polymers.* 2017 May;163:280–91.
225. De Silva RT, Pasbakhsh P, Goh K-L, Chai S-P, Ismail H. Physico-chemical characterisation of chitosan/halloysite composite membranes. *Polymer Testing.* 2013;32(2):265–71.
226. Zhou WY, Guo B, Liu M, Liao R, Rabie ABM, Jia D. Poly (vinyl alcohol)/halloysite nanotubes bionanocomposite films: Properties and in vitro osteoblasts and fibroblasts response. *J Biomed Mat Res.* 2010;93(4):1574–87.
227. Shuai C, Peng B, Liu M, Peng S, Feng P. A self-assembled montmorillonite-carbon nanotube hybrid nanoreinforcement for poly-l-lactic acid bone scaffold. *Mat Today Adv.* 2021 Sep;11: 100158.
228. Ambre AH, Katti KS, Katti DR. Nanoclay based composite scaffolds for bone tissue engineering applications. *J Nanotechnol Engg Med.* 2010 Aug 1;1(3):031013.
229. Kumar S, Sarita, Nehra M, Dilbaghi N, Tankeshwar K, Kim K-H. Recent advances and remaining challenges for polymeric nanocomposites in healthcare applications. *Prog Polymer Sci.* 2018 May;80:1–38.
230. Lin L-S, Cong Z-X, Cao J-B, Ke K-M, Peng Q-L, Gao J. Multifunctional Fe_3O_4 @polydopamine core–shell nanocomposites for intracellular mRNA detection and imaging-guided photothermal therapy. *ACS Nano.* 2014 Apr 22;8(4):3876–83.

231. Bahadır EB, Sezgintürk MK. Applications of commercial biosensors in clinical, food, environmental, and biothreat/biowarfare analyses. *Anal Biochem.* 2015 Jun;478:107–20.
232. Barsan MM, Ghica ME, Brett CMA. Electrochemical sensors and biosensors based on redox polymer/carbon nanotube modified electrodes: A review. *Analytica Chimica Acta.* 2015 Jun;881:1–23.
233. Knopfmacher O, Hammock ML, Appleton AL, Schwartz G, Mei J, Lei T. Highly stable organic polymer field-effect transistor sensor for selective detection in the marine environment. *Nat Commun.* 2014 May;5(1):2954.
234. Krishnamoorthy M, Hakobyan S, Ramstedt M, Gautrot JE. Surface-initiated polymer brushes in the biomedical field: Applications in membrane science, biosensing, cell culture, regenerative medicine and antibacterial coatings. *Chem Rev.* 2014 Nov 12;114(21):10976–1026.
235. Hu W, Chen S, Yang J, Li Z, Wang H. Functionalized bacterial cellulose derivatives and nanocomposites. *Carbohydrate Polymers.* 2014 Jan;101:1043–60.
236. Huang J, Zhu Y, Jiang W, Yin J, Tang Q, Yang X. Parallel carbon nanotube stripes in polymer thin film with remarkable conductive anisotropy. *ACS Appl Mater Interfaces.* 2014 Feb 12;6(3):1754–8.
237. Tang L-C, Wang X, Gong L-X, Peng K, Zhao L, Chen Q. Creep and recovery of polystyrene composites filled with graphene additives. *Composites Sci Technol.* 2014 Jan;91:63–70.
238. Turkmen E, Bas SZ, Gulce H, Yildiz S. Glucose biosensor based on immobilization of glucose oxidase in electropolymerized poly(o-phenylenediamine) film on platinum nanoparticles-polyvinylferrocenium modified electrode. *Electrochimica Acta.* 2014 Mar 20;123:93–102.
239. Bugatti V, Sorrentino A, Gorrasi G. Encapsulation of lysozyme into halloysite nanotubes and dispersion in PLA: Structural and physical properties and controlled release analysis. *Euro Polymer J.* 2017;93:495–506.
240. Oliveira GC, Moccelini SK, Castilho M, Terezo AJ, Possavatz J, Magalhães MRL. Biosensor based on atemoya peroxidase immobilised on modified nanoclay for glyphosate biomonitoring. *Talanta.* 2012 Aug 30;98:130–6.
241. Tzialla AA, Pavlidis IV, Felicissimo MP, Rudolf P, Gournis D, Stamatis H. Lipase immobilization on smectite nanoclays: Characterization and application to the epoxidation of α-pinene. *Bioresource Technol.* 2010 Mar;101(6):1587–94.
242. Menezes-Blackburn D, Jorquera M, Gianfreda L, Rao M, Greiner R, Garrido E. Activity stabilization of Aspergillus niger and Escherichia coli phytases immobilized on allophanic synthetic compounds and montmorillonite nanoclays. *Bioresource Technol.* 2011 Oct;102(20):9360–7.
243. Garripelli VK, Jo S. Nanocomposite thermogel for controlled release of small proteins. *J Bioactive Compatible Polymers.* 2012 May;27(3):198–209.
244. Haraguchi K, Li H-J, Matsuda K, Takehisa T, Elliott E. Mechanism of forming organic/inorganic network structures during in-situ free-radical polymerization in PNIPA–clay nanocomposite hydrogels. *Macromolecules.* 2005 Apr 1;38(8):3482–90.
245. Haraguchi K, Takehisa T. Nanocomposite hydrogels: A unique organic–inorganic network structure with extraordinary mechanical, optical, and swelling/de-swelling properties. *Adv Mater.* 2002 Aug 16;14(16):1120.
246. Gaharwar AK, Schexnailder PJ, Kline BP, Schmidt G. Assessment of using Laponite® cross-linked poly(ethylene oxide) for controlled cell adhesion and mineralization. *Acta Biomaterialia.* 2011 Feb;7(2):568–77.
247. Gaharwar AK, Rivera CP, Wu C-J, Schmidt G. Transparent, elastomeric and tough hydrogels from poly(ethylene glycol) and silicate nanoparticles. *Acta Biomaterialia.* 2011 Dec;7(12):4139–48.

248. Haraguchi K, Murata K, Takehisa T. Stimuli-responsive nanocomposite gels and soft nanocomposites consisting of inorganic clays and copolymers with different chemical affinities. *Macromolecules.* 2012 Jan 10;45(1):385–91.
249. Haroun AA, Gamal-Eldeen A, Harding DRK. Preparation, characterization and in vitro biological study of biomimetic three-dimensional gelatin–montmorillonite/cellulose scaffold for tissue engineering. *J Mater Sci: Mater Med.* 2009 Dec;20(12):2527–40.
250. Depan D, Kumar A, Singh R. Cell proliferation and controlled drug release studies of nanohybrids based on chitosan-g-lactic acid and montmorillonite. *Acta Biomaterialia.* 2009 Jan;5(1):93–100.
251. Zheng JP, Wang CZ, Wang XX, Wang HY, Zhuang H, Yao KD. Preparation of biomimetic three-dimensional gelatin/montmorillonite–chitosan scaffold for tissue engineering. *Reactive and Functional Polymers.* 2007 Sep;67(9):780–8.
252. Lee YH, Lee JH, An I-G, Kim C, Lee DS, Lee YK. Electrospun dual-porosity structure and biodegradation morphology of Montmorillonite reinforced PLLA nanocomposite scaffolds. *Biomaterials.* 2005 Jun;26(16):3165–72.
253. Liu H-J, Chu H-C, Lin L-H, Hsu S-Y. Preparation and drug release of aspirin-loaded PLGA-PEG-PLGA/montmorillonite microparticles. *Int J Polymeric Mat Polymeric Biomat.* 2015 Jul 3;64(1):7–14.
254. Jain S, Datta M. Montmorillonite-PLGA nanocomposites as an oral extended drug delivery vehicle for venlafaxine hydrochloride. *Appl Clay Sci.* 2014 Sep;99:42–7.
255. Ambre AH, Katti KS, Katti DR. Nanoclay based composite scaffolds for bone tissue engineering applications. *J Nanotechnol Engg Med.* 2010 Aug 1;1(3):031013.
256. Pinto FCH, Silva-Cunha A, Pianetti GA, Ayres E, Oréfice RL, Da Silva GR. Montmorillonite clay-based polyurethane nanocomposite as local triamcinolone acetonide delivery system. *J Nanomater.* 2011;2011:1–11.
257. Feng S-S, Mei L, Anitha P, Gan CW, Zhou W. Poly(lactide)–vitamin E derivative/montmorillonite nanoparticle formulations for the oral delivery of Docetaxel. *Biomaterials.* 2009 Jul;30(19):3297–306.
258. da Silva GR, da Silva-Cunha A, Behar-Cohen F, Ayres E, Oréfice RL. Biodegradable polyurethane nanocomposites containing dexamethasone for ocular route. *Mat Sci Engg.* 2011 Mar;31(2):414–22.
259. Kelly HM, Deasy PB, Ziaka E, Claffey N. Formulation and preliminary in vivo dog studies of a novel drug delivery system for the treatment of periodontitis. *Int J Pharm.* 2004 Apr;274(1–2):167–83.
260. Liu M, Chang Y, Yang J, You Y, He R, Chen T. Functionalized halloysite nanotube by chitosan grafting for drug delivery of curcumin to achieve enhanced anticancer efficacy. *J Mater Chem B.* 2016;4(13):2253–63.
261. Koosha M, Mirzadeh H, Shokrgozar MA, Farokhi M. Nanoclay-reinforced electrospun chitosan/PVA nanocomposite nanofibers for biomedical applications. *RSC Adv.* 2015;5(14):10479–87.
262. Xu R, Manias E, Snyder AJ, Runt J. New biomedical poly(urethane urea)–layered silicate nanocomposites. *Macromolecules.* 2001 Jan 1;34(2):337–9.

7 Metal Matrix Nanocomposites (MMNCs) in Engineering Applications

Abhishek Tevatia
Netaji Subhash University of Technology

CONTENTS

7.1 INTRODUCTION

Nanocrystalline materials have much potential due to the fact that bulk solids with a size of grain less than 100 nm have better properties than microcrystalline forms (Koch, 2007; Meyers et al., 2006; Zhang et al., 2002). Also, nanostructured materials have lower ductility and hardness than microcrystalline materials (Suryanarayana, 2005). A nanosize particle-reinforced metal matrix can increase the mechanical properties such as stiffness, hardness, ductility, and tensile strength (He et al., 2008). These unique features of the metal matrix nanocomposite (MMNC) offer a plethora of industrial applications for the production of weight-sensitive and stiffness-critical components such as connecting rod, crankshaft, driving shaft, and brake rotor disk

DOI: 10.1201/9781003279389-7

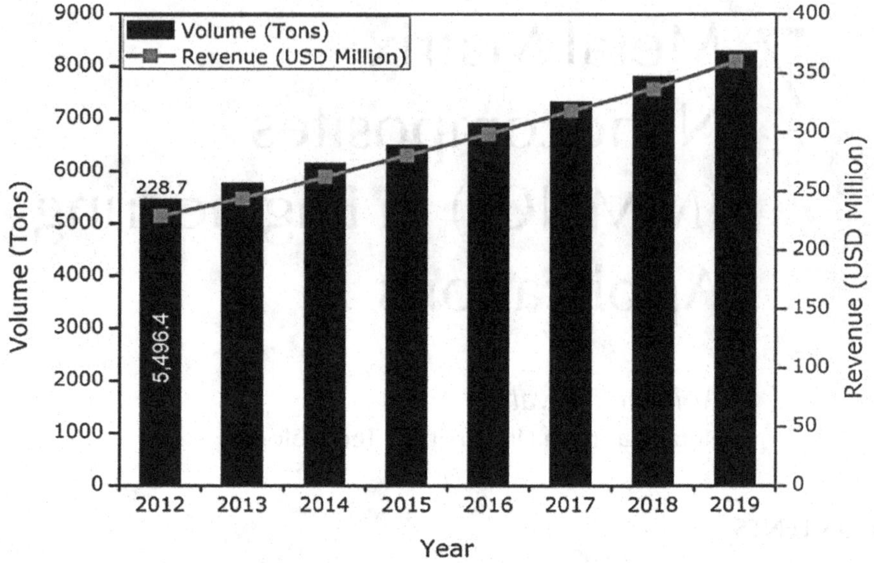

FIGURE 7.1 Global demand of MMNCs in manufacturing industries (Kumar et al., 2020).

(Ceschini et al., 2017; Kaczmar et al., 2000; Malaki et al., 2019; Pan, Jin, et al., 2022; Pan, Wang, et al., 2022; Rawal, 2001). Generally, the MMNC is a fine amalgamation of various matrices and reinforcements (Zhang et al., 2002). A market review by Kumar et al. (2020) estimated the demand of the metal matrix composite at the global level to increase from 5496 tons to 8000 tons during 2012–2019 (Figure 7.1).

Aluminum-based nanocomposites have nanosize particles homogeneously spread throughout the matrix. As a result, Al nanocomposites have enhanced attributes including superior specific stiffness (Akbari et al., 2013; Lim et al., 2012), improved ductility (Kang and Chan, 2004), significant strength (Goh et al., 2008), fatigue resistance (Ghasemi Yazdabadi et al., 2013; Goh et al., 2008; Gupta et al., 2022), stability in wear (Deuis et al., 1997), creep (Choi and Bae, 2011), and thermal properties (Čadek et al., 1999). Several studies have clearly demonstrated the advantages of employing nanoparticles, rather than microsize particles. Sajjadi et al. (2012) compared the strength of A356 reinforced with Al_2O_3 micro-particles (20 μm, 10 wt.%) and A356 reinforced with Al_2O_3 nanoparticles (50 nm, 3 wt.%), which are 453 MPa and 610 MPa, respectively. This study showed that the nanometric reinforcement phase resulted in about a 1.35 times increase in strength. Similarly, Ma et al. (1996) observed that the smaller percentages of nanoparticle reinforcements in nanocomposites (1 vol.% nano-Si3N4/Al) result in improved strength compared with the higher percentages of micro-particles in microcomposites (15 vol.% micro-SiC-Al). Several studies (Choi et al., 2012; El-Mahallawi et al., 2012; Mazahery et al., 2009) have shown that nanocomposites have higher strength than microcomposites. The tensile strength and ductility of Mg-based nanocomposites are also significantly higher than those of their microcomposite counterparts (Hassan et al., 2008; Paramsothy et al.,

2011; Radi and Mahmudi, 2010; Sun et al., 2012). This is particularly significant for structural applications that require high tensile strength and ductility.

The manufacturing aspect of nanocomposites might open technical possibilities of mass production-advanced materials in various industries (Ceschini et al., 2017; Ibrahim et al., 1991). The MMNCs have been manufactured by a variety of techniques and can be classified into three groups: (i) liquid-phase (Abdizadeh et al., 2014; Beygi et al., 2011; Lai and Chung, 1994; Tahamtan et al., 2013), (ii) solid-phase (Abdizadeh et al., 2014; Liu et al., 1994; Tun and Gupta, 2007), and (iii) two-phase processing techniques (Ceschini et al., 2017). The liquid-phase technique is more time-efficient than the solid-phase processing technique (Ceschini et al., 2017). Reinforcing nanoparticles must be distributed uniformly in the metal matrix of nanocomposites to obtain improved mechanical properties. However, it is difficult to achieve the required distribution of ceramic nanoparticles in metals. Nanosize particles have a tendency to cluster due to their inclination to agglomerate and limited wettability inside the molten matrix. One of the prominent solutions to this mixing problem has been applying high-frequency ultrasonic waves to evenly disseminate ceramic nanoparticles in molten metals (Ceschini et al., 2017).

Various approaches have been proposed for predicting the yield strength of MMNCs. Ramakrishnan (1996) has developed a mathematical model to predict the yield strength of discontinuous MMNCs and to consolidate two strengthening mechanisms related to the load-bearing effect of reinforcement and the enhanced dislocation density mechanism. Mirza and Chen (2012, 2015) considered the effect of porosity to develop the yield strength model. Zhang and Chen (2006, 2008) developed a mathematical model to predict the yield strength of MMNCs, taking into account the effect of load-bearing capacity of discontinuous reinforcement particles, the Orowan strengthening factor, and the enhanced dislocation density. The other parameters influencing the mechanical properties of MMNCs are the distribution of nanoparticles and volume fraction (Ceschini et al., 2017). In this way, various investigations had been carried out to understand these variables to improve the efficient properties of MMNCs (Ceschini et al., 2017; Gupta et al., 2022; Tevatia and Srivastava, 2015)

This chapter provides an overview of current advancements in processing and mechanical properties of MMNCs. Liquid and semi-solid methods have been discussed to explore the latest fabrication routes and properties of Mg- and Al-based MMNCs. Based on the manufacturing processes employed, the mechanical properties of composite materials with various matrix and reinforcement phases are compared. The strengthening mechanism of MMNCs has been discussed, and the yield strength model has been proposed and compared with other available models in the literature. Finally, the engineering applications and future scope of MMNCs are presented.

7.2 PROCESSING TECHNIQUES FOR MMNCs

The manufacturing process of MMNCs is classified into two routes: ex situ and in situ. Ex situ approaches involve a reinforcing phase previously manufactured and then added to the matrix (Ceschini et al., 2017), whereas in situ routes involve the reinforcement formed during the composite synthesis, often by controlled reactions (Ceschini et al., 2017). Ex situ processing is further divided into three groups, viz., liquid-phase (Abdizadeh et al., 2014; Beygi et al., 2011; Lai and Chung, 1994; Tahamtan et al., 2013), solid-phase (Abdizadeh et al., 2014; Liu et al., 1994; Tun and Gupta, 2007), and two-phase processing techniques (Ceschini et al., 2017). The liquid-phase technique is more time-efficient than the solid-phase processing technique. In liquid-phase processing, the liquid matrix is used for the fabrication of MMNCs using the processes such as liquid–metal infiltration (stir casting), squeeze casting, pressure casting, spray co-deposition, and compo-casting (Ceschini et al., 2017). The basic principle of this technique is to permeate the reinforcement material with the molten matrix to produce MMNCs. In the solid-phase processing technique, the powder matrix material is used for the fabrication of MMNCs using the processes such as powder metallurgy, hot rolling, diffusion bonding, drawing, pneumatic impaction, and explosive welding (Ceschini et al., 2017).

In powder metallurgy, the mixing process homogeneously distributes the reinforcement, followed by compaction of the mixture under high pressure. Next, the compact mixture is heated below the melting temperature to obtain the solid-state diffusion, which develops the bonding between the matrix and ceramic reinforcement (Liu et al., 1994; Tun and Gupta, 2007). There are several studies (Ceschini et al., 2017; Deuis et al., 1997; He et al., 2008; Ibrahim et al., 1991) that focus on different liquid- and solid-phase processing techniques. Table 7.1 describes various processing techniques used in the fabrication of MMNCs.

7.2.1 Reinforcement Material and Structures

The reinforcement in MMNCs is selected based on the specific application and can be categorized based on the shape and the size. The reinforcement of nanoparticles into the matrix significantly improves the stiffness and strength of MMNCs (Ma et al., 1996; Zhong et al., 2007). The most common nanoreinforcement is alumina (Al_2O_3), silicon carbide (SiC), titanium carbide (TiC), boron nitride (BN), magnesium oxide (MgO), boron carbide (B_4C), yttrium oxide (Y_2O_3), zirconium dioxide (ZrO_2), zinc oxide (ZnO), carbon nanotube, and graphite (Ceschini et al., 2017; He et al., 2008). The particle diameter varies from 1 nm to 100 nm (Ceschini et al., 2017). SiC and Al_2O_3 are mostly used as reinforcement in aluminum-based matrix materials. The reinforcement of Al_2O_3 ceramic particles enhances their strength and modulus about twice that of pure aluminum (El-Mahallawi et al., 2012; Radi and Mahmudi, 2010; Sajjadi et al., 2012). On the other hand, the reinforcement of SiC particles increases the modulus of pure aluminum approximately equal to that of steel (Lianxi and Erde, 2000; Sun et al., 2012).

7.2.2 Matrix Material Selection

Matrix materials such as aluminum, magnesium, titanium, nickel, and copper alloy develop MMNCs (Ceschini et al., 2017; He et al., 2008; Ibrahim et al., 1991). The matrix material is selected based on its mechanical and electrical properties such

TABLE 7.1
Processing Techniques Used in the Development of MMNCs

Processing Technique	Cost of Fabrication	Description	Applications	Reference
Stir casting	Least expensive	The process is suited for particulate reinforcement in Al matrix composite. It depends on material properties and process parameters.	This process is a commercial method for producing Al-based MMNCs and is applicable for mass production.	Abdizadeh et al. (2014); Beygi et al., 2011; Ceschini et al., 2017; Lai and Chung, 1994; Sajjadi et al., 2012; Tahamtan et al., 2013)
Compo-casting	Low	The process is suited for discontinuous fibers (particulate reinforcement).	The application is in aerospace and automotive industries.	(Ceschini et al., 2017; Sajjadi et al., 2012)
Squeeze casting	Medium	The process is suited for any type of reinforcement and may be used for the mass production.	The application in automotive industries for producing cylinder heads, pistons, rocker arms, and connecting rods.	(Ceschini et al., 2017; Lianxi and Erde, 2000)
Spray casting	Medium	The process is suited for particulate reinforcement and produce full density materials.	Material is used in the production of electrical brushes and contacts, friction materials, cutting, and grinding tools.	(Ceschini et al., 2017; Lianxi and Erde, 2000)
In situ (reactive) processing	Expensive	This process is used for homogeneous distribution of reinforcement.	The application is in automotive industries.	(Ceschini et al., 2017)
Liquid–metal infiltration	Low/ medium	The process is suited for the production of structural shapes like tubes, rods, and beams having the properties in the uniaxial direction.	Reinforcement filaments are used.	(Ceschini et al., 2017; Lai and Chung, 1994)
Powder metallurgy	Medium	In this process, the powder of matrix and reinforcements are used to develop the composite. The process is suited for particulate reinforcement. High-strength MMNCs are obtained by this process.	Application in producing valves, pistons, bolts, heat-resistant, and high-strength materials.	(Abdizadeh et al., 2014; Ceschini et al., 2017; Liu et al., 1994; Tun and Gupta, 2007)
Ultrasonic-assisted casting	Expensive	The process has good dispersion and uniform distribution of reinforcement.	The application in the development for net shape fabrication of complex structural components.	(Ceschini et al., 2017)
Diffusion bonding	High	Used to produce monofilament reinforced AMCs.	Used to make blades, sheets, vane shafts, and some other structural components.	(Ceschini et al., 2017)
Friction stir welding	Moderate/ expensive	This process is used to increase the microhardness of the surface.	Used in aerospace and automotive applications.	(Ceschini et al., 2017)

TABLE 7.2
Aluminum Alloy Groups Accepted as Matrix Materials Designated by Aluminum Association System (Ceschini et al., 2017; Ibrahim et al., 1991)

S. No.	Aluminum Alloy Group	AAS Designation	Superior Characteristics
1.	Al-Cu-Mg (2xxx)	AA2009, AA2014, AA2048, AA2124	Provide excellent combination of strength and damage tolerance.
2.	Al-Mg-Si-Cu (6xxx)	AA6061, AA6082, AA6090	Improve corrosion resistance for severe environments.
3.	Al-Zn-Mg-Cu (7xxx)	AA7075, AA7090, AA7091	Offer high-strength potential, which suited for aerospace applications.
4.	Al-Fe-Li (8xxx)	AA8090	Show good wettability characteristics and provide the opportunity for higher temperature tolerance.

as good corrosion resistance, high thermal conductivity, low density, low electrical resistance, and heat treatment capability. These MMNCs have low cost, wide acceptance in the aerospace, automobile and hydrospace industries, and overall versatility processing (Ibrahim et al., 1991). Aluminum alloys are most widely accepted as matrix material in the commercial applications of MMNCs and have high market value (Deuis et al., 1997; Lai and Chung, 1994; Tahamtan et al., 2013). Table 7.2 presents the characteristics of aluminum alloy as a matrix material (Ceschini et al., 2017; Ibrahim et al., 1991).

7.3 STRENGTHENING MECHANISMS

There are generally two methodologies related to the prediction of the theoretical strengthening effect in MMNCs (Goh et al., 2007; Ibrahim et al., 1991; Mirza and Chen, 2012, 2015; Ramakrishnan, 1996; Zhang and Chen, 2006, 2008; Zhong et al., 2007). The first methodology depends on the dislocation loop mechanism, counting different dislocation line blocking systems that may add to the strengthening impact because of the proximity of the "rigid" discontinuous reinforced nanoscale particles. The second methodology depends on nanomechanics with regard to the Eshelby consideration hypothesis (Ceschini et al., 2017).

7.3.1 Load-Bearing Effect

The load-bearing effect is defined as the shear load shift from the soft metal matrix to the nanoparticles. The modified shear lag model is applicable when a cohesive bond is established between nanoparticles and the metal matrix (Ceschini et al., 2017; Ramakrishnan, 1996; Tevatia and Srivastava, 2015, Mirza and Chen, 2015). At this nanoscale point, some of the shear load transfers are carried out by metal matrix reinforcement particles, which depend on the volume fraction of the discontinuous reinforcement particles of MMNCs. Mathematically, the modified shear lag model is expressed as follows (Ramakrishnan, 1996):

$$\Delta\sigma_{MSL} = \frac{1}{2}V_{np}V_m \tag{7.1}$$

where V_{np} and V_m are the volume fractions of nanoparticles and the metal matrix material, respectively. The strengthening mechanism in the presence of porosity can be expressed as follows (Mirza and Chen, 2015):

$$\Delta\sigma_{MSL} = \sigma_{ym}\left(\frac{1}{2}V_f - P\right) \tag{7.2}$$

where σ_{ym} is the yield strength of the matrix material (MPa) and P is the volume fraction of porosity.

7.3.2 Enhanced Dislocation Density Mechanism

The enhanced dislocation density mechanism, also known as Taylor strengthening mechanism, represents an increase in yield stress in MMNCs due to the presence of the geometrical dislocation. At the point where MMNCs are cooled from the working temperature to room temperature, the volumetric strain mismatch between the metal matrix and the discontinuous reinforcement components in MMNCs can occur due to the difference in the coefficient of thermal expansion (CTE). The strain mismatch creates a geometrically necessary dislocation in the range of gain components to the force CTE dissimilarity (Ramakrishnan, 1996). Mathematically, the enhanced dislocation density model is expressed as follows (Ramakrishnan, 1996):

$$\Delta\sigma_{CTE} = G_m bK\sqrt{\frac{12\Delta\alpha V_f \Delta T}{(1-V_f)bd_p}} \tag{7.3}$$

where G_m is the shear modulus of the matrix (MPa); K is a constant ($K \approx 1.25$), based on theoretical estimates (Mirza and Chen, 2015); b is the Burgers vector of the matrix (nm); d_p is the nanoparticle size (nm); $\Delta T = T_{\text{process}} - T_{\text{test}}$ is the temperature change; and $\Delta\alpha = \alpha_m - \alpha_p$ is the coefficient of the thermal expansion difference between the reinforcement phase and the matrix.

7.3.3 Orowan Strengthening

In MMNCs, a barrier is built by widely spaced rigid particles against dislocation motion during discontinuous enhanced MMNCs and is referred to the Orowan enhancement effect. Since the size of discontinuous reinforcement components is large and the distance between the particles is even larger, it has consequences. The movement of the dislocation lines occurs by temporarily restraining these nanoscale particles by initially bending, then reconnecting, and finally forming a dislocation line loop in the region of the discontinuous reinforcement particles. By adding nanoscale and strong discontinuous reinforced particles, it is reasonable to accept that Orowan strengthening can occur (Mirza and Chen, 2012, 2015; Zhang and Chen, 2006, 2008). Mathematically, the Orowan strengthening model is expressed as follows (Zhang and Chen, 2006):

$$\Delta\sigma_{\text{Orowan}} = \frac{0.13G_m b}{\lambda}\ln\left(\frac{d_p}{2b}\right) \tag{7.4}$$

where $\lambda = d_p\left[\left(\frac{1}{2V_f}\right)^{\frac{1}{3}} - 1\right]$.

7.3.4 Modulus Mismatch Strengthening

The modulus mismatch enhances the yield strength of MMNCs. The elastic modulus mismatch strengthening mechanism portrays the development of geometrically necessary dislocation once an MMNC is exposed to compressive load conditions. Many geometrically necessary offsets must be made to compensate for the modulus dissimilarity between the matrix and nanoparticles due to the presence of discontinuous nanoscale reinforcing particles, and in this way, distortion deformation arises during post-processing (Mirza and Chen, 2012):

$$\Delta\sigma_{EM} = \sqrt{3}\beta G_m b\sqrt{\frac{6V_f}{bd_p}\varepsilon} \tag{7.5}$$

where β is material specific coefficient and ε is the bulk strain of MMNCs.

7.3.5 Yield Strength Model

Considering the load-bearing effect of the reinforcement (MSL model), enhanced dislocation density strengthening mechanism (EDD model), modulus mismatches strengthening mechanism (EM model), and the Orowan strengthening effect, the yield strength model can be expressed as follows:

$$\sigma_{yc} = \left(1 + \frac{1}{2}V_{np} - P\right)\left(1 + \left(\frac{\sqrt{3}\beta G_m b\sqrt{\frac{6V_{np}}{bd_p}\varepsilon}}{\sigma_{ym}}\right)\right)\left(\sigma_{ym} + C + D + \frac{CD}{\sigma_{ym}}\right) \tag{7.6}$$

where $C = 1.25G_m b\sqrt{\frac{12\Delta T\Delta\alpha V_{np}}{(1 - V_{np})bd_p}}$, $D = \frac{0.13G_m b}{d_p\left[\left(\frac{1}{2V_{np}}\right)^{\frac{1}{3}} - 1\right]}\ln\left(\frac{d_p}{2b}\right)$.

TABLE 7.3
Mechanical Properties for the Different Types of MMNCs

Parameter of MMNCs	Mg-SiC (Mirza and Chen, 2015)	Mg carbon nanotube(Mirza and Chen, 2015)	Mg-Al_2O_3 (Mirza and Chen, 2015)	Mg-Y_2O_3 (Goh et al., 2007; Mirza and Chen, 2015)
σ_{ym} (MPa)	97	97	97	97
E_m (MPa)	42.8×103	42.8×103	42.8×103	42.8×103
v	0.3	0.3	0.3	0.3
G_m (MPa)	16.5×103	16.5×103	16.5×103	16.5×103
b (nm)	0.32	0.32	0.32	0.32
α_m (oC)-1	28.4×10–6	28.4×10–6	28.4×10–6	28.4×10–6
α_p (oC)-1	4.5×10–6	–1.52×10–6	9×10–6	7.5×10–6
$T_{process}$ (oC)	300	350	300	250
T_{test} (oC)	20	20	20	20
d_p (nm)	50	50	50	50
n	1.94	1.94	1.94	1.94
K	0.05	0.05	0.05	0.05
$V_{porosity}$(%)	1	1	1	1
ε	0.0741	0.0823	0.084	0.0789

The values of modeling parameters used in the calculation of yield strength (Equation 7.6) for different MMNCs are presented in Table 7.3.

Figure 7.2 depicts the effect of agreement between the yield strength (Equation 7.6) and experimental data available for different MMNCs. Figure 7.2 suggests that the present model estimations are off the experimental data set. For instance, at $V_f = 0.002$, the present model underestimates the experimental data (Zhong et al., 2007) (experimental $\sigma_{yc} = 144.38$ MPa and predicted $\sigma_{yc} = 132.97$ MPa); at $V_f = 0.0065$, the present model gives results approximately equal to the experimental data (experimental $\sigma_{yc} = 170.09$ MPa and predicted $\sigma_{yc} = 166.32$ MPa); at $V_f = 0.011$, the present model also gives results approximately equal to the experimental data (experimental $\sigma_{yc} = 194.28$ MPa and predicted $\sigma_{yc} = 189.91$ MPa); and at $V_f = 0.0115$, the present model overestimates the experimental data (experimental $\sigma_{yc} = 176.67$ MPa and predicted $\sigma_{yc} = 191.42$ MPa). These skewed estimations are due to the presence of porosity in the MMNCs (Mirza and Chen, 2015).

Figure 7.3 shows the effectiveness of the yield strength model (Equation 7.6) is comparable to that of different mathematical models found in the literature (Zhang and Chen, 2006; Mirza and Chen, 2015; Ramakrishnan, 1996; Zhong et al., 2007; Goh et al., 2007). Figure 7.3 depicts the comparison of fitness of present and literature-cited analytical models with available experimental data points (Zhong et al., 2007). Thus, by considering the effect of porosity and modulus mismatch, prediction can be improved.

Figure 7.4 shows the effect of nanoparticle size (d_p) on the yield strength (σ_{yc}) for different values of volume fraction (V_f) using Equation (7.6). The porosity volume

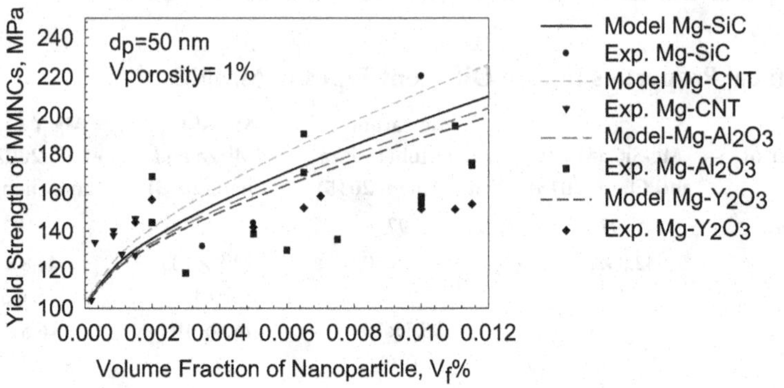

FIGURE 7.2 Comparison of analytical yield strength with experimental data for different MMNCs with varied volume fractions of nanoparticles.

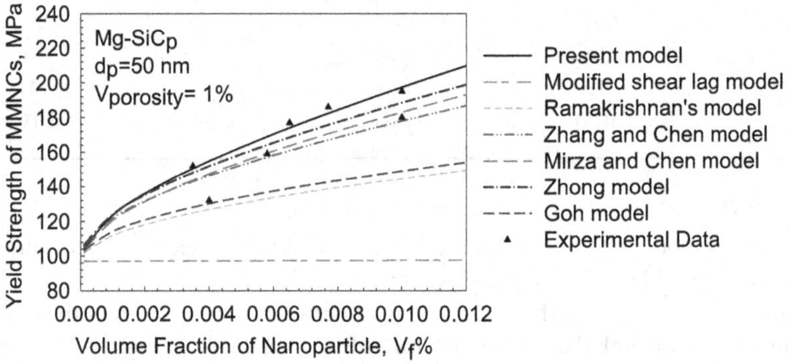

FIGURE 7.3 Comparison of the present model with other models fitting the experimental data.

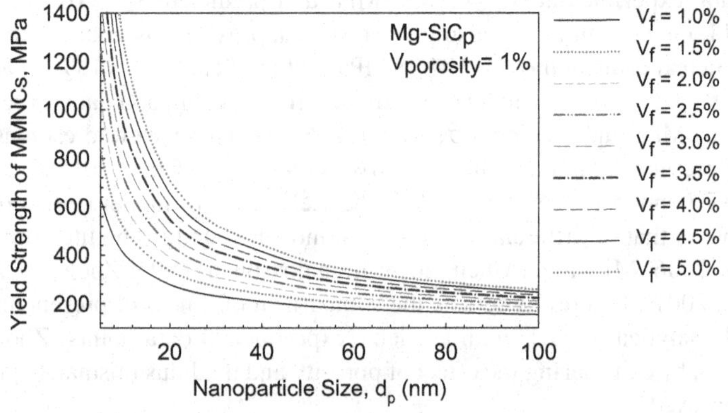

FIGURE 7.4 Yield strength as a function of nanoparticle size for different volume fractions of nanoparticles in DRMMNCs.

fraction ($V_{Porosity}$) is fixed at 1%. For a given range of volume fraction (0.001–0.014), higher yield strength is observed at lower nanoparticle size levels. The yield strength of MMNCs increases with an increase in the volume fraction value, regardless of the nanoparticle size levels. On the other hand, the nanoparticle size estimates strongly affect the yield strength. A small volume fraction of under 0.05 can improve the yield strength of MMNCs.

7.4 MECHANICAL AND THERMAL PROPERTIES

The nanosize particles can have improved Young's modulus, yield quality, extreme rigidity, and hardness of ductile-based lattice (Goh et al., 2008; Hassan et al., 2008; Lim et al., 2012). Hassan et al. (2008) revealed that the modulus is expanded by 22% when 1.11 vol.% Al_2O_3 nanoparticles of ~50 nm are reinforced into the Mg grid. Goh et al. (2008) uncovered that the modulus is expanded by 32% by reinforcing 1.0 vol.% MgO particles of ~36 nm into Mg-based nanocomposites. The modeling parameters used in the calculation of the elastic modulus and CTE for different Mg-based nanoparticle-reinforced MMNCs are presented in Table 7.3.

7.4.1 Elastic Modulus

For MMNCs, the dependence of the composite elastic modulus E_C on the volume fraction V_f of nanoparticle can be expressed as follows (Ceschini et al., 2017):

$$E_c = \overline{E}\, V_{np}^{\frac{2}{3}} + E_m\left(1 - V_{np}^{\frac{2}{3}}\right) \tag{7.7}$$

where $\overline{E} = \dfrac{E_m E_{np}}{E_{np} + V_{np}^{\frac{1}{3}}\left(E_m - E_{np}\right)}$, and E_m E_{np} are the elastic modulus of matrix and reinforcement particles (MPa), respectively. The present model Equation (7.7) can determine the elastic modulus of all the conservative composites without considering the size effect of reinforcement particle of MMNCs. To take the effect of nanoparticle size for predicting the elastic modulus of MMNCs, a size efficiency factor k_f is introduced in Equation (7.7), which takes care of different sizes of the reinforcements (E_{nc}), as follows:

$$E_{nc} = k_f \overline{E}\, V_f^{\frac{2}{3}} + E_m\left(1 - V_f^{\frac{2}{3}}\right) \tag{7.8}$$

where $k_f = 1 + k_0 e^{-\left(sd_p{*}10^9\right)}$, k_o and s are constants.

Figure 7.5 depicts the effect of particle size (d_p) on the elastic modulus (Equation 7.8) for different volume fractions (V_f), keeping other modeling parameters constant. The constraint k_o and s are fixed for nanoparticle-reinforced Mg-based MMNCs to the values of 25 and 0.20, respectively. For $0 \le d_p \le 600$ nm in MMNCs, a higher

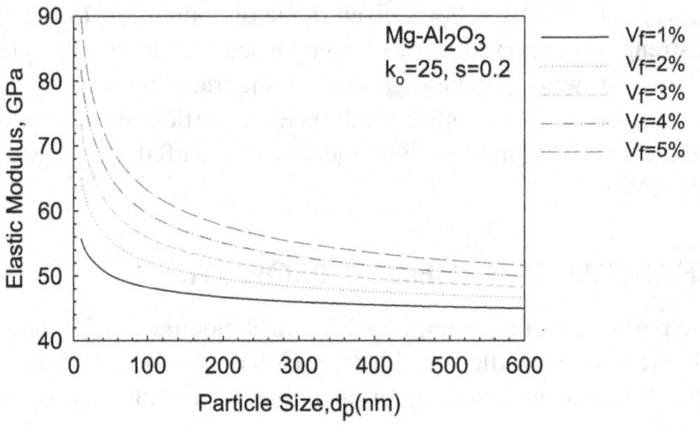

FIGURE 7.5 Predicted elastic modulus of the present model (Equation 7.8) as a function of nanoparticle size for different volume fractions of nanoparticles for Mg-based MMNCs.

volume fraction led to a higher elastic modulus (Figure 7.5), and a higher value of particle size results in a lower elastic modulus. These results present an inverse relation between the elastic modulus and particle size.

7.4.2 Coefficient of Thermal Expansion

The CTE of particulate-reinforced MMNCs with nanosize particles α_{nc} can be specified by the following relation (Ceschini et al., 2017):

$$\alpha_{nc} = \frac{\bar{\alpha} V_f^{\frac{2}{3}} \bar{E} + \alpha_m E_m \left(1 - V_f^{\frac{2}{3}}\right)}{E_{nc}} \tag{7.9}$$

in which

$$\bar{\alpha} = \alpha_m - V_f^{\frac{1}{3}} \left(\alpha_m - \alpha_p\right) \tag{7.10}$$

where α_m and α_p denote the CTE of the matrix and reinforcement particles, respectively.

Figure 7.6 shows the effect of particle size (d_p) on the CTE (Equation 7.9) for different volume fractions (V_f), keeping other modeling parameters constant. The constraints and k_o are assumed to be fixed at 0.02 and 25, respectively, for particulate-reinforced Mg-based MMNCs. For $0 \leq d_p \leq 600\,\text{nm}$ in MMNCs, a higher volume fraction always led to a lower CTE (Figure 7.6), and a higher value of particle size resulted in a higher CTE. These results comply with the contrary relation between CTE and particle size.

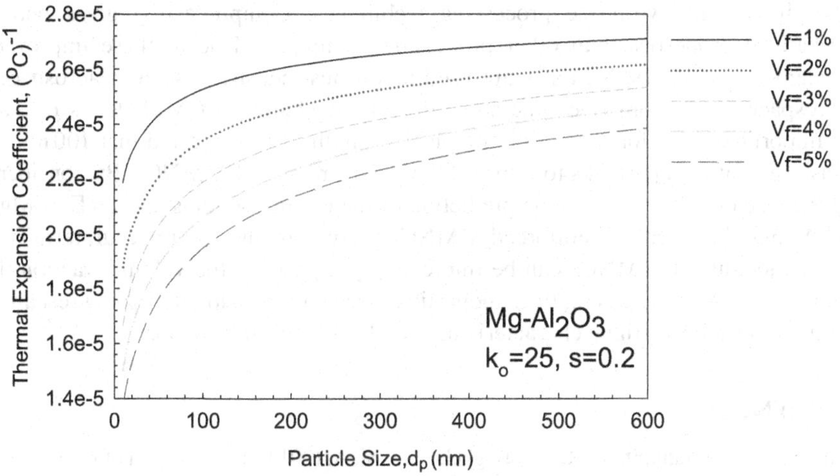

FIGURE 7.6 Predicted CTE of the present model (Equation 7.9) as a function of nanoparticle size for different volume fractions of nanoparticles for MMNCs. CTE, coefficient of thermal expansion.

7.5 APPLICATIONS OF MMNCs

MMNCs have a plethora of application in the aerospace industry (Ceschini et al., 2017) as it focuses on the improvement of properties including low density, tailored thermal expansion coefficient, thermal conductivity, and mechanical property. Aircraft organizations focus on the development and improvement of the above-listed properties, rather than on its manufacturing cost. On the other hand, high strength, high wear resistance, and creep resistance, as well as low thermal expansion coefficient, are major properties required in the automotive industry. Materials such as Al-SiC, Al-Al_2O_3, Mg-SiC, and Mg-Al_2O_3 and their combinations are best suited for manufacturing piston rods, frames, valve spring caps, etc. (He et al., 2008).

Products such as turbine blades and bullet-proof armors demand high specific strength, impact strength, and, most importantly, fatigue resistance are preferred in military applications. Therefore, materials such as Al-B, Al-SiC, Al-C, Ti-SiC, Mg-C, Al-B_4C, and Mg-B_4C are preferred (He et al., 2008; Ibrahim et al., 1991). Energy industry products such as superconductors, semiconductors, and bushes have properties like high thermal and electrical conductivities, wear resistivity, and anti-corrosiveness. Therefore, Cu-C, Al-Al_2O_3, Al-C, and Ag-Ni are favored in the energy industry (Ceschini et al., 2017).

7.6 CONCLUDING REMARKS

This chapter focuses on the current advances in the processing and mechanical properties of MMNCs. MMNCs can be manufactured by a variety of techniques. Furthermore, MMNCs can be divided into three groups, i.e., liquid-phase, solid-phase, and two-phase processing techniques. Compo-casting gives better mechanical properties than other processing techniques. Due to these improved mechanical properties, MMNCs are used in various industries, such as aerospace, hydrospace, and automotive. The strengthening mechanism of MMNCs is one of the important bases for predicting the yield strength. It is suggested that 100 nm is a basic size of nanoparticles to improve the yield strength of MMNCs. By considering the size efficiency factor, the prediction of the elastic modulus and CTE of Mg- and Al-based particulate-reinforced MMNCs can be obtained more accurately. The elastic modulus of MMNCs can be improved by increasing the volume fraction of nanoparticles and decreasing the nanoparticle size. There is a lot of ongoing research in this area such as fatigue characterization and tribological behavior.

REFERENCES

Abdizadeh, H., Ebrahimifard, R., and Baghchesara, M. A. (2014). Investigation of microstructure and mechanical properties of nano MgO reinforced Al composites manufactured by stir casting and powder metallurgy methods: A comparative study. *Composites Part B: Engineering*, *56*, 217–221.

Akbari, M. K., Mirzaee, O., and Baharvandi, H. R. (2013). Fabrication and study on mechanical properties and fracture behavior of nanometric Al_2O_3 particle-reinforced A356 composites focusing on the parameters of vortex method. *Materials and Design*, *46*, 199–205.

Beygi, H., Ezatpour, H. R., Sajjadi, S. A., and Zebarjad, S. M. (2011). Microstructure evolution of Al-Al2O3 micro and nano composites fabricated by a modified stir casting route. *18th Int. Conf. Compos. Mater*, 1–6.

Čadek, J., Kuchařová, K., and Šustek, V. (1999). A PM 2124Al-20SiCp composite: disappearance of true threshold creep behaviour at high testing temperatures. *Scripta Materialia*, *40*(11), 1269–1275.

Ceschini, L., Dahle, A., Gupta, M., Jarfors, A. E. W., Jayalakshmi, S., Morri, A., Rotundo, F., and Toschi, S. (2017). *Aluminum and magnesium metal matrix nanocomposites*. Springer.

Choi, H. J., and Bae, D. H. (2011). Creep properties of aluminum-based composite containing multi-walled carbon nanotubes. *Scripta Materialia*, *65*(3), 194–197.

Choi, H., Jones, M., Konishi, H., and Li, X. (2012). Effect of combined addition of Cu and aluminum oxide nanoparticles on mechanical properties and microstructure of Al-7Si-0.3 Mg alloy. *Metallurgical and Materials Transactions A*, *43*(2), 738–746.

Deuis, R. L., Subramanian, C., and Yellup, J. M. (1997). Dry sliding wear of aluminium composites—a review. *Composites Science and Technology*, *57*(4), 415–435.

El-Mahallawi, I., Abdelkader, H., Yousef, L., Amer, A., Mayer, J., and Schwedt, A. (2012). Influence of Al2O3 nano-dispersions on microstructure features and mechanical properties of cast and T6 heat-treated Al Si hypoeutectic Alloys. *Materials Science and Engineering: A*, *556*, 76–87.

Ghasemi Yazdabadi, H., Ekrami, A., Kim, H. S., and Simchi, A. (2013). An investigation on the fatigue fracture of P/M Al-SiC nanocomposites. *Metallurgical and Materials Transactions A*, *44*(6), 2662–2671.

Goh, C. S., Wei, J., Lee, L. C., and Gupta, M. (2007). Properties and deformation behaviour of $Mg–Y_2O_3$ nanocomposites. *Acta Materialia*, *55*(15), 5115–5121.

Goh, C. S., Wei, J., Lee, L. C., and Gupta, M. (2008). Ductility improvement and fatigue studies in Mg-CNT nanocomposites. *Composites Science and Technology*, *68*(6), 1432–1439.

Gupta, S., Tevatia, A., and Rana, K. P. S. (2022). A fatigue crack growth life prediction model for metal matrix nanocomposites: contribution of strengthening factors. *Advanced Composite Materials*, *31*(1), 43–61.

Hassan, S. F., Tan, M. J., and Gupta, M. (2008). High-temperature tensile properties of Mg/Al_2O_3 nanocomposite. *Materials Science and Engineering: A*, *486*(1–2), 56–62.

He, F., Han, Q., and Jackson, M. J. (2008). Nanoparticulate reinforced metal matrix nanocomposites-a review. *International Journal of Nanoparticles*, *1*(4), 301.

Ibrahim, I. A., Mohamed, F. A., and Lavernia, E. J. (1991). Particulate reinforced metal matrix composites—a review. *Journal of Materials Science*, *26*(5), 1137–1156.

Kaczmar, J. W., Pietrzak, K., and Włosiński, W. (2000). The production and application of metal matrix composite materials. *Journal of Materials Processing Technology*, *106*(1–3), 58–67.

Kang, Y.-C., and Chan, S. L.-I. (2004). Tensile properties of nanometric Al_2O_3 particulate-reinforced aluminum matrix composites. *Materials Chemistry and Physics*, *85*(2–3), 438–443.

Koch, C. C. (2007). Structural nanocrystalline materials: an overview. *Journal of Materials Science*, *42*(5), 1403–1414.

Kumar, P. A., Rohatgi, P., and Weiss, D. (2020). 50 years of foundry-produced metal matrix composites and future opportunities. *International Journal of Metalcasting*, *14*(2), 291–317.

Lai, S. W., and Chung, D. D. L. (1994). Fabrication of particulate aluminium-matrix composites by liquid metal infiltration. *Journal of Materials Science*, *29*(12), 3128–3150.

Lianxi, H., and Erde, W. (2000). Fabrication and mechanical properties of SiCw/ZK51A magnesium matrix composite by two-step squeeze casting. *Materials Science and Engineering: A*, *278*(1–2), 267–271.

Lim, J.-Y., Oh, S.-I., Kim, Y.-C., Jee, K.-K., Sung, Y.-M., and Han, J. H. (2012). Effects of CNF dispersion on mechanical properties of CNF reinforced A7 nanocomposites. *Materials Science and Engineering: A*, *556*, 337–342.

Liu, Y. B., Lim, S. C., Lu, L., and Lai, M. O. (1994). Recent development in the fabrication of metal matrix-particulate composites using powder metallurgy techniques. *Journal of Materials Science*, *29*(8), 1999–2007.

Ma, Z. Y., Li, Y. L., Liang, Y., Zheng, F., Bi, J., and Tjong, S. C. (1996). Nanometric Si_3N_4 particulate-reinforced aluminum composite. *Materials Science and Engineering: A*, *219*(1–2), 229–231.

Malaki, M., Xu, W., Kasar, A. K., Menezes, P. L., Dieringa, H., Varma, R. S., and Gupta, M. (2019). Advanced metal matrix nanocomposites. *Metals*, *9*(3), 330.

Mazahery, A., Abdizadeh, H., and Baharvandi, H. R. (2009). Development of high-performance A356/nano-Al_2O_3 composites. *Materials Science and Engineering: A*, *518*(1–2), 61–64.

Meyers, M. A., Mishra, A., and Benson, D. J. (2006). Mechanical properties of nanocrystalline materials. *Progress in Materials Science*, *51*(4), 427–556.

Mirza, F. A., and Chen, D. L. (2012). An analytical model for predicting the yield strength of particulate-reinforced metal matrix nanocomposites with consideration of porosity. *Nanoscience and Nanotechnology Letters*, *4*(8), 794–800.

Mirza, F. A., and Chen, D. L. (2015). A unified model for the prediction of yield strength in particulate-reinforced metal matrix nanocomposites. *Materials*, *8*(8), 5138–5153.

Pan, S., Jin, K., Wang, T., Zhang, Z., Zheng, L., and Umehara, N. (2022). Metal matrix nanocomposites in tribology: Manufacturing, performance, and mechanisms. *Friction*, 1–39.

Pan, S., Wang, T., Jin, K., and Cai, X. (2022). Understanding and designing metal matrix nanocomposites with high electrical conductivity: A review. *Journal of Materials Science*, 1–37.

Paramsothy, M., Chan, J., Kwok, R., and Gupta, M. (2011). Adding TiC nanoparticles to magnesium alloy ZK60A for strength/ductility enhancement. *Journal of Nanomaterials*, *2011*.

Radi, Y., and Mahmudi, R. (2010). Effect of Al2O3 nano-particles on the microstructural stability of AZ31 Mg alloy after equal channel angular pressing. *Materials Science and Engineering: A*, *527*(10–11), 2764–2771.

Ramakrishnan, N. (1996). An analytical study on strengthening of particulate reinforced metal matrix composites. *Acta Materialia*, *44*(1), 69–77.

Rawal, S. P. (2001). Metal-matrix composites for space applications. *Jom*, *53*(4), 14–17.

Sajjadi, S. A., Ezatpour, H. R., and Parizi, M. T. (2012). Comparison of microstructure and mechanical properties of A356 aluminum alloy/Al_2O_3 composites fabricated by stir and compo-casting processes. *Materials and Design*, *34*, 106–111.

Sun, K., Shi, Q. Y., Sun, Y. J., and Chen, G. Q. (2012). Microstructure and mechanical property of nano-SiCp reinforced high strength Mg bulk composites produced by friction stir processing. *Materials Science and Engineering: A*, *547*, 32–37.

Suryanarayana, C. (2005). Recent developments in nanostructured materials. *Advanced Engineering Materials*, *7*(11), 983–992.

Tahamtan, S., Halvaee, A., Emamy, M., and Zabihi, M. S. (2013). Fabrication of Al/A206–Al_2O_3 nano/micro composite by combining ball milling and stir casting technology. *Materials and Design*, *49*, 347–359.

Tevatia, A., and Srivastava, S. K. (2015). Modified shear lag theory based fatigue crack growth life prediction model for short-fiber reinforced metal matrix composites. *International Journal of Fatigue*, *70*, 123–129.

Tun, K. S., and Gupta, M. (2007). Improving mechanical properties of magnesium using nano-yttria reinforcement and microwave assisted powder metallurgy method. *Composites Science and Technology*, *67*(13), 2657–2664.

Zhang, H., Maljkovic, N., and Mitchell, B. S. (2002). Structure and interfacial properties of nanocrystalline aluminum/mullite composites. *Materials Science and Engineering: A*, *326*(2), 317–323.

Zhang, Z., and Chen, D. L. (2006). Consideration of Orowan strengthening effect in particulate-reinforced metal matrix nanocomposites: A model for predicting their yield strength. *Scripta Materialia*, *54*(7), 1321–1326.

Zhang, Z., and Chen, D. L. (2008). Contribution of Orowan strengthening effect in particulate-reinforced metal matrix nanocomposites. *Materials Science and Engineering: A*, *483*, 148–152.

Zhong, X. L., Wong, W. L. E., and Gupta, M. (2007). Enhancing strength and ductility of magnesium by integrating it with aluminum nanoparticles. *Acta Materialia*, *55*(18), 6338–6344.

8 Environmental Impact in Terms of Nanotoxicity and Limitations of Employing Inorganic Nanofillers in Polymers

Feba Anna John, Ajith James Jose, and Litty Theresa Biju
St. Berchmans College

CONTENTS

DOI: 10.1201/9781003279389-8

8.1 INTRODUCTION

The growth of nanotechnology in all fields of research, as well as its use in diverse areas of the environment, has been lauded throughout the cosmos. However, the consequences after the invasion of the materials with measurement below 100 nm at least in one dimension in the environment are seen more important to address worldwide and find solutions to. Nanoparticles are wildly blended with other materials to bring about enhanced coatings, composites, quantum effects, electronics, spintronics, molecular magnets, etc., even though the role of inorganic nanomaterials is more significant and a matter for discussion. Some common examples of inorganic nanoparticles are gold nanoparticles, titanium dioxide nanoparticles, silicon dioxide nanoparticles, etc.

Nanoparticles can be released into the environment in a variety of ways during their manufacturing, use, and recycling. The release may be either intentional or non-intentional. Nanoparticles that are found to be non-toxic during pre-application analysis have the potential to become harmful through chemical or physical mechanisms. These nanotoxic components will easily spread through the air, water, and soil. The ingestion of harmful nanocomponents will disrupt the environment's overall systems (Soenen et al., 2011).

8.2 VARIOUS TYPES OF NANOMATERIALS THAT PROLIFERATED IN THE ENVIRONMENT

The various types of nanomaterials can be grouped and listed under different titles according to its origin (organic and inorganic), or mode of origin (natural, synthetic, and accidental), or dimension (1D, 2D, 3D), and so on. Nowadays, nanoparticles are used in various fields such as photonics, catalysis, magnetics, and biotechnology including cosmetics, pharmaceutics, and medical purposes. This rapid development of use of nanoparticles causes exposure to environmental systems including human beings. Figure 8.1 depicts probable classification of nanomaterials which are closer to the environmental systems.

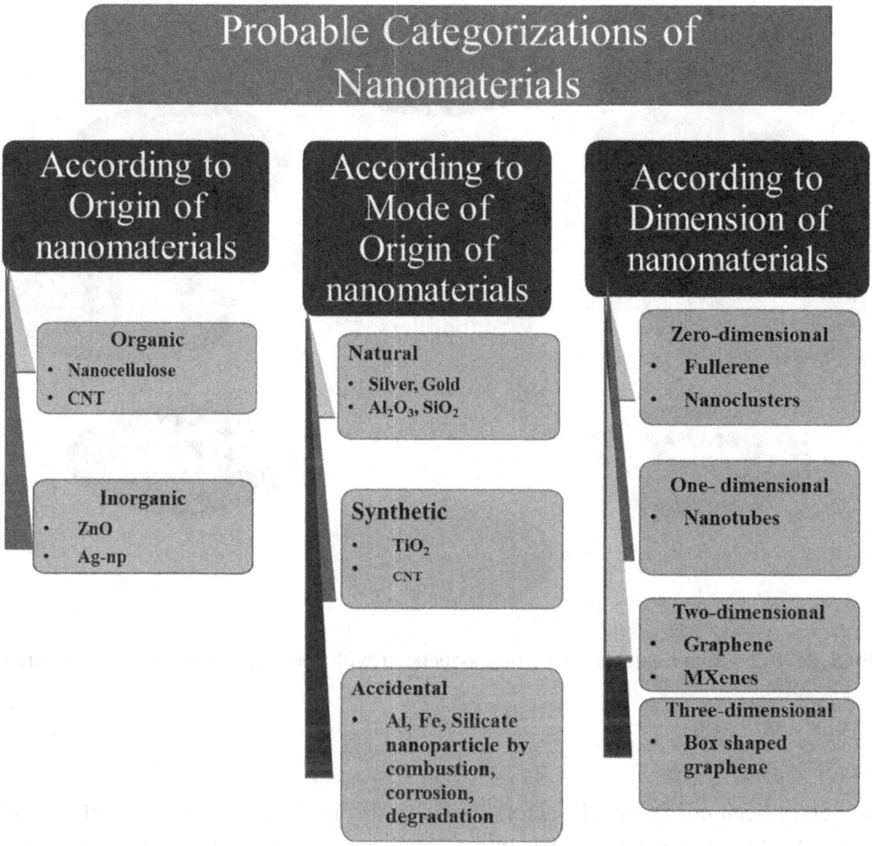

FIGURE 8.1 Probable categorization of nanomaterials.

8.2.1 Organic Nanomaterials

Organic nanomaterials are a mixture of aliphatic and aromatic compounds and considered most abundant nanomaterials in the earth. Organic nanomaterials are either released naturally from biomass or synthesized. These nanomaterials have been extensively studied, with liposomes, polymersomes, polymer constructions, and micelles being used for imaging, drug delivery, and other applications. For example, proteins, peptides, polyethylenimine, dendrimers, polyphosphoesters, etc. (Virlan et al., 2016; Khalid et al., 2020).

8.2.2 Inorganic Nanomaterials

Inorganic nanomaterials include metal nanoparticles (Cu, Au, Ag), metal oxides (SiO_2, TiO_2, Al_2O_3, Fe_3O_4), nanolayers (layered silicates), nanosheets (2D), nanotubes, mesoporous silica, semiconductors (CdS, PbS), etc. Yusuke Imai (2014) are

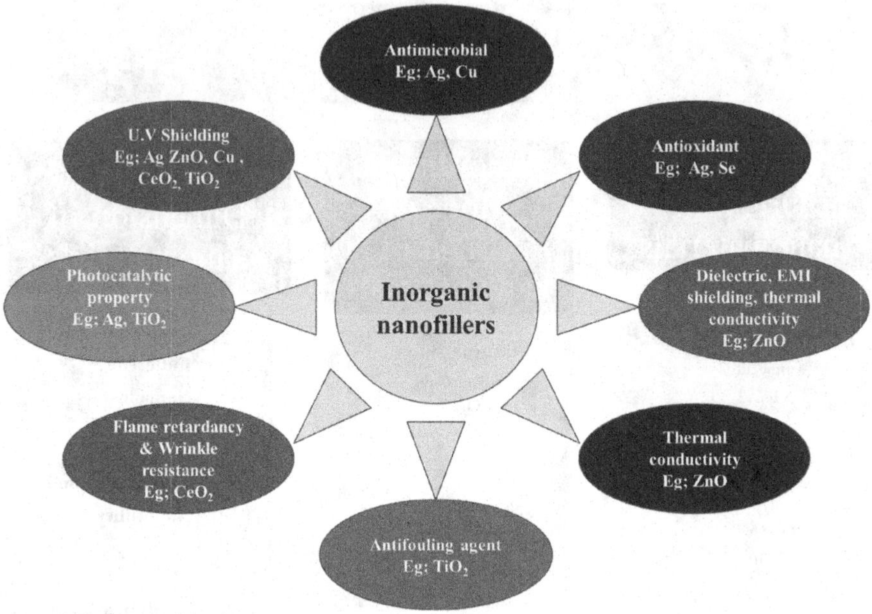

FIGURE 8.2 Some notable properties associated with inorganic nanomaterials (Jamróz et al., 2019).

somehow been utilized in all sectors of day-to-day needs (Nworie et al., 2016; Imai, 2020). Copper nanoparticles have antibacterial action against a wide range of micro-organisms. Copper ions released into the environment have a high redox potential and can damage or trigger apoptosis in microbial cell components (Sengan, 2019). Silver nanoparticles have attracted increasing attention for use in wound healing, food packaging, and medical applications, due to their superior physicochemical and biological properties. When compared to selenium, selenium nanoparticles have a substantially lower toxicity and can also be used as a platform for conveying various medications to their intended destinations (Khurana,2019; Jamróz et al., 2019).

Zinc oxide, titanium oxide, silica, aluminum oxide, cerium oxide, iron oxide, and copper oxide have all been used as active nanofillers in polymer matrices. Figure 8.2 shows some notable properties associated with inorganic nanomaterials (Jamróz et al., 2019).

8.2.3 Natural Nanomaterials

The nanomaterials that formed in the Earth's crust through biogeochemical processes are coined as natural nanomaterials. The natural nanomaterials are varied in their morphology as well as compositions, which includes organic nanomaterials, inorganic nanomaterials, layered nanomaterials, and so on (Malakar et al., 2021; Griffin et al., 2018; Westerhoff et al., 2018; Taghavi et al., 2013).

8.2.4 Synthetic Nanomaterials

The nanomaterials that are manufactured purposefully for instance research, commercial utility, drug delivery, etc., are termed synthetic nanomaterials. The yearly flow of synthetic nanomaterials is less than that of natural nanomaterials, even though the synthetic nanomaterials are considered contaminants and pose harm to the environment (Bundschuh et al., 2018; Malakar et al., 2021).

8.2.5 Accidental Nanomaterials

Automobile exhaust, industrial waste, mining waste, combustion processes, wear, corrosion processes, and degradation process are all sources of accidental nanoparticles (Malakar et al., 2021). In addition, the accidental nanomaterials may sometimes release during the commercial or laboratory synthesis of nanomaterials. The accidental nanoparticles include mostly carbon-based nanomaterials (combustion), nanoplastics (thermal degradation), metal, or metal oxide nanomaterials (corrosion) (Martínez et al., 2021).

8.2.6 Zero-, One-, Two-, and Three-Dimensional Nanomaterials

All the above-categorized nanomaterials can be classified again according to number of nanodimensions such as zero-dimensional(all the three dimensions are in nanoscale), one dimensional (only one dimension in nanoscale), two dimensional (two dimensions are in nanoscale), and three dimensional (do not possess nanoscale measurement in any dimension). Nanotoxicology currently monitors the safety of the use of nanomaterials in various fields that include in situ, in vivo, and environmental studies (Singh et al., 2020).

8.3 HOW THE ENVIRONMENTAL SYSTEMS GET CLOSER TO VARIOUS TYPES OF NANOMATERIALS

Nanomaterials enter into various systems of the environment directly either by chemical reactions like oxidation and reduction or by physical aggregation (Louie et al., 2014). The oxidation and reduction reactions upon nanomaterials cause the release of reactive oxygen species (ROS) and respective ions. Even if the nanomaterial is categorized as non-toxic, the ions generated from it by chemical change may be a toxic one. These ions happen to be aggregated. The physical aggregation takes place in two modes: homogenic aggregation of same material or heterogenic aggregation of different types of materials. The surface-area-to-volume effects on nanomaterial reactivity are reduced when nanoparticles are aggregated. This rise in aggregate size has an impact on their performance sedimentation, uptake by organisms, reactivity, transport in porous media, and sometimes toxicity (Martínez et al., 2021).

On exposure to sunlight, nanoparticles such as TiO_2 and ZnO generate ROS through photocatalyst reaction mechanism which increases the level of toxicity in organism. In photochemical reaction, holes are generated on ZnO surfaces which attack the Zn-O bond and lead to dissociation of Zn^{2+} from ZnO surface. The toxicity of zinc oxide nanoparticles is enhanced by the presence of sunlight; other environmental conditions such as natural organic matter, ionic strength pH, electrolyte concentration temperature, and light affect the level of toxicity. Nanoparticles could act as carriers of various contaminants to the cells of aquatic and terrestrial organisms. Certain organism in aquatic and terrestrial habitat could be used as monitors of environmental toxicity of engineered nanoparticles. The presence of titanium dioxide, silicon dioxide, and zinc oxide nanoparticles is important in determining the level of toxicity of nanoparticles in environment (Abbas et al., 2020).

8.4 TOXICOLOGY OF NANOMATERIALS

Toxicity concerns about the nanomaterials associated with their aggregation, physical and chemical nature, solubility, absorbability, geometry as well as its circumambient atmosphere. The too-small materials may have been absorbed by the various systems easily; this is the primary concern about the Nanos. And the effect is mostly depending on the surface with which it will be adhered. The term nanotoxicity has been expressed in different contexts as follows: (i) nanocytotoxicity: the nanotoxicity on living cells; (ii) nanogenotoxicity: the nanotoxicity causing damage to genetic materials of biological systems; and (iii) nanoecotoxicity: the nanotoxicity which affects ecosystems. The examination of the toxicological qualities of nanosized materials with the goal of establishing whether they may constitute an environmental or social threat and the extent of the threat is known as nanotoxicology (Chavon walters 2016).

The behavior of nanoparticles in a variety of environmental matrices is complicated and involves a number of processes. The properties of nanomaterials are distinct from those of ordinary materials. Particle size, surface area and charge, shape/structure, solubility, and surface coatings have been shown to have an impact on nanotoxicity. Large-scale comparative studies surely play important role in determining the toxicological effects of nanoparticles.

To assess nanotoxicology, there are in vitro and in vivo methods established. But, the in vivo methods increase risk to animals and raise some ethical issues. In order to avoid the issues, it is necessary to study, establish, and promote novel in vitro methods for the assessment of nanotoxicity. The in vitro methods include in vitro cell-based assays, tissue engineering, in silico structure-based techniques, and quantitative structure–activity relationship (QSAR) studies (Huang et al., 2021) (Figure 8.3).

8.4.1 Nanotoxicity Impact on Soil and Associated Microbes

Toxicity of engineered nanomaterials has considerable effect on the property of soil and organisms in the soil. Many useful bacteria and other microorganisms are badly affected by toxicity of nanoparticles. Among them, an important class is the symbiotic nitrogen-fixing bacteria. Different studies are conducted to determine the effect of toxicity of microscale aggregates of zinc oxide and cerium dioxide engineered nanomaterials. It was found that nanomaterials of ZnO completely inhibit the growth of bacteria. This is due to unique delivery of Zn^{2+} to the cell surface, whereas dissolved usual-sized Zn^{2+} are less bioavailable through complexation and possible precipitation. The nanoparticles alter the structure of proteins and polysaccharides of

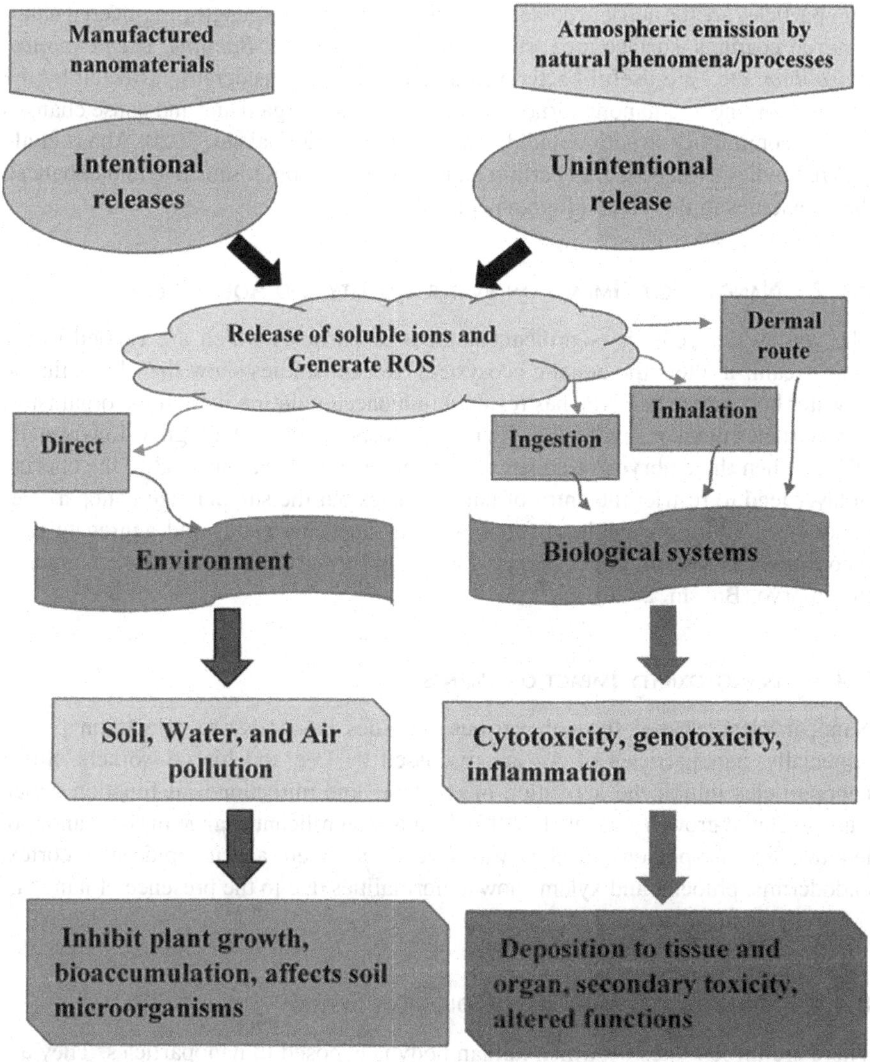

FIGURE 8.3 The toxicity pathway of nanomaterials/nanofillers to the environment and biological systems at a glance (Armstead & Li, 2016; Mohammad pour, 2019).

these bacteria. It was found that soil enriched with nanoparticles especially those of silver, zinc oxide, and titanium oxide has reduced number of root noodles. The nanoparticles disrupt the process of nodulation and negatively affect the plant health. As a result, the amount of nitrogen fixed to the soil reduces and did considerably affect the fertility of soil and thus plant growth (Bakshi, 2020; Abbas et al., 2020).

The engineered nanoparticles interact with different forms of bacteria found in the soil. The bacterium may respond to the presence of nanomaterials in different ways. The different parameters that influence the interaction of bacterium and nanoparticles are the nutrient content, organic matter, ionic strength, engineered nanomaterial coatings, surface area, shape integration pH, etc. *Bacillus, Pseudomonas, Rhizobium,* etc., are useful bacteria in soil which are considerably affected by the toxicity. Engineered nanomaterials interact with microorganisms and cause changes in their community structure and dynamic functioning (Bakshi, 2020; Abbas et al., 2020). Studies evidenced that certain nanoparticles are photosensitive and enhanced their toxicities in this route (Taghavi et al., 2013).

8.4.2 Nanotoxicity Impact on Water Quality and Aquatic Life

The wastewater contains significant number of particles which are carried to the mainstream. It alters the aquatic ecosystem. Recent studies show that the pollution of water bodies due to ENPs has resulted in bioaccumulation. In aquatic organisms, nanoparticles uptake is by feeding, dermal adsorption, or through gills (Lewis et al., 2019). When the embryos of aquatic organism encounter nanoparticles, the chorine in layer tend to restrict the entry of nanoparticles but the smaller size nanoparticles make entry and are capable of functioning as nucleation sites and aggregate with incoming ones to form larger particles which in turn affect the transport channel of the embryo (Bakshi, 2020).

8.4.3 Nanotoxicity Impact on Plants

Nanoparticles released through various activities have adverse effects on plants. Especially, nanoparticles of Ag are discussed by Lee and his co-workers. Silver nanoparticles inhibit the activation of ethylene, and mitochondrial function which causes stunted growth (Lee et al., 2012). It causes significant change in shoot and root lengths. The morphology of plant was severely affected, and its epidermis, cortex, endodermis, phloem, and xylem show abnormalities due to the presence of nanoparticles (Lei et al., 2016).

8.4.4 Nanotoxicity Impact on Biological Systems

There are three routes by which human body is exposed to nanoparticles. They are inhalation of airborne nanoparticles, ingestion of drinking or food additives, and dermal penetration (Choi et al., 2009). Toxicity may be instant or delayed based on the time required for its manifestation (Mohammadpour et al., 2019; Buzea et al., 2007).

8.4.4.1 Ingestion

We humans uptake directly a lot of food ingredients, additives and supplements containing nanofillers. Inhaled nanoparticles can also reach the stomach with the help of mucous ciliary cells. Nanofillers can move to the blood and can reach each organ on crossing the epithelium (Mohammadpour et al., 2019; Buzea et al., 2007).

Gastrointestinal tract comes across both exogenous and endogenous nanoparticles. Titanium dioxide, silicate, and aluminosilicate are classified as exogenous inorganic particles which are important in food and pharmaceutical industries as additives.

The most commonly used nanoparticles in food industry are silver, titanium oxide, zinc oxide, silicon dioxide, etc. (Mohammadpour et al., 2019; Buzea et al., 2007).

8.4.4.2 Inhalation

Inorganic nanocomposites may reach lungs by inhalation. Furthermore, inhalation is due to occupational exposure. Size is the main factor that determines the entry of nanoparticles to the respiratory tract.

Nanoparticles with diameter between 5 and 30 μm remain in nasopharyngeal region, and those with size 0.1–1 μm can reach the alveolar region by gravitational sedimentation and Brownian diffusion. Translocation in blood capillary is very easy for nanoparticles having diameter less than 0.5 μm (De Matteis, 2017).

8.4.4.3 Dermal Route

Nanomaterials are widely used in cosmetic field. Studies show that nanofillers such as iron can penetrate through hair follicle reaching the basal and spinous layer. Here also size is the main factor that determines the entry. Nanoparticles with diameter around 4 nm can penetrate intact skin, and those with size up to 45 nm can only penetrate through damaged skin (Mohammadpour et al., 2019; Buzea et al., 2007).

8.4.5 Nanotoxicity Impact of Nanofillers on Human Body

Nanoparticles reaching gastrointestinal tract are considerably affected by the pH and the peristalsis. It causes a change in physiochemical properties. As the stomach has an acidic environment, dissolution at lower pH promotes degradation of nanofillers by digestive fluids. It triggers oxidative stress, DNA damage, and inflammation. Accumulation of nanoparticles causes depolarization of mitochondria membrane causing cytotoxicity. Titanium dioxide causes alteration of microvilli on the apical surface which in turn causes increase in calcium levels.

Activities such as coughing accelerate the motion of nanoparticles deposited in the respiratory tract. Studies showed that these inhaled nanoparticles play a crucial role in inducing cardiovascular diseases, and inhalation of nanoparticles also increased the risk of lung cancer (Choi et al., 2009).

Severe inflammatory response followed by microgranulomatous changes, increase in B&T cell counts along with histopathological changes in lungs were reported on exposure of inorganic nanofillers (Mohammadpour et al., 2019). Surveys showed that exposure to gold nanoparticles resulted in detection of nanoparticles even after three months in the blood and urine samples. Titanium dioxide nanoparticles are the most exclusively used ones in food and medical industry, and its prolonged exposure can result in liver damage since it is the site where major accumulation of these particles occurs (Singh et al., 2019).

Inhalation of nanoparticles causes lung disorders such as fibrosis pneumoconiosis and even asthma. Kidney can filter small-sized nanofillers whereas large particles are captured by Kupffer cells and splenic macrophages. Nanofillers may accumulate in renal blood vessels. It causes chronic toxicity in kidneys including glomerular adhesion to Bowman's capsule, membranes thickening, and inflammation (Figure 8.4).

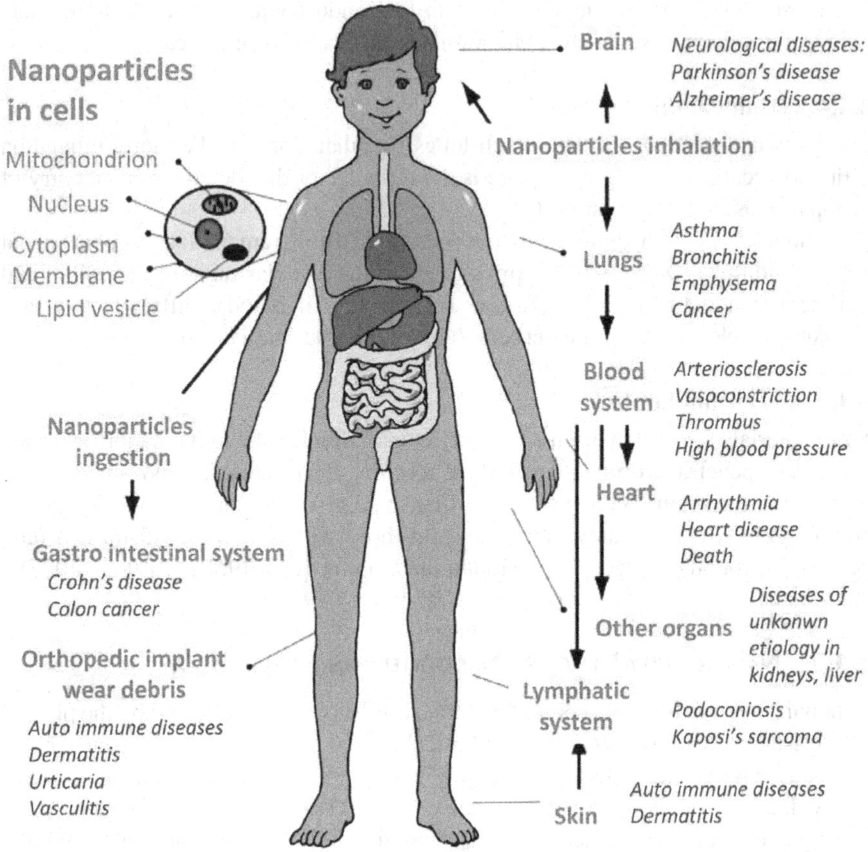

FIGURE 8.4 Nanoparticle exposure and detrimental effects on human health (Buzea et al., 2007).

8.4.6 Nanotoxicity Impact on Nano Medicine Delivery Applications

One of the chief parameters in nanotoxicity is determining the appropriate dose for administration. In delivery applications of nanomaterials, the relative dose of cargo and carrier is important based on the efficacy and toxicity of both. Classic toxicology studies use particle mass, but modern studies consider surface area and number of particles to predict the harmful effect (Mohammadpour et al., 2019).

According to studies conducted by Watson and co-workers, percentage of nanofillers with the host matrix will influence dose–response (Watson-Wright, 2017) relationships. A prominent view in the field of nanotoxicology is that surface area is an important factor in determining toxicity (Fadeel & Garcia-Bennett, 2010). Studies reported that designing of nanofillers above certain limit tend to aggregate and agglomerate rapidly to form layer particles with dimensions often exceeding the nanometric range, his father enhances the level of toxicity song (Mohammadpour et al., 2019).

The physicochemical properties of nanofillers such as shape and composition can considerably affect the toxicity (Watson-Wright et al., 2017). Size is also important in determining the toxicity of nanofillers. Gold nanoparticles having size 35 nm were non-toxic to murine in macrophage like cell line. It was reported that gold nanoparticles having size about 5 nm have effects on primary human umbilical vein endothelial cells and on gene expression. The gold clusters' compound with a distinct particle size of 1.4 nm has shown to interact with major grooves of DNA, which could account for the remarkable toxicity of nanostructure (Fadeel & Garcia-Bennett, 2010).

8.5 INFLUENCE OF INORGANIC NANOFILLERS IN VARIOUS POLYMER NANOCOMPOSITES

The properties of polymers can be effectively augmented by adding nanofillers, and the resultant materials are termed polymer nanocomposites. Numerous studies had reported that nanosized fillers have a lager surface area than the conventional microsized fillers, thus enhancing the properties due to better interfacial interactions with the polymer matrix (Bioplastic & Ngoh, 2022). So, the nanofillers are more effective than conventional macro- or microfillers. Inorganic nanoparticles are promising options for extending the life of polymer coatings and lowering repair costs (Pourhashem et al., 2020). The inorganic nanofillers are commonly used to change the mechanical characteristics, viz. mechanical strength, heat resistance, modulus, and stiffness of polymers (Taghizadeh & Nasirianfar, 2020; Afzal et al., 2016). The cost of polymer composites has decreased due to the development of fillers and their qualities (Afzal et al., 2016; Fu et al., 2019; Abarca, 2021).

ZnO, TiO_2, $CaCO_3$, $Mg(OH)_2$, talc, and mica are examples of inorganic fillers that are frequently employed in polymeric nanocomposites. Zinc oxide, titanium oxide, silica, aluminum oxide, cerium oxide, iron oxide, and copper oxide have all been used as active nanofillers in polymer matrices (Jamróz et al., 2019). The nanoclays are utilized with polymers specially to enhance its stiffness (Jagadeesh et al., 2021; Das et al., 2021). The most commonly used, less toxic, and biocompatible inorganic nanofillers are ZnO, TiO_2, silver nanoparticle, and gold nanoparticle (Figure 8.5, Tables 8.1 and 8.2).

8.5.1 INORGANIC NANOFILLERS WITH BIOPLASTICS

The traditional polymers are mostly made from non-renewable petrochemical sources that cannot be destroyed by sun radiation or microbial decomposers. As a result, plastic trash continues to accumulate in the environment. This could have a significant negative influence on the ecosystem. Thus, the demand of biocompatible plastics has elevated in recent years.

Bioplastics are bio-based polymers that are biodegradable and/or compostable by microorganisms and are derived from renewable resources. Nanoclays, nanosilica, metal, and metal oxide nanofillers are added to starch-based bioplastics particularly to enhance its antimicrobial and mechanical properties. Table 8.3 portrays certain inorganic nanofillers and its effects on bioplastics. Starch-based bioplastics are thermoplastics (Bioplastic & Ngoh, 2022; Jamróz et al., 2019; Jagadeesh et al., 2021) (Table 8.3).

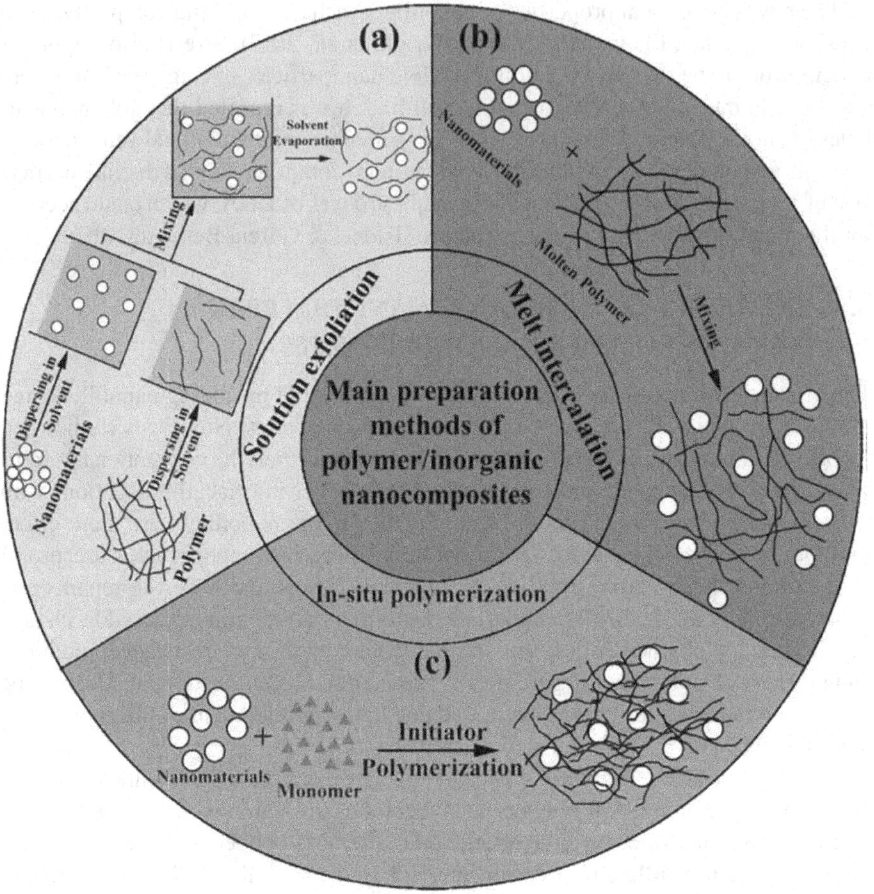

FIGURE 8.5 The main methods for preparation of polymer/inorganic nanocomposites: (a) solution exfoliation, (b) melt intercalation, and (c) in situ polymerization (Pourhashem et al., 2020).

TABLE 8.1
Salient Factors that Influence the Properties of Polymer Nanocomposites

Size of the Fillers
Morphology of fillers and matrix
Percentage of composition (usually wt% of fillers w.r.t. matrix)
Number of fillers and matrix
Dispersion of filler/s in the matrix/ces
Solvent
Method of preparation
Atmospheric condition (temperature & pressure)

TABLE 8.2
Frequently Employed Inorganic Nanofillers in Polymer Nanocomposites (Afzal et al., 2016; Fu et al., 2019; Kausar, 2019; Imai, 2020; Abarca, 2021)

Type	Examples
Oxides	ZnO, TiO_2, Glass, MgO, Sb_2O_3, SiO_2
Hydroxides	$Al(OH)_2$, $Mg(OH)_2$,
Salts	$CaCO_3$, $BaSO_4$, $CaSO_4$, $Ca_3(PO_4)_2$
Silicates	Talc, mica, montmorillonite, nanoclay, kaolin, zeolite
Metals and metal oxides	Boron, steel, gold, silver and its oxides, aluminum oxides, iron oxides, zirconium, platinum, palladium, rhodium
Semiconductors	CdS, CdSe
2D	MXenes, metal chalcogenides

8.5.2 Photopolymers

The polymers which alter their properties in the presence of light are termed as photopolymers. The inorganic nanofillers nanoprecipitated calcium carbonate and nanoclay when incorporated in a 3D-printed photopolymer and assessed its effect on the mechanical and morphological properties. A stress transfer can be determined by examining the morphological features. Furthermore, the stress transmission mechanism was substantiated by the improved mechanical properties of the composite filled with CNC and inorganic nanofillers (Bae & Kim, 2021).

8.5.3 Polymer Electrolytes

Polymer electrolytes are the polymer matrix with conducting ions which finds extensive applications in electrochemistry. Polymer electrolytes are divided into two categories based on their compositions: gel polymer electrolytes, in which the liquid electrolyte is incorporated into the matrix of the swollen polymer host, and solid or dry polymer electrolytes, in which the electrolyte salt is blended with the polymer of choice without the use of an organic solvent after processing. Polyvinylidene fluoride (PVDF), Poly methyl methacrylate (PMMA), and Poly (vinyl alcohol) (PVA) are some important examples for polymer electrolytes. Inorganic clay (montmorillonite, bentonite) is a form of phyllosilicate that has an octahedral magnesium or aluminum sheet sandwiched between one or two tetrahedral silicon oxide sheets, depending on whether it is 1:1 or 2:1. Each layer is 1 nm thick, resulting in large surface areas, aspect ratios of over 1000, and tremendous mechanical strength.

Negative charges are produced via isomorphic substitution of cations on the tetrahedral or octahedral sheets, which are counterbalanced by Na^+ or Ca^{2+} present in the interlayer gaps known as galleries. Clay becomes hydrophilic when these cations and water molecules are present, making it compatible with hydrophilic polymers, while hydrophobic polymers can be accommodated with prior surface preparation (Chua et al., 2018).

TABLE 8.3
Bioplastics and Inorganic Nanofillers

Bioplastics	Inorganic Nanofillers	Remarkable Effects
Starch-based bioplastics	Nanoclay	• Reduction of water vapor permeability • Reduction of microbial growth • Improvement of tensile strength • Increase in moisture absorption
	montmorillonites (MMTs)	• Improvement of tensile strength • Improvement of Young's modulus • Improvement of surface hydrophobicity of film • Reduction of moisture uptake
	Zinc oxide (ZnO)	• Improvement of tensile strength • Inhibition of microbial growth
	Ag	• Improvement of tensile strength • Improvement of antimicrobial activity
Gelatin	Ag	• Improvement of antimicrobial activity
	TiO_2	• Improvement of tensile strength • Photocatalytic microbial inhibition
	ZnO	• Enhancement of UV screening effect • Improvement of antimicrobial activity
Gelatin/Agar	TiO_2	Improvement of tensile strength Showed UV block effect
Starch/gelatin	ZnO	Improvement of tensile strength Enhancement of UV barrier property
Chitosan/gelatin	Ag	Increase of the shelf life of grapes wrapped with this film
Chitosan/cellulose	Ag	Improvement of antimicrobial activity
Chitosan/PVA	Ag	Improvement of antioxidant activity Improvement of antimicrobial activity
Chitosan	Au	Improvement of antimicrobial activity
	Ag	Improvement of antimicrobial activity
	ZnO	Biocompatibility, non-toxicity
Furcellaran	Ag	Improved UV blocking activity
	Se	Improvement of antimicrobial activity
PLA	ZnO	Improvement of tensile strength Improvement of antimicrobial activity

8.6 VARIOUS STRATEGIES TO REDUCE NANOTOXICITY IN THE ENVIRONMENTS

We should carry out more researches in the field of nanotoxicology and social awareness regarding manufacturing, usage, and disposal of nano-contained materials. There are some national and international moves to achieve proper usage of nanomaterials. In addition, certain research-based suggestions have been put forward to increase the nanoscale of the materials, assure the purity, etc.

8.6.1 National and International Level Regulations Implemented for Effective Use of Nanomaterials (Schwirn et al., 2020)

1. The European regulation on the Registration, Evaluation, Authorisation and Restriction of Chemicals (REACH, EC No 1907/2006).

   In 2018, REACH implemented various nanospecific amendments for a proper assessment of nanomaterials (EC2018). EC 2018 came in effect from January 1, 2020.

   European Chemical Agency (ECHA).
2. Under REACH, the ECHA came up with nanospecific annexes to guide registration of nanomaterials.

   Organization for Economic Co-operation and Development (OECD).
3. The OECD organized a special conference on the human health risk and environmental safety of produced nanoparticles for the first time in 2005. The construction of databases describing the benefits and toxicity of nanomaterials, as well as highlighting research gaps, was the result of this work. This group also approved methods for nanomaterial toxicity testing standards, such as subacute and subchronic toxicity studies.
4. Some research initiatives as realistic nanoexposure estimation.

   EU Framework Programme (FP) 7 SUN, EU FP 7 NanoFATE, Horizon 2020 NanoFaSe, or EU LIFE Nano MONITOR (NanoSafety Cluster 2020) was focused with exposure assessment and applicability of nanotechnology (Schwirn et al., 2020).

8.6.2 Increasing the Size of Nanofillers

Nanoparticles with size less than 5 nm are considered to be the most hazardous one due to its ability of nuclear penetration and high-surface-area-to-volume ratio. So, increasing size of nanofiller to 40 nm could diminish its cytotoxic effects. Studies showed that nanofillers with size in the range 10–30 nm are considered to be optimal choice. The surface properties of fillers must be designed carefully to avoid particle aggregation.

8.6.3 Assure the Purity of Nanofillers

Purity of the nanofillers should be maintained in order to reduce its toxic effects. Any impurities such as metal or metal oxides must be removed before the use of fillers. Leached ions and stabilizers can be removed by dialysis against physiological buffer (Soenen et al., 2011).

8.6.4 Biocompatible Coating on Free Nanoparticle

It is better to coat the reactive, free nanoparticle with biocompatible materials which are about to contact with biological systems, so as to prevent interaction of the potential nanoparticles with foreign materials (nano or non-nano) and avoid the generation of toxic materials (Bakshi, 2020).

8.6.5 Effective Categorization of Nanowaste/Nanofillers

The use of nanoparticles in various fields has developed another category of waste materials known as nanowaste. For effective management of nanowaste, categorization is an important step. Categorization is done on the basis of level of toxicity and exposure potency. Physical, chemical, and biological methods are used for the treatment of nanowaste. Natural clay plays an important role in removal of engineered nanoparticles; clay particles can coagulate engineered nanoparticles in aqueous environment and thus play a role in the removal of nanocontaminants. Biochar which is a carbon-rich product formed from different waste biomass materials during pyrolysis has variety of functional groups, and it helps in reducing toxicity of nanowaste by immobilizing it in the soil by surface interactions (Abbas et al., 2020).

8.6.6 Effective Characterization of Nanomaterials

Characterization of nanoparticle is necessary to determine its extent of toxicity. Shown are physicochemical properties such as chemical composition size and surface topography of nanofillers. For a type of nanoparticle, the concentration and incubation time should be accurately defined. Nanomaterials must be characterized in order to understand how their physical and chemical properties connect to chemical, ecological, and biological reactions. The bulk (shape, size, phase, electronic structure, and crystallinity) and surface (surface area, arrangement of surface atoms, surface electronic structure, surface composition, and functionality) features of nanomaterials are both determined during the characterization process. Environmental conditions (such as temperature, pH, ionic strength, salinity, and organic matter) can also influence NP toxicity and behavior (Fadeel & Garcia-Bennett, 2010; Chavon Walters, 2016).

The OECD developed standardized tests to summarize some of the analytical procedures to characterize nanomaterials. Scanning electron microscopy (SEM) and transmission electron microscopy (TEM) are widely used to study the morphology. The combination of energy-dispersive X-ray (EDX) spectrometry with SEM is a typical method for characterizing elemental analysis of NMs. Dynamic light scattering (DLS), which analyzes the Brownian movement of the NPs, or electrophoretic light scattering spectroscopy (ELS), which uses an oscillating electric field, can also be used to estimate particle size in the aqueous phase. The Scherrer method is used to quantify particle size in the dry state using X-ray powder diffraction (XRD).

Large-scale comparative studies surely play an important role in determining the toxicological effects of nanoparticles. Comparison of different nanoparticles in a single step is to assess the effect of size, surface chemistry, etc., on toxicity (Soenen et al., 2011; Chavon Walters, 2016) (Table 8.4).

8.7 CONCLUSION

Nanomaterials are pertinent to all domains of environment and human life. Almost all the consumer goods from food packages to energy materials consist of nanomaterials or its composites. According to literature survey, most of the studies concerned about the cost, availability, and processability of the inorganic nanofillers. They seldom concerned

TABLE 8.4
The Characteristic Techniques Defined by the OECD

Characteristics	Techniques
Morphology	SEM TEM
Characterizing elemental analysis	The combination of EDX spectrometry with SEM
Brownian movement of the NPs	DLS
Used to estimate particle size in the aqueous phase	ELS
The Scherrer method— to quantify particle size in the dry state	XRD

about the toxic factors of the applied nanofillers/nanomaterials. A unified guideline that is universally accepted is to be implemented for the proper utilization nanomaterials. The guideline should portray the probable toxic pathways and consequences associated with each type of nanomaterials. So the ongoing as well as forthcoming research works must focus on the nanotoxicity. It is imperative for a healthy future.

REFERENCES

Abarca, R. M. (2021). Functional ingredients from algae for foods and nutraceuticals. *Nuevos sistemas de comunicación e información*, 2013–2015.

Afzal, A., Kausar, A., & Siddiq, M. (2016). A Review on polymer/cement composite with carbon nanofiller and inorganic filler. *Polymer - Plastics Technology and Engineering*, *55*(12), 1299–1323. https://doi.org/10.1080/03602559.2016.1163594

Armstead, A. L., & Li, B. (2016). Nanotoxicity: Emerging concerns regarding nanomaterial safety and occupational hard metal (WC-Co) nanoparticle exposure. *International Journal of Nanomedicine*, *11*, 6121–6133. https://doi.org/10.2147/IJN.S121238

Bae, S. U., & Kim, B. J. (2021). Effects of cellulose nanocrystal and inorganic nanofillers on the morphological and mechanical properties of digital light processing (DLP) 3D-Printed photopolymer composites. *Applied Sciences (Switzerland)*, *11*(15). https://doi.org/10.3390/app11156835

Bakshi, M. S. (2020). Impact of nanomaterials on ecosystems: Mechanistic aspects in vivo. *Environmental Research*, *182* (October 2019), 109099. https://doi.org/10.1016/j.envres.2019.109099

Chua, S., Fang, R., Sun, Z., Wu, M., Gu, Z., Wang, Y., Hart, J. N., Sharma, N., Li, F., & Wang, D. W. (2018). Hybrid solid polymer electrolytes with two-dimensional inorganic nanofillers. *Chemistry - A European Journal*, *24*(69), 18180–18203. https://doi.org/10.1002/chem.201804781

Das, P. P., Chaudhary, V., Ahmad, F., & Manral, A. (2021). Effect of nanotoxicity and enhancement in performance of polymer composites using nanofillers: A state-of-the-art review. *Polymer Composites*, *42*(5), 2152–2170. https://doi.org/10.1002/pc.25968

De Matteis, V. (2017). Exposure to inorganic nanoparticles: Routes of entry, immune response, biodistribution and in vitro/In vivo toxicity evaluation. *Toxics*, *5*(4), 1–21. https://doi.org/10.3390/toxics5040029

Fu, S., Sun, Z., Huang, P., Li, Y., & Hu, N. (2019). Some basic aspects of polymer nanocomposites: A critical review. *Nano Materials Science*, *1*(1), 2–30. https://doi.org/10.1016/j.nanoms.2019.02.006

Huang, H. J., Lee, Y. H., Hsu, Y. H., Liao, C. Te, Lin, Y. F., & Chiu, H. W. (2021). Current strategies in assessment of nanotoxicity: Alternatives to in vivo animal testing. *International Journal of Molecular Sciences*, *22*(8), 1–14. https://doi.org/10.3390/ijms22084216

Imai, Y. (2020). Encyclopedia of polymeric nanomaterials. *Encyclopedia of Polymeric Nanomaterials*, 3–9. https://doi.org/10.1007/978-3-642-36199-9

Jagadeesh, P., Puttegowda, M., Mavinkere Rangappa, S., & Siengchin, S. (2021). Influence of nanofillers on biodegradable composites: A comprehensive review. *Polymer Composites*, *42*(11), 5691–5711. https://doi.org/10.1002/pc.26291

Jamróz, E., Kulawik, P., & Kopel, P. (2019). The effect of nanofillers on the functional properties of biopolymer-based films: A review. *Polymers*, *11*(4), 1–43. https://doi.org/10.3390/polym11040675

Kausar, A. (2019). Inorganic nanomaterials in polymeric water decontamination membranes. *International Journal of Plastics Technology*, *23*(1), 1–11. https://doi.org/10.1007/s12588-019-09230-x

Khalid, K., Tan, X., Mohd Zaid, H. F., Tao, Y., Lye Chew, C., Chu, D. T., Lam, M. K., Ho, Y. C., Lim, J. W., & Chin Wei, L. (2020). Advanced in developmental organic and inorganic nanomaterial: a review. *Bioengineered*, *11*(1), 328–355. https://doi.org/10.1080/21655979.2020.1736240

Louie, S. M., Ma, R., & Lowry, G. V. (2014). Transformations of nanomaterials in the environment. *Frontiers of Nanoscience*, *7*, 55–87. https://doi.org/10.1016/B978-0-08-099408-6.00002-5

Malakar, A., Kanel, S. R., Ray, C., Snow, D. D., & Nadagouda, M. N. (2021). Nanomaterials in the environment, human exposure pathway, and health effects: A review. *Science of the Total Environment*, *759*, 143470. https://doi.org/10.1016/j.scitotenv.2020.143470

Martínez, G., Merinero, M., Pérez-Aranda, M., Pérez-Soriano, E. M., Ortiz, T., Begines, B., & Alcudia, A. (2021). Environmental impact of nanoparticles' application as an emerging technology: A review. *Materials*, *14*(1), 1–26. https://doi.org/10.3390/ma14010166

Mohammadpour, R., Dobrovolskaia, M. A., Cheney, D. L., Greish, K. F., & Ghandehari, H. (2019). Subchronic and chronic toxicity evaluation of inorganic nanoparticles for delivery applications. *Advanced Drug Delivery Reviews*, *144*, 112–132. https://doi.org/10.1016/j.addr.2019.07.006

Nworie, F. S., Ugwu, M. E., Eluu, S. O., & Ezekwena, C. (2016). Recent advances and applications of cyclic and polymeric inorganic nanocomposites. *Journal of Chemistry and Materials Research*, *5*(4), 58–67. www.oricpub.com%5Cnwww.oricpub.com/jcmr

Pourhashem, S., Saba, F., Duan, J., Rashidi, A., Guan, F., Nezhad, E. G., & Hou, B. (2020). Polymer/Inorganic nanocomposite coatings with superior corrosion protection performance: A review. *Journal of Industrial and Engineering Chemistry*, *88*, 29–57. https://doi.org/10.1016/j.jiec.2020.04.029

Schwirn, K., Voelker, D., Galert, W., Quik, J., & Tietjen, L. (2020). Environmental risk assessment of nanomaterials in the light of new obligations under the reach regulation: which challenges remain and how to approach them? *Integrated Environmental Assessment and Management*, *16*(5), 706–717. https://doi.org/10.1002/ieam.4267

Singh, V., Yadav, P., & Mishra, V. (2020). Recent advances on classification, properties, synthesis, and characterization of nanomaterials. *Green Synthesis of Nanomaterials for Bioenergy Applications, September*, 83–97. https://doi.org/10.1002/9781119576785.ch3

Soenen, S. J., Rivera-Gil, P., Montenegro, J. M., Parak, W. J., De Smedt, S. C., & Braeckmans, K. (2011). Cellular toxicity of inorganic nanoparticles: Common aspects and guidelines for improved nanotoxicity evaluation. *Nano Today*, *6*(5), 446–465. https://doi.org/10.1016/j.nantod.2011.08.001

Taghavi, S. M., Momenpour, M., Azarian, M., Ahmadian, M., Souri, F., Taghavi, S. A., Sadeghain, M., & Karchani, M. (2013). Effects of nanoparticles on the environment and outdoor workplaces. *Electronic Physician*, *5*(4), 706–712. https://doi.org/10.14661/2013.706-712

Taghizadeh, M. T., & Nasirianfar, S. (2020). Mechanical, rheological and computational study of pvp/pani with additives. *Iranian Journal of Chemistry and Chemical Engineering*, *39*(1), 281–296. https://doi.org/10.30492/IJCCE.2020.32959

Tan, S. X., Andriyana, A., Ong, H. C. , Lim, S., Pang, Y. L. , Ngoh, G. C. (2022). A Comprehensive review on the emerging roles of nanofillers fabrication. *Polymer Review*, 14(664), 1–27.

Virlan, M. J. R., Miricescu, D., Radulescu, R., Sabliov, C. M., Totan, A., Calenic, B., & Greabu, M. (2016). Organic nanomaterials and their applications in the treatment of oral diseases. *Molecules*, *21*(2), 1–23. https://doi.org/10.3390/molecules21020207

9 Inorganic Nanofillers-Derived Polymers in Energy Storage Devices

B. C. Bhadrapriya, Bosely Anne Bose, and Nandakumar Kalarikkal
Mahatma Gandhi University

CONTENTS

9.1 INTRODUCTION

Energy is the capacity to do work. There are mainly two different forms of energy – potential energy and kinetic energy. Potential energy is the energy stored in an object due to its position or structure, whereas kinetic energy is the energy of motion. Energy resources are of two types – renewable and non-renewable energy resources. Renewable sources can be regenerated by natural processes and are continuously available without depletion. Solar, wind, hydro, geothermal, and biomass energy are renewable energy resources. Non-renewable resources can't be regenerated by natural processes and hence get depleted with usage. Fossil fuels like oil, natural gas, and coal are examples. Energy is the most crucial resource for human life and society's progress and sustainability. Researchers around the globe are striving to come up with green and sustainable energy sources as well as more sustainable and cost-effective ways to store energy. The energy storage sector is evolving to meet the ever-increasing energy demands of the world. It is high time we start depending on renewable energy resources. The intermittent nature of renewable resources makes it challenging to use them as a reliable energy source (Owusu & Asumadu-Sarkodie, 2016). But by using advanced energy storage devices, we can improve reliability and ensure the timely delivery of electrical

DOI: 10.1201/9781003279389-9

power to a great extent. Research and development in clean energy and its integration with enhanced energy storage methods can tackle global concerns like depletion of energy resources, environmental pollution, etc.

Because of the tremendous increase in electrical energy consumption, the development of efficient electrical energy storage systems has become increasingly crucial. Over the last few years, significant progress has been achieved in improving the energy densities of polymer nanocomposites by tuning the chemical structures of ceramic fillers and polymer matrices and by tailoring the polymer–ceramic interfaces. Electrical energy storage systems convert electrical energy into other forms of energy and store it for future use. According to the converted form of energy, they are classified into chemical, mechanical, electrical, thermal, and electrochemical energy storage systems. When chemical bonds are formed or broken, energy is absorbed or released. Coal, hydrogen, gasoline, liquefied petroleum gas, butane, etc., store energy in their chemical bonds.

Electrochemical energy storage systems store electrical energy generated as chemical energy, and at any point, this energy can be turned back into electrical energy. Batteries, fuel cells, and electrochemical capacitors are electrochemical energy storage systems. There are mainly two different types of batteries: rechargeable (or primary) and non-rechargeable (or secondary) batteries (Faria et al., 2014; Yoshino, 2012). Due to the irreversible reactions inside a primary battery, they are meant for a single-time use only (Schumm, 2000). Hence, they are used in low-power devices such as watches and calculators. They are also low cost (Xianfeng Hu et al., 2021). On the other hand, secondary batteries can be used multiple times by recharging and are used in high-power systems (Yoshino, 2012; Kim et al., 2020). Lithium-ion batteries are electrochemical energy storage devices and are the most commonly used rechargeable batteries. A schematic diagram of a basic Li-ion cell is shown in Figure 9.1. The working principle of LIBs is the exchange of lithium ions between a cathode and an anode (Owusu & Asumadu-Sarkodie, 2016). Generally, carbonaceous materials such as graphite and transition metal oxides are chosen as the anode, and lithium salts such as lithium cobalt oxide, lithium nickel manganese cobalt, and lithium iron phosphate are selected as cathode material (Roy & Srivastava, 2015). Commonly used electrolyte materials are $LiPF_6$, $LiBF_4$, etc. (Li et al., 2018).

LIBs have high efficiency, long shelf life, minimal self-discharge rate, and higher specific energies than batteries made from other materials, such as zinc and lead, due to their low density and relatively lightweight (Kaifeng Yu et al., 2020). Lead–acid batteries, even though heavy, are highly beneficial for high-power applications, especially in the automotive industry. They are comparably cheaper and have an efficiency of 70%–80% (Hannan et al., 2017). Supercapacitors possess good stability, high power, excellent storage efficiency, and a longer lifetime than batteries, but are not suitable for long-term energy requirement applications due to their low energy density and specific capacitance values (Raza et al., 2018). Capacitors store electrical energy as an electrostatic charge and are capable of storing energy with very minimal loss. A dielectric capacitor consists of a dielectric material separating two conducting electrodes. Figure 9.2 shows the schematic diagram of a typical dielectric capacitor (Zha et al., 2021). The capacitance of a dielectric capacitor depends upon the distance between electrically conductive plates and the area.

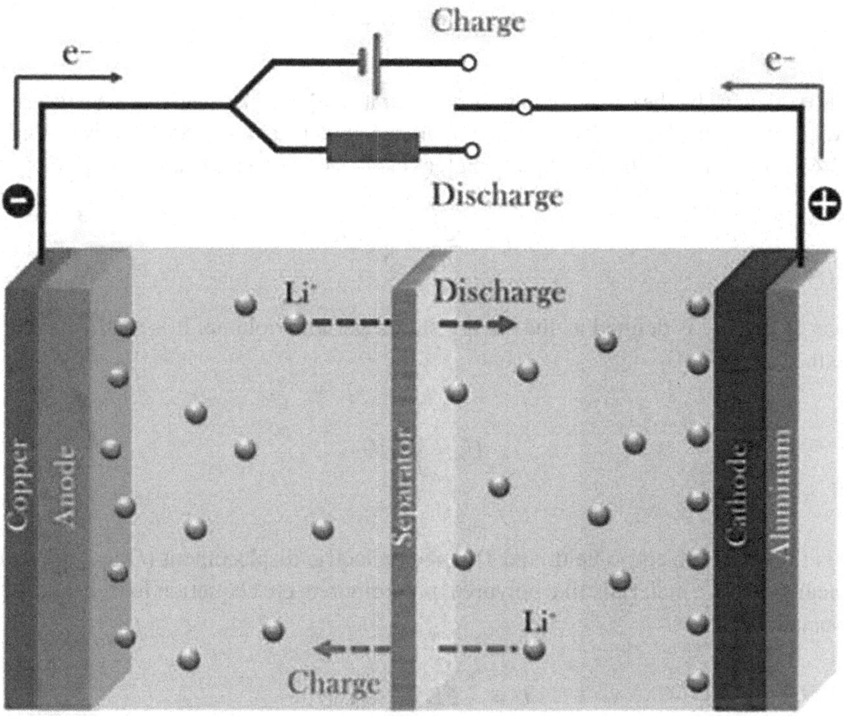

FIGURE 9.1 Schematic diagram of lithium-ion cell.

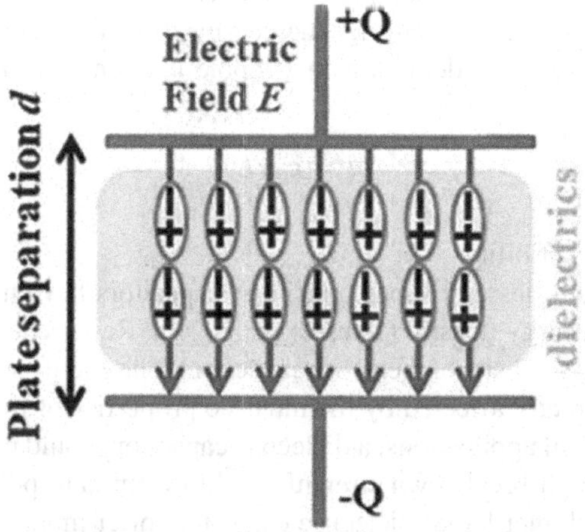

FIGURE 9.2 Schematic diagram of a dielectric capacitor.

$$\text{Capacitance, } C = \frac{\varepsilon_0 A}{d}$$

The work done by the capacitor to transport a unit charge from a negative plate to a positive plate is equivalent to energy stored in the capacitor (Purcell, 1985; Serway & Vuille, 2014; Yang et al., 2019). The expression for energy storage in a capacitor is given by

$$U = \frac{Q^2}{2C} = \frac{1}{2}CV^2$$

Energy density is defined as the energy stored per unit volume. It is mathematically defined as

$$U_e = \int EdD$$

E is the applied electric field, and D is the dielectric displacement ($D = \varepsilon_0\varepsilon_r E$). For linear dielectric materials like polyurea, polythiourea, etc., equation for the dielectric constant becomes

$$U = \frac{1}{2}ED = \frac{1}{2}\varepsilon_0\varepsilon_r E^2$$

Dielectric breakdown strength refers to the ability of a material to withstand applied electric field without losing its insulating nature (Guo et al., 2019). In the case of polymer nanocomposites, the addition of nanofillers up to threshold volume fraction enhances the dielectric breakdown strength of the material. Beyond this limit, conductive pathways are created in the composite material due to nanoparticle aggregation, particle–particle interactions, defect formation, etc. (Feng et al., 2019; Shen et al., 2017).

Another parameter that determines the energy storage property of a material is dielectric polarisation. It is defined as net dipole moments per unit volume of the material.

$$P = \varepsilon_0 \chi E$$

where χ is the susceptibility.

Compared to batteries and supercapacitors, capacitors have high power density, low loss, and low energy density (Jiang et al., 2021b). Researchers are focussing on improving the energy density of dielectric capacitors. The energy storage performance of a capacitor is directly affected by the intrinsic properties of the chosen dielectric material. For practical applications, a dielectric capacitor should possess high saturation polarisation, high breakdown strength, and low remnant polarisation (Kaifeng Yu et al., 2020). Polymer-based dielectric capacitors offer more processability, high

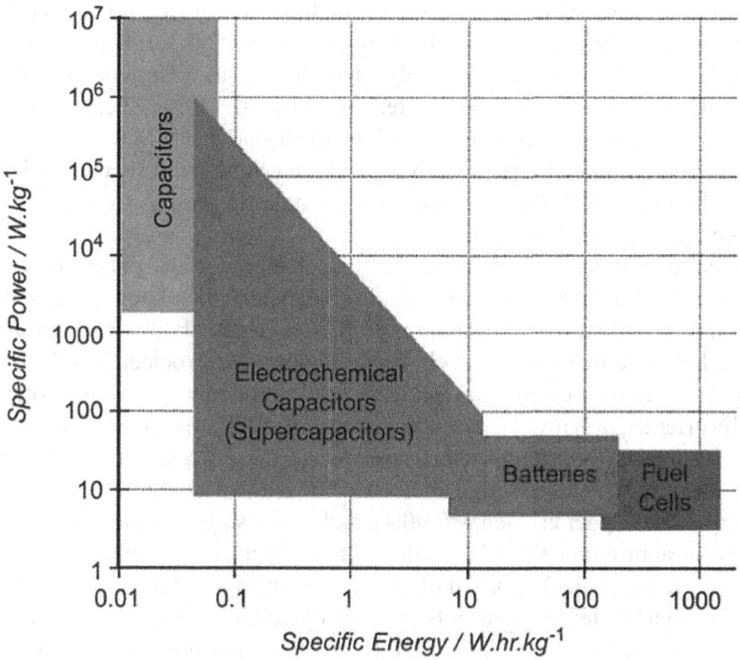

FIGURE 9.3 Ragone plot.

breakdown strength, and low dielectric loss compared to ceramic-based dielectric capacitors, and find attractive applications in the field of electronics. A Ragone plot helps to compare the energy storage performance of various energy storage devices (conventionally used for batteries, capacitors, and fuel cells) and choose the appropriate energy storage method for a specific application (Christen & Carlen, 2000; Hall & Bain, 2008) (Figure 9.3).

9.2 INORGANIC NANOFILLERS-DERIVED POLYMERS IN ENERGY STORAGE AND DEVICES

Polymer nanocomposites with inorganic fillers have been used in energy storage devices. Nanoparticles have a larger interacting area with the polymer matrix than microparticles and therefore can achieve significantly higher polarisation levels, breakdown strengths, and mechanical enhancements, providing more options for tailoring and optimising the properties of polymer nanocomposites. Furthermore, the dimensions of polymer nanocomposite-based devices can be reduced by introducing nano-sized fillers instead of micro-sized ones, favouring the device miniaturisation. M. Armand et al. (Armand, 1983; Armand et al., 1979b, 1979a) suggested the use of a solid electrolyte like a poly(ethylene oxide)

(PEO) salt complex, which has good ion conductivity, in electrochemical energy storage devices. This has paved the way for to develop solid-state lithium-ion batteries containing polymer electrolytes with various inorganic fillers, which are safer than typical lithium-ion batteries (Liu et al., 2019; Zheng et al., 2018). Nanocomposites based on poly(methyl methacrylate) (PMMA) (Zhang et al., 2015), poly(vinylidene fluoride) (PVDF), poly(acrylonitrile) (Sivaraj et al., 2021), PEO (Mohanta et al., 2018), etc., have been commonly used as solid electrolyte in solid-state lithium-ion batteries.

Compared to conventional inorganic dielectric materials, polymer dielectrics exhibit excellent mechanical properties, high dielectric breakdown strength, good processability, and electrical insulation properties (Hu et al., 2013a). Four different types of dielectric materials, namely linear dielectric, ferroelectric, relaxor ferroelectric, and anti-ferroelectric, are widely explored for energy storage applications. Biaxially oriented polypropylene, abbreviated as BOPP, is one of the widely explored and commercially important dielectric materials for energy storage applications. It is a linear polymer with high dielectric breakdown strength (~700 kV mm^{-1}), low energy loss, higher efficiency (~90%), and low cost. But, the dielectric constant of BOPP is approximately 2.2, resulting in a lower energy density (Xin Zhang et al., 2016; Dang et al., 2013). Nanofillers can be introduced into polymer matrices to improve their dielectric constants, thereby enhancing energy density values and breakdown strength (Guo et al., 2019). By synthesising polymer nanocomposites by incorporating nanofillers into the polymer matrix, it is expected to have synergistic properties for both the nanoparticles and the polymer (Jiang et al., 2021b). Introducing ceramic nanoparticles with higher dielectric constants improves the polymer nanocomposites' dielectric constant values but not necessarily the energy density values (Sun et al., 2017). Filler morphology, polymer–filler interface compatibility, etc., also play a role in the energy storage performance of the material. Surface functionalisation helps to reduce the nanofiller aggregation and interfacial energy barriers. Incorporating nanomaterials with high aspect ratios such as nanorods, nanowires, and nanosheets was found to enhance the breakdown strength of polymer nanocomposites (Zhang et al., 2020; Xinxin Zhang et al., 2020; Jiang et al., 2021a). Hu et al. studied the dielectric performance of polyimide–$BaTiO_3$ nanocomposites with different $BaTiO_3$ morphologies (0D nanoparticles and 1D nanofibres). Higher breakdown strength values were obtained for nanocomposites with 1D filler (BTNF) than those obtained for composites with 0D filler (BTNP) for the same volume fraction of nanofiller (Figure 9.4) (Hu et al., 2018; Sun et al., 2017).

Composites having divinyltetramethyldisiloxane-bis(benzocyclobutene) polymer matrix with Al_2O_3 nanoparticles of different morphologies (nanoparticles, nanowires, and nanoplates) as filler were synthesised, and their dielectric properties were studied. Among the nanocomposites, the one with 2D nanoplates as filler showed more significant breakdown strengths than others (Li et al., 2019). PMMA is another linear dielectric having high breakdown strength, high glass transition temperature (T_g ~ 160°C), and better dielectric performance (Lu et al., 2020). B. Xie et al. reported that PMMA-based nanocomposites with 5.0 wt% BT@SiO_2@dopa nanowires showed energy density values 213% higher than those of pristine PMMA polymer (Xie et al., 2021).

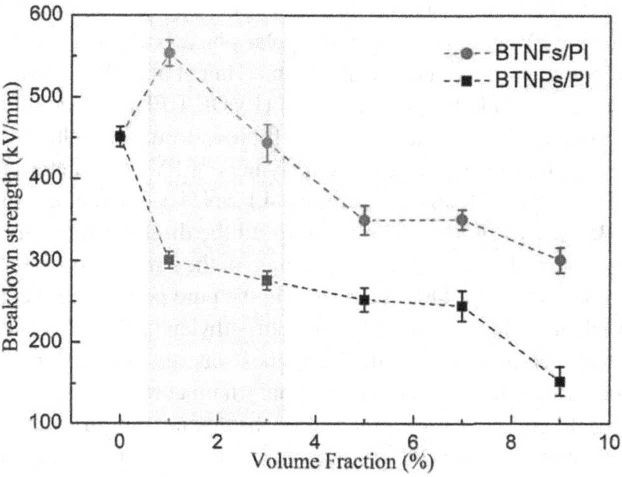

FIGURE 9.4 Breakdown strength of BT/PI nanocomposites with varying filler morphology (nanoparticles and nanofibres) as function of filler volume fraction.

9.3 ENERGY STORAGE PERFORMANCE OF FERROELECTRIC POLYMERS WITH VARIOUS INORGANIC FILLERS

Ferroelectric polymers have attracted attention in polymer dielectrics because of their excellent dielectric performance and hence higher energy density values (Jiang et al., 2021b). PVDF and its co-/terpolymers are among the widely explored ferroelectric polymers due to their superior dielectric constant values, excellent chemical and physical stabilities, high breakdown strength, and better mechanical flexibility (Xinping Hu et al., 2018; Prateek et al., 2016). Further improvement in the energy storage performance of ferroelectric polymers can be achieved by incorporating ceramic fillers with high values of dielectric constant such as $BaTiO_3$ (BTO) (Xie et al., 2017) and $BaTiZrO_3$ (Mayeen et al., 2020), or conductive fillers such as graphene (Mishra et al., 2020), carbon nanotubes (Begum et al., 2018), and metal nanoparticles (Issa et al., 2017; Gayen et al., 2017). Surface modification of fillers was also found to be an effective way to improve the dielectric properties and breakdown strength of polymer nanocomposites. Surface hydroxylated BTO nanoparticles in PVDF polymer matrix enhanced the breakdown strength of the composite compared to that of bare BTO nanoparticles (Zhou et al., 2011). PVDF nanocomposites containing BTO nanoparticles functionalised with polyvinylpyrrolidone as filler showed dielectric constant and electric breakdown strength values of 77 and 336 MV/m, respectively, and an energy density value of 6.8 J/cm^3 (Yu et al., 2013).

PVDF is a semicrystalline polymer with polar as well as non-polar crystalline phases. Out of the five different crystalline phases of PVDF, the β-phase has the highest polarity with an all-trans (TTTT) chain conformation (Martins et al., 2014). But, the non-polar α-phase is the dominant phase often. Researchers have improved the polar phase content through several methods such as electrospinning, melt crystallisation, using polar solvents, and incorporating nanofillers. Another technique to favour the PVDF crystallise in the polar phase is containing monomers and using them in copolymer and terpolymer forms (Han et al., 2012; Lu et al., 2006).

Poly(vinylidene fluoride-trifluoroethylene) (PVDF-TrFE) is one of the copolymers of PVDF that directly crystallises in the β-phase from the melt. PVDF-HFP and PVDF-CTFE are other widely explored copolymers of PVDF. Hu et al. reported that the PVDF-TrFE polymer matrix containing 4.1 vol.% of dopamine-functionalised Bi_2O_3-doped Ba0.3Sr0.7TiO_3 nanofibres improved the discharged energy density values 300%. Also, the dielectric breakdown strength of the nanocomposite containing 4.1 vol.% filler was enhanced by 140% compared to the pure polymer (Hu et al., 2013b).

Poly (vinylidene difluoride-co-chlorotrifluoroethylene), P(VDF-CTFE), is a ferroelectric copolymer of PVDF with the highest energy density among the polymers (Lovinger, 1983). Li et al. prepared ternary nanocomposites of P(VDF-CTFE) containing two fillers, viz. two-dimensional hexagonal boron nitride nanosheets (BNNSs) and barium titanate (BT) nanoparticles, with improved dielectric constant and breakdown strength values via solution-casting method. BNNSs not only prevented the aggregation of BT nanoparticles but also reduced leakage current and space-charge conduction. The highest energy density value was obtained for the nanocomposite with 12 wt% BNNSs and 15 wt% BT nanoparticles (Li et al., 2014).

Energy storage performance of dopamine-functionalised ST nanowires incorporated P(VDF-CTFE) nanocomposite was studied by Zhang et al. Nanocomposites with paraelectric 1D $SrTiO_3$ fillers exhibited better charge–discharge efficiency and discharge energy density than nanocomposites with ferroelectric 1D $BaTiO_3$ fillers. Since $SrTiO_3$ has a lower dielectric constant, inhomogeneous electric fields are reduced in nanocomposites, resulting in improved dielectric breakdown strength (Zhang et al., 2018).

The alignment of 1D fillers in the matrix also plays a significant role in the energy storage performance of polymer nanocomposites. Dopamine-functionalised $BaTiO_3$ nanowires aligned in the X–Y direction and Z direction in the P(VDF-CTFE) matrix were synthesised using a physical-assisted casting technique. The Z-aligned nanocomposite exhibited better energy density and charge–discharge efficiency than the X–Y aligned nanocomposite. The Z-aligned nanocomposite showed an energy density of 10.8 J/cm^3 at 2400 kV/cm (Xie et al., 2017).

D. Kang et al. reported that nanocomposites synthesised by dispersing core–shell nanowires with TiO_2 core and $BaTiO_3$ shell into poly(vinylidene fluoride-co-hexafluoropropylene) (PVDF-HFP) matrix possess better energy storage performance than nanocomposites with either TiO_2 or $BaTiO_3$ nanowire fillers. The core–shell nanowire fillers also reduced the dielectric loss at low- and high-frequency regions. Approximately a 35% increase in charge–discharge efficiency was observed when core–shell nanowire fillers were used instead of individual nanowire fillers. Even low filler content (~5% core–shell nanowires) showed a higher energy density value than the commercial BOPP (Kang et al., 2018).

Modifications of the interface between filler and polymer can enhance the dielectric performance of polymer nanocomposites. Luo et al. introduced a liquid crystalline polymer (poly {2, 5-bis [(4-methoxyphenyl)oxycarbonyl] styrene} (PMPCS)) in the interface between PVDF-HFP and $Na_2Ti_3O_7$ nanofibres. The presence of PMPCS results in strong interfacial polarisation and hence an overall increase in the dielectric constant of the nanocomposite. When the interfacial layer thickness varied from 9 to 25 nm, the dielectric constant varied from 16.9 to 69.6 (Luo et al., 2017).

Poly (vinylidene fluoride-trifluoroethylene chlorotrifluoroethylene) (PVDF-TrFE-CTFE) and poly (vinylidene fluoride-trifluoroethylene-chlorofluoroethylene) P(VDF-TrFE-CFE) are the terpolymers of PVDF and are relaxor ferroelectrics. The advantage of relaxor ferroelectrics over ferroelectrics is that they possess narrow hysteresis loops implying lesser hysteresis loss. They possess high dielectric constant values at room temperature compared to other dielectric polymers (Xu & Cheng, 2001). However, the incorporation of a third bulky monomer degrades the mechanical strength and, in turn, the dielectric breakdown strength of the terpolymers (Zhou et al., 2009). Adopting a nanocomposite approach can effectively improve their properties (Takala et al., 2010).

H. Tang et al. reported that ethylenediamine-functionalised $BaTiO_3$ nanowires incorporated P(VDF-TrFE-CFE) polymer nanocomposites showed a significant increase in energy density compared to the pristine polymer. The discharge energy density value (4.8 J/cc) and power density value (1.2 MW/cc) are, respectively, 17.5 times and 14 times larger than commercial BOPP capacitors (Tang et al., 2013). Incorporating BNNSs in the P(VDF-TrFE-CFE) terpolymer matrix improves the charge–discharge efficiency, dielectric breakdown strength, and energy storage performance of the nanocomposites. The solution-casting nanocomposite films exhibited improved properties compared to dielectric films obtained via traditional melt extrusion methods (Li et al., 2015).

9.4 CONCLUSION

Dielectric polymers are widely explored for their energy storage performances and are better candidates for developing flexible and efficient energy storage devices. Incorporating nanofillers onto polymer matrix enhances the dielectric performance of polymers. In addition to that, surface functionalisation of the fillers helps in improving the filler–matrix interactions and reduces the filler aggregation in the polymer matrix. The morphology and the alignment of the fillers also affect the dielectric and energy storage performance of the polymer nanocomposites. Two-dimensional fillers were found to enhance the breakdown strength of polymer nanocomposites.

REFERENCES

Armand, M. (1983). Polymer solid electrolytes-an overview. *Solid State Ionics*, *9*, 745–754.

Armand, M. B., Chabagno, J. M., & Duclot, M. J. (1979a). Fast ion transport in solids. *Electrodes and Electrolytes*, *131*, 2944–2955.

Armand, M. B., Chabagno, J. M., & Duclot, M. J. (1979b). Poly-ethers as solid electrolytes. *Fast Ion Transport in Solids*, *131*.

Begum, S., Ullah, H., Kausar, A., Siddiq, M., & Aleem, M. A. (2018). Fabrication of epoxy functionalised MWCNTs reinforced PVDF nanocomposites with high dielectric permittivity, low dielectric loss and high electrical conductivity. *Composites Science and Technology*, *167*, 497–506.

Christen, T., & Carlen, M. W. (2000). Theory of Ragone plots. *Journal of Power Sources*, *91*(2), 210–216.

Dang, Z. M., Yuan, J. K., Yao, S. H., & Liao, R. J. (2013). Flexible nanodielectric materials with high permittivity for power energy storage. *Advanced Materials*, *25*(44), 6334–6365. https://doi.org/10.1002/adma.201301752

Faria, R., Marques, P., Garcia, R., Moura, P., Freire, F., Delgado, J., & de Almeida, A. T. (2014). Primary and secondary use of electric mobility batteries from a life cycle perspective. *Journal of Power Sources*, *262*, 169–177.

Feng, Y., Wu, Q., Deng, Q., Peng, C., Hu, J., & Xu, Z. (2019). High dielectric and breakdown properties obtained in a PVDF based nanocomposite with sandwich structure at high temperature via all-2D design. *Journal of Materials Chemistry C*, *7*(22), 6744–6751.

Gayen, A. L., Mondal, D., Roy, D., Bandyopadhyay, P., Manna, S., Basu, R., Das, S., Bhar, D. S., Paul, B. K., & Nandy, P. (2017). Improvisation of electrical properties of PVDF-HFP: use of novel metallic nanoparticles. *Journal of Materials Science: Materials in Electronics*, *28*(19), 14798–14808. https://doi.org/10.1007/s10854-017-7349-9

Guo, M. F., Jiang, J. Y., Shen, Z. H., Lin, Y. H., Nan, C. W., & Shen, Y. (2019). Nanocomposites for capacitive energy storage: Enhanced breakdown strength and improved discharged efficiency. *Mater. Today*, *29*, 49–67.

Guo, M., Jiang, J., Shen, Z., Lin, Y., Nan, C. W., & Shen, Y. (2019). High-energy-density ferroelectric polymer nanocomposites for capacitive energy storage: Enhanced breakdown strength and improved discharge efficiency. *Materials Today*, *29*(xx), 49–67. https://doi.org/10.1016/j.mattod.2019.04.015

Hall, P. J., & Bain, E. J. (2008). Energy-storage technologies and electricity generation. *Energy Policy*, *36*(12), 4352–4355. https://doi.org/10.1016/j.enpol.2008.09.037

Han, R., Jin, J., Khanchaitit, P., Wang, J., & Wang, Q. (2012). Effect of crystal structure on polarization reversal and energy storage of ferroelectric poly (vinylidene fluoride-co-chlorotrifluoroethylene) thin films. *Polymer*, *53*(6), 1277–1281.

Hannan, M. A., Hoque, M. M., Mohamed, A., & Ayob, A. (2017). Review of energy storage systems for electric vehicle applications: Issues and challenges. *Renewable and Sustainable Energy Reviews*, *69*, 771–789.

Hu, P., Song, Y., Liu, H., Shen, Y., Lin, Y., & Nan, C.-W. (2013a). Largely enhanced energy density in flexible P (VDF-TrFE) nanocomposites by surface-modified electrospun BaSrTiO3 fibers. *Journal of Materials Chemistry A*, *1*(5), 1688–1693.

Hu, P., Song, Y., Liu, H., Shen, Y., Lin, Y., & Nan, C. W. (2013b). Largely enhanced energy density in flexible P(VDF-TrFE) nanocomposites by surface-modified electrospun $BaSrTiO_3$ fibers. *Journal of Materials Chemistry A*, *1*(5), 1688–1693. https://doi.org/10.1039/c2ta00948j

Hu, P., Sun, W., Fan, M., Qian, J., Jiang, J., Dan, Z., Lin, Y., Nan, C.-W., Li, M., & Shen, Y. (2018). Large energy density at high-temperature and excellent thermal stability in polyimide nanocomposite contained with small loading of $BaTiO_3$ nanofibers. *Applied Surface Science*, *458*, 743–750.

Hu, Xianfeng, Robles, A., Vikström, T., Väänänen, P., Zackrisson, M., & Ye, G. (2021). A novel process on the recovery of zinc and manganese from spent alkaline and zinc-carbon batteries. *Journal of Hazardous Materials*, *411*, 124928.

Hu, Xinping, Yi, K., Liu, J., & Chu, B. (2018). High energy density dielectrics based on PVDF-based polymers. *Energy Technology*, *6*(5), 849–864. https://doi.org/10.1002/ente.201700901

Issa, A., Al-Maadeed, M., Luyt, A., Ponnamma, D., & Hassan, M. (2017). Physico-mechanical, dielectric, and piezoelectric properties of PVDF electrospun mats containing silver nanoparticles. *APL Materials*, *3*(4), 30. https://doi.org/10.3390/c3040030

Jiang, Y., Zhou, M., Shen, Z., Zhang, X., Pan, H., & Lin, Y.-H. (2021a). Ferroelectric polymers and their nanocomposites for dielectric energy storage applications. *APL Materials*, *9*(2), 20905.

Jiang, Y., Zhou, M., Shen, Z., Zhang, X., Pan, H., & Lin, Y. H. (2021b). Ferroelectric polymers and their nanocomposites for dielectric energy storage applications. *APL Materials*, *9*(2). https://doi.org/10.1063/5.0039126

Kang, D., Wang, G., Huang, Y., Jiang, P., & Huang, X. (2018). Decorating TiO_2 nanowires with $BaTiO_3$ nanoparticles: A new approach leading to substantially enhanced energy storage capability of high-k polymer nanocomposites. *ACS Applied Materials and Interfaces*, *10*(4), 4077–4085. https://doi.org/10.1021/acsami.7b16409

Kim, J., Krüger, L., & Kowal, J. (2020). On-line state-of-health estimation of lithium-ion battery cells using frequency excitation. *Journal of Energy Storage*, *32*, 101841.

Li, H., Ai, D., Ren, L., Yao, B., Han, Z., Shen, Z., Wang, J., Chen, L., & Wang, Q. (2019). Scalable polymer nanocomposites with record high-temperature capacitive performance enabled by rationally designed nanostructured inorganic fillers. *Advanced Materials*, *31*(23), 1900875.

Li, M., Lu, J., Chen, Z., & Amine, K. (2018). 30 Years of lithium-ion batteries. *Advanced Materials*, *30*(33), 1800561. https://doi.org/10.1002/adma.201800561

Li, Q., Han, K., Gadinski, M. R., Zhang, G., & Wang, Q. (2014). High energy and power density capacitors from solution-processed ternary ferroelectric polymer nanocomposites. *Advanced Materials*, *26*(36), 6244–6249. https://doi.org/10.1002/adma.201402106

Li, Q., Zhang, G., Liu, F., Han, K., Gadinski, M. R., Xiong, C., & Wang, Q. (2015). Solution-processed ferroelectric terpolymer nanocomposites with high breakdown strength and energy density utilizing boron nitride nanosheets. *Energy and Environmental Science*, *8*(3), 922–931. https://doi.org/10.1039/c4ee02962c

Liu, H., Cheng, X., Xu, R., Zhang, X., Yan, C., Huang, J., & Zhang, Q. (2019). Plating/stripping behavior of actual lithium metal anode. *Advanced Energy Materials*, *9*(44), 1902254.

Lovinger, A. J. (1983). Ferroelectric polymers. *Science*, *220*(4602), 1115–1121.

Lu, X., Zou, X., Shen, J., Zhang, L., Jin, L., & Cheng, Z.-Y. (2020). High energy density with ultrahigh discharging efficiency obtained in ceramic-polymer nanocomposites using a non-ferroelectric polar polymer as matrix. *Nano Energy*, *70*, 104551.

Lu, Y., Claude, J., Zhang, Q., & Wang, Q. (2006). Microstructures and dielectric properties of the ferroelectric fluoropolymers synthesized via reductive dechlorination of poly (vinylidene fluoride-co-chlorotrifluoroethylene) s. *Macromolecules*, *39*(20), 6962–6968.

Luo, H., Ma, C., Zhou, X., Chen, S., & Zhang, D. (2017). Interfacial design in dielectric nanocomposite using liquid-crystalline polymers. *Macromolecules*, *50*(13), 5132–5137. https://doi.org/10.1021/acs.macromol.7b00792

Martins, P., Lopes, A. C., & Lanceros-Mendez, S. (2014). Electroactive phases of poly(vinylidene fluoride): Determination, processing and applications. *Progress in Polymer Science*, *39*(4), 683–706. https://doi.org/10.1016/j.progpolymsci.2013.07.006

Mayeen, A., Kala, M. S., Sunija, S., Rouxel, D., Bhowmik, R. N., Thomas, S., & Kalarikkal, N. (2020). Flexible dopamine-functionalized $BaTiO_3$/$BaTiZrO_3$/$BaZrO_3$-PVDF ferroelectric nanofibers for electrical energy storage. *Journal of Alloys and Compounds*, *837* (20), 155492. https://doi.org/10.1016/j.jallcom.2020.155492

Mishra, S., Sahoo, R., Unnikrishnan, L., Ramadoss, A., Mohanty, S., & Nayak, S. K. (2020). Investigation of the electroactive phase content and dielectric behaviour of mechanically stretched PVDF-GO and PVDF-rGO composites. *Materials Research Bulletin*, *124*(August), 110732. https://doi.org/10.1016/j.materresbull.2019.110732

Mohanta, J., Padhi, D. K., & Si, S. (2018). Li-ion conductivity in PEO-graphene oxide nanocomposite polymer electrolytes: A study on effect of the counter anion. *Journal of Applied Polymer Science*, *135*(22), 46336.

Owusu, P. A., & Asumadu-Sarkodie, S. (2016). A review of renewable energy sources, sustainability issues and climate change mitigation. *Cogent Engineering*, *3*(1), 1167990.

Prateek, Thakur, V. K., & Gupta, R. K. (2016). Recent progress on ferroelectric polymer-based nanocomposites for high energy density capacitors: Synthesis, dielectric properties, and future aspects. *Chemical Reviews*, *116*(7), 4260–4317. https://doi.org/10.1021/acs.chemrev.5b00495

Purcell, E. M. (1985). *Berkeley Physics Course: Electricity and Magnetism*. McGraw-Hill.

Raza, W., Ali, F., Raza, N., Luo, Y., Kim, K.-H., Yang, J., Kumar, S., Mehmood, A., & Kwon, E. E. (2018). Recent advancements in supercapacitor technology. *Nano Energy*, *52*, 441–473.

Roy, P., & Srivastava, S. K. (2015). Nanostructured anode materials for lithium ion batteries. *Journal of Materials Chemistry A*, *3*(6), 2454–2484. https://doi.org/10.1039/C4TA04980B

Schumm, B. (2000). Advances and trends in primary and small secondary batteries with zinc anodes and manganese dioxide and/or air cathodes. *Fifteenth Annual Battery Conference on Applications and Advances (Cat. No. 00TH8490)*, Long Beach, CA, USA, 89–94. IEEE.

Serway, R. A., & Vuille, C. (2014). *College Physics*. Cengage Learning.

Shen, Z.-H., Wang, J.-J., Zhang, X., Lin, Y., Nan, C.-W., Chen, L.-Q., & Shen, Y. (2017). Space charge effects on the dielectric response of polymer nanocomposites. *Applied Physics Letters*, *111*(9), 92901.

Sivaraj, P., Abhilash, K. P., Nalini, B., Perumal, P., & Selvin, P. C. (2021). Free-standing, high Li-ion conducting hybrid PAN/PVdF/LiClO4/Li0. 5La0. 5TiO_3 nanocomposite solid polymer electrolytes for all-solid-state batteries. *Journal of Solid State Electrochemistry*, *25*(3), 905–917.

Sun, W., Lu, X., Jiang, J., Zhang, X., Hu, P., Li, M., Lin, Y., Nan, C.-W., & Shen, Y. (2017). Dielectric and energy storage performances of polyimide/$BaTiO_3$ nanocomposites at elevated temperatures. *Journal of Applied Physics*, *121*(24), 244101.

Takala, M., Ranta, H., Nevalainen, P., Pakonen, P., Pelto, J., Karttunen, M., Virtanen, S., Koivu, V., Pettersson, M., Sonerud, B., & Kannus, K. (2010). Dielectric properties and partial discharge endurance of polypropylene-silica nanocomposite. *IEEE Transactions on Dielectrics and Electrical Insulation*, *17*(4), 1259–1267. https://doi.org/10.1109/TDEI.2010.5539698

Tang, H., Lin, Y., & Sodano, H. A. (2013). Synthesis of high aspect ratio batio3 nanowires for high energy density nanocomposite capacitors. *Advanced Energy Materials*, *3*(4), 451–456. https://doi.org/10.1002/aenm.201200808

Xie, B., Wang, Q., Zhang, Q., Liu, Z., Lu, J., Zhang, H., & Jiang, S. (2021). High energy storage performance of PMMA nanocomposites utilizing hierarchically structured nanowires based on interface engineering. *ACS Applied Materials and Interfaces*, *13*(23), 27382–27391. https://doi.org/10.1021/acsami.1c03835

Xie, B., Zhang, H., Zhang, Q., Zang, J., Yang, C., Wang, Q., Li, M.-Y., & Jiang, S. (2017). Enhanced energy density of polymer nanocomposites at a low electric field through aligned BaTiO3 nanowires. *Journal of Materials Chemistry A*, *5*(13), 6070–6078. https://doi.org/10.1039/C7TA00513J

Xie, Y., Yu, Y., Feng, Y., Jiang, W., & Zhang, Z. (2017). Fabrication of stretchable nanocomposites with high energy density and low loss from cross-linked PVDF filled with poly (dopamine) encapsulated BaTiO3. *ACS Applied Materials & Interfaces*, *9*(3), 2995–3005.

Xu, H., Cheng, Z. Y., Olson, D., Mai, T., Zhang, Q. M., & Kavarnos, G. (2001). Ferroelectric and electromechanical properties of poly (vinylidene-fluoride–trifluoroethylene–chlorotrifluoroethylene) terpolymer. *Applied Physics Letters*, 78(16), 2360–2362.

Yang, L., Kong, X., Li, F., Hao, H., Cheng, Z., Liu, H., Li, J.-F., & Zhang, S. (2019). Perovskite lead-free dielectrics for energy storage applications. *Progress in Materials Science*, *102*, 72–108.

Yoshino, A. (2012). The birth of the lithium-ion battery. *Angewandte Chemie International Edition*, *51*(24), 5798–5800. https://doi.org/10.1002/anie.201105006

Yu, Kaifeng, Wang, J., Wang, X., Li, Y., & Liang, C. (2020). Zinc–cobalt bimetallic sulfide anchored on the surface of reduced graphene oxide used as anode for lithium ion battery. *Journal of Solid State Chemistry*, *290*, 121619.

Yu, Ke, Niu, Y., Zhou, Y., Bai, Y., & Wang, H. (2013). Nanocomposites of surface-modified $BaTiO_3$ nanoparticles filled ferroelectric polymer with enhanced energy density. *Journal of the American Ceramic Society*, *96*(8), 2519–2524. https://doi.org/10.1111/jace.12338

Zha, J. W., Zheng, M. S., Fan, B. H., & Dang, Z. M. (2021). Polymer-based dielectrics with high permittivity for electric energy storage: A review. *Nano Energy*, *89*(PB), 106438. https://doi.org/10.1016/j.nanoen.2021.106438

Zhang, H., Marwat, M. A., Xie, B., Ashtar, M., Liu, K., Zhu, Y., Zhang, L., Fan, P., Samart, C., & Ye, Z. (2020). Polymer matrix nanocomposites with 1D ceramic nanofillers for energy storage capacitor applications. *ACS Applied Materials & Interfaces*, *12*(1), 1–37. https://doi.org/10.1021/acsami.9b15005

Zhang, H., Zhu, Y., Li, Z., Fan, P., Ma, W., & Xie, B. (2018). High discharged energy density of polymer nanocomposites containing paraelectric $SrTiO_3$ nanowires for flexible energy storage device. *Journal of Alloys and Compounds*, *744*, 116–123. https://doi.org/10.1016/j.jallcom.2018.02.052

Zhang, S., Cao, J., Shang, Y., Wang, L., He, X., Li, J., Zhao, P., & Wang, Y. (2015). Nanocomposite polymer membrane derived from nano TiO 2-PMMA and glass fiber nonwoven: High thermal endurance and cycle stability in lithium ion battery applications. *Journal of Materials Chemistry A*, *3*(34), 17697–17703.

Zhang, Xin, Shen, Y., Xu, B., Zhang, Q., Gu, L., Jiang, J., Ma, J., Lin, Y., & Nan, C.-W. (2016). Giant energy density and improved discharge efficiency of solution-processed polymer nanocomposites for dielectric energy storage. *Advanced Materials*, *28*(10), 2055–2061. https://doi.org/10.1002/adma.201503881

Zhang, Xinxin, Wang, F., Dou, L., Cheng, X., Si, Y., Yu, J., & Ding, B. (2020). Ultrastrong, superelastic, and lamellar multiarch structured ZrO_2–Al_2O_3 nanofibrous aerogels with high-temperature resistance over 1300°C. *ACS Nanoparticles*, *14*(11), 15616–15625.

Zheng, F., Kotobuki, M., Song, S., Lai, M. O., & Lu, L. (2018). Review on solid electrolytes for all-solid-state lithium-ion batteries. *Journal of Power Sources*, *389*, 198–213.

Zhou, T., Zha, J.-W., Cui, R.-Y., Fan, B.-H., Yuan, J.-K., & Dang, Z.-M. (2011). Improving dielectric properties of $BaTiO_3$/ferroelectric polymer composites by employing surface hydroxylated $BaTiO_3$ nanoparticles. *ACS Applied Materials & Interfaces*, *3*(7), 2184–2188.

Zhou, X., Zhao, X., Suo, Z., Zou, C., Runt, J., Liu, S., Zhang, S., & Zhang, Q. M. (2009). Electrical breakdown and ultrahigh electrical energy density in poly (vinylidene fluoride-hexafluoropropylene) copolymer. *Applied Physics Letters*, *94*(16), 162901.

10 Fundamental Applications of Inorganic Nanofillers for Water Purification Using Polymers

P. A. Nizam and Sabu Thomas
Mahatma Gandhi University

CONTENTS

10.1 INTRODUCTION

The scarcity of water as a result of technological advancements is a key concern that must be handled as a priority. Water resources are being contaminated by industrial effluents and other pollutants such as dyes, bacteria, heavy metals, and so on. These waters must be treated before they may be reused for home and technological applications. To illustrate, the water resource utilized in an industry may be processed, treated, and reused for its intended purpose, hence reducing the usage of surplus resources.

DOI: 10.1201/9781003279389-10

Membrane technologies are the most promising and efficient techniques of treatment in which particles of various sizes may be sieved using various morphological membranes such as ultrafiltration membrane, microfiltration membrane, nanofiltration membrane, and so on. Prior to designing each membrane, the intended purpose is determined, such as the pollutants that must be separated, such as germs, dyes, or heavy particles, and then the porosities and other qualities are adjusted. Polymers are the most common materials utilized in the fabrication of membranes because they can be tuned using various modifications to meet their end needs. Furthermore, modifying the pores and reinforcing them with various inorganic fillers improve their features such as mechanical strength, antibacterial property, antifouling, rejection efficiency, and so on. Metal nanoparticles, carbon-based nanoparticles, metal-organic frameworks, and other inorganic nanoparticles are among the most common.

This chapter surely offers light on numerous membrane technologies, their synthesis process, their mode of reinforcing with inorganic nanoparticles, the various polymers used, and the changes in their features with the addition of various inorganic fillers.

10.2 MEMBRANE TECHNOLOGY

Membrane technology is an effective solution to the increasing need for water treatment. These are pressure-driven filtering systems in which membranes are utilized as a barrier to stop water toxins or pollutants such as nutrients, bacteria, dyes, organic particles, inorganic particles, metal ions, turbidity, and so on. The semipermeable membrane facilitates the movement of smaller molecules while retaining the larger ones. Molecule's natural movement from high concentration to lower concentration can be altered by inducing pressure on which molecules move from low concentration to higher concentration (Shon *et al.*, 2013). The pressure difference between both sides of the membrane enhances a steady flow of permeate. The filtration process is of two types, (i) conventional or dead-end filtration and (ii) cross-flow filtration. Figure 10.1 depicts (a) dead-end filtration and (b) cross-flow filtration processes.

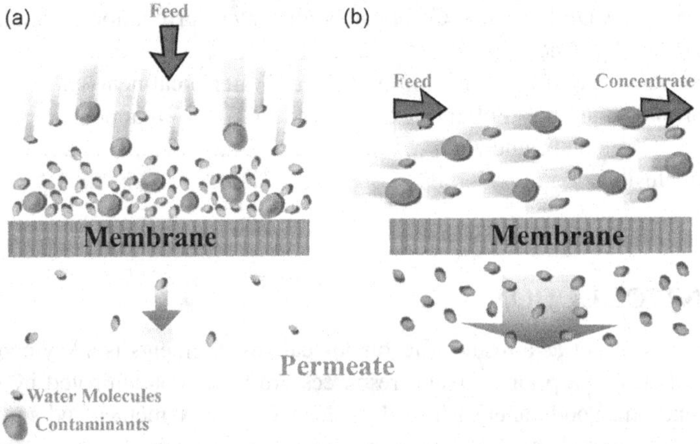

FIGURE 10.1 (a) dead-end filtration and (b) cross-flow filtration processes.

Dead-end filtration is a straightforward batch-filtering process in which the input water is driven through the membrane by applying pressure. There are two streamlines retentate the source line and permeate the filtered stream. The retentate flowed through the membrane, leaving the trapped particles on the membrane's surface, which is referred to as a cake. As the process progresses, the size of the cake grows, resulting in a low flux for the permeate. They have a high filtering capacity and are often used in medical sectors and labs. The final membrane cannot be cleaned or backwashed since particles clog the interior pores. The cross-flow filtration technique, also known as tangential flow, is one in which the water flows parallel to the membrane. The continual turbulent flow prevents pollutants from accumulating on the membrane's surface and transporting them along with the feed. The pressure differential permits the retentate to pass through the membrane while washing away the particles that have remained on the surface. Unlike the dead end, which has two streamlines, this procedure has three: one for the feed, one for the permeate, and one for the water with retained particles carrying out the concentrate. This process is called cross-flow because the feed and concentrate are flowing 90 degrees to permeate. The method is used to filter water with a high concentration of contaminant particles. As the feed and concentrate continue to flow, these help the membrane to stay clean. The membrane process is divided into four basic categories based on pore size: microfiltration (MF), ultrafiltration (UF), nanofiltration (NF), and reverse osmosis (RO). Let's have a quick peek into these processes.

10.3 TYPES OF MEMBRANES

Ultrafiltration (UF) is a treatment process in which water is passed over a semi-permeable membrane, where high-molecular-weight solutes and particles become entrapped on the retentate side while water and low-molecular-weight solutes filter to the permeate side. Ultrafiltration membrane pore sizes range from 0.1 to 0.01 microns. This technique can remove viruses, organic molecules, and a range of salts. Their applications include paper pulp effluent treatment, food, and beverage industry, treatment of wastewater from industries, etc. Ultrafiltration membranes must be cleaned regularly to avoid fouling by scaling, solids, and microbiological agents such as bacteria and algae. Liu et al. modified the antifouling characteristics of the (polyvinylidene fluoride) PVDF membrane using quaternized graphene oxide (QGO). QGO was synthesized by growing ammonium salt on the surface of GO in situ. This modification improves the membrane's hydrophilicity and mechanical and antibacterial properties. QGO has strong bonding with the membrane and barely departs, giving the membrane a prolonged antibacterial function (Liu *et al.*, 2020).

Microfiltration is similar to UF in that the polluted liquid is passed over a semi-permeable membrane, and solid particles big enough to pass through the pores are visible on one side of the membrane. The primary distinction between these two processes is the membrane pore size. While UF pore diameters vary from 0.1 to 0.01 m, MF pore sizes range from 0.1 to 10 microns. To further understand where to apply this process, consider milk. If you need to remove colloidal particles while keeping the proteins in the permeate, MF is the ideal solution. Because MF has wider pore sizes than UF, most multivalent ions, viruses, and water flow through but solid pollutants and bacteria are restricted. They are frequently used in petroleum refining, oil/water separation, and wastewater treatment. They are also employed as a prior treatment before RO. Along with PVDF, another extensively utilized polymer for membrane synthesis is polyether sulfone (PES). These membranes were modified utilizing polyethylenimine (PEI) and graphene oxide in a layer-by-layer construction (GO). PEI's positive charge strengthens their bonds to PES as well as the incoming GO. The change significantly enhanced the dye adsorption capacity of the membrane while maintaining excellent flux, implying that GO, an inorganic nanofiller, is a viable material for the water purification process (Homem *et al.*, 2019).

The most efficient water treatment procedures are RO and nanofiltration (NF). All of these approaches are similar to the preceding one, but the pore size of the membrane differs. NF can separate particles with sizes ranging from 0.002 to 0.005 m, which are used to clean portable water. They can remove large divalent ions, pesticides, and other contaminants. Osmosis is a process in which a low concentration liquid flows to a high concentration liquid. With the application of pressure to a highly concentrated liquid, the flow is reversed to a low concentrated liquid across a semipermeable membrane in RO. RO is capable of eliminating all tiny and big ions, and it has a narrow pore that can remove particles as small as 0.01 nm. Using a pre-seeding interfacial polymerization approach, Lakhotia et al. investigated the effect of iron oxide (FeO) on PES membranes. The synthesized membrane demonstrated a direct influence of FeO, with a rise in FeO concentration facilitating an increase in flux, hydrophilicity, and surface charge. This membrane has a high fouling resistance as well as a saltwater rejection rate of greater than 90% (Lakhotia *et al.*, 2019). To create a thin layer of RO membrane, graphitic carbon nitride sheets, which approximate the 2D structure of graphene, triangular nanopores, and defects such as lamellar networks were acidified and integrated with polyamide (PA). These modifications resulted in negatively charged, hydrophilic membranes with better antifouling properties, an increase in permeate flow of 79.3%, and a NaCl rejection of 98.6% (Gao *et al.*, 2017).

10.4 COMMON METHOD OF FABRICATION OF MEMBRANES

Membrane techniques are developed with the construction material, type of membrane, end use, and membrane geometry in mind. The pores are regulated throughout their manufacture, and post-treatments are frequently performed for the specific activity. Organic polymers are commonly used in the fabrication of membranes since they are extremely permeable, flexible, and processable. Inorganic materials are also used where thermal stability and chemical resistance are required. To improve their performance, these membranes are further modified using inorganic and organic fillers, as well as metal-organic frameworks (MOF). Some of the common techniques used in the fabrication of the membranes are discussed.

10.4.1 Phase Inversion Technique

The phase inversion (PI) approach is a widely acknowledged technique for creating asymmetric membranes that involve a de-mixing process in which the polymer solution is turned into a solid state using a coagulation bath. When the polymer solution is in a non-solvent coagulation bath, a solvent exchange occurs. The key elements influencing the PI are solution composition, solvent selection, coagulation bath composition, non-solvent system, and lastly film casting conditions (Zahid *et al.*, 2018). The approach is frequently used in both laboratory and industrial settings. The PI process is classified into four categories. Non-solvent induced phase inversion (NIPS), thermally induced phase inversion (TIPS), evaporation-induced phase inversion (EIPS), and vapor-induced phase inversion (VIPS) are the four types of phase inversion (VIPS). The polymer solution in a solvent is cast onto a flat surface in the NIPS process, followed by inversion into a non-solvent. The solvent–non-solvent exchange promotes the formation of polymer membranes, wherein morphology is essentially determined by the mutual interchange of two solvents. Controlling the solvent exchange rate by altering the coagulation medium, dope composition, quenching bath temperature, and evaporation time is required to boost the effectiveness and structure formed by NIPS.

When polymers are insoluble at room temperature, the TIPS approach is used. In this case, polymers are dissolved in solvents at elevated ambient temperatures before being cast into the film at the same temperature. They are then cooled to a temperature suitable for precipitation, at which point they generate regulated gelation. The solvent is evaporated, removed, or freeze-dried when the precipitation is complete. Thermodynamic parameters such as solvent system, polymer concentration, temperature, and cooling rate decide the pore size of the symmetric membranes (Gohil and Choudhury, 2018). VIPS emphasizes the preparation of highly porous membranes, mostly MF membranes. This process includes introducing a casted film to a vapor chamber's environment of non-solvent (typically water) vapors. The influx of non-solvent vapors into the polymeric-casted film and the outflow of solvent from the casted film cause phase inversion. Because of their low volatility, non-solvent vapors dominate mass transfer as compared to solvents (Zahid *et al.*, 2018). In solvent EIPS, a homogenous polymer solution is produced from two or more volatile solvents with distinct boiling temperatures and dissolving characteristics and then distributed as a thin layer. Evaporation of a more volatile solvent from the cast polymer solution sheet causes polymer precipitation resulting in the formation of the membrane.

10.4.2 Electrospinning

Electrospinning techniques are used for the fabrication of fibers in the range of submicron to nanometer range. Many polymers, ceramics, and composites can be spun via this technique. The main advantage of the process is that they only require a dissolving solvent, and do not require any functionalization or complex processes. Solution viscosity, flow rate, the concentration of the solution, and electric field are some of the governing factors of this process (Jose Varghese *et al.*, 2019).

This electrodynamic technique entails electrifying a liquid droplet to create a jet, which is subsequently stretched and expanded to produce the fiber. A typical electrospinning setup includes a spinneret, a syringe pump, a high-voltage power source, and a conductive collector. The process begins with a droplet extruded from the end of the spinneret due to surface tension, forms as a Taylor cone, and ejects a charged jet upon electrification. This jet is collected at the collector surface initially traveling a straight path and followed by a whipping path due to the bending instabilities. The materials solidify quickly because of the stretching action of the jet into fibers. In a nutshell, the approach is broken into four phases: (i) Formation of Taylor cone droplets after charging a liquid droplet, (ii) jet movement along a straight line (iii) electric field thinning of jet fibers, followed by whipping, (iv) solidified fiber collection on the collector (Xue *et al.*, 2019).

10.4.3 Track Etching

When compared to other traditional methods, track membranes provide great structural control. This technique generates membranes with retention and transport properties. Track and etching procedures govern the density, shape, and size of this process. Many polymers are being widely employed for this process study. Track etching is an industrial technique for the manufacturing of isoporous membranes in which polyethylene naphthalate or polycarbonate films of a particular thickness, typically 6–35 μm, are exposed to accelerated heavy ions. When these ions move through the film, their journey creates a distorted trajectory. The density of the pores is determined by the duration spent in the reactor, with a longer time exposure resulting in larger holes on the film. To induce porosity, the film is rinsed with caustic etchants such as sodium hydroxide to remove damaged zones from trajectories. Etchants are required to complete the course's full-fledged opening. The porosity may also be controlled by adjusting the etchant's exposure temperature, and concentration (Apel, 2001).

10.4.4 Photolithography

Microfabrication is used in lithography to create a precise pattern on a film. For this procedure, a source light, generally optical, UV, is employed, and the substrate is covered with a photoresist substance. These are often extremely light-sensitive materials. Following that, a pattern mask is applied to prevent parts of the film from being exposed to light. Only the uncovered portions are exposed to light in this way. Depending on the photoresist used, the unmasked or masked portions are removed using a chemical called a developer after the light explosion. For a positive photoresist, light deteriorates the unmasked portions exposed to light, which are then dissolved by the chemical treatment. In this situation, a pattern resembling the masking pattern is achieved. When a negative photoresist is exposed to light, the uncovered portions polymerize, while the masked areas are removed during chemical treatment. A pattern comparable to uncovered areas is generated in this kind. Other methods, such as ion-beam lithography and electron-beam lithography, use a similar idea, but instead of light, bursts of charged particles or electrons are used (Gohil and Choudhury, 2018).

10.4.5 Sintering and Stretching

Sintering is a process used to produce porous membranes without liquefaction from inorganic and organic powdered materials with high melting points. Powdered materials are sintered using heat and pressure slightly below their melting point after being formed into films via tape casting, pressing, extrusion, or slip casting. The particle size and distribution control the porosity and pore distribution. The material used in manufacturing determines the mechanical and thermomechanical properties. Organic materials often used in membranes include polytetrafluoroethylene (PTFE), whereas inorganic membranes commonly include ceramic, graphite, metal powders, and other materials. This approach enables the creation of various membrane configurations such as tubes, cartridges, disks, and so on (Gohil and Choudhury, 2018).

By physically rupturing hollow fiber precursors and partly crystalline films, stretching creates pores. This method extrudes film/hollow fibers with crystallites aligned along the drawing direction. They are then annealed and cooled further to produce pores mechanically rupturing by cold stretching. At a faster strain rate, the films are ruptured perpendicular to the direction of drawing. Hot stretching is sometimes used with a decreased strain rate to increase pore size. Controlling the annealing, stretching, cold and hot factors like crystallinity, temperature, and strain rate allow for the fabrication of asymmetric membranes with holes ranging from 0.2 to 20m.

10.5 POLYMER INORGANIC COMPOSITES

Polymers are widely employed for the fabrication of membranes for water treatment due to their variable properties and ease of modification. The incorporation of inorganic fillers aids their performance such as antibacterial, antifouling, rejection of heavy metals dyes, organic particles, etc. Some of the polymer-based inorganic nanocomposites are briefly discussed.

10.5.1 PVDF Inorganic Composites for Water Purification

Due to its outstanding thermal stability, mechanical properties, superhydrophobicity, and chemical resistance, polyvinylidene fluoride (PVDF) is a key contributing polymer for water treatment membranes. This polymer is commonly used in UF, MF, and NF applications, and it is now being investigated as a contender for membrane contactor and distillation (Liu *et al.*, 2011). PVDF is more hydrophobic than other membrane polymers such as PES, polysulfone (PSF), and polyimide (PI), which are typically connected to surface tension. Other polymers, such as polytetrafluoroethylene (PTFE) and polypropylene (PP), frequently need the use of complicated solvents in the fabrication of membranes. In contrast to these polymers, PVDF is sensitive to practically all organic solvents and dissolves easily, allowing it to be cast into membranes using a phase inversion approach. Their high chemical resistance and mechanical capabilities increase their usage in wastewater treatment; moreover, their thermal stability trait is used in industrial applications. Because of their thermodynamic compatibility with other polymers and low level of extractability, they can be used to make membranes for biomedical and other purposes.

PVDF being a semicrystalline polymer contains almost 60% of which imparts hydrophobicity to the polymer. They are typically synthesized using suspension or emulsion polymerization with –CH2–CF2– repeating unit. Fluorocarbons are having high thermal stability than hydrocarbons due to the high bond dissociation energy of the C–F bond and the high electronegativity of fluorine atoms. PVDF also possesses excellent chemical resistance to many harsh chemicals like inorganic acids, chlorinated aromatic solvents, halogens, and oxidants. These characteristics of PVDF escalate their use in membrane technology.

Oil spills and oily wastewaters endanger both environment and humans by causing health concerns. PVDF is commonly used to solve these challenges, and in one investigation, PVDF and PS reinforced with halloysite nanoclay and $CoFe_2O_4$ were synthesized as sorbent materials using electrospinning. The electrospinning method is carried out by creating a homogeneous dispersion of polymer and nanofiller. The solution is placed in a syringe, and the tip of the syringe is connected to a strong electric discharge. The samples are being collected at a collector surface covered with aluminum foil. The fluid in the syringe is released at a certain flow rate, and the high discharge drives the samples to deflect, which are collected at the collector surface. The spectroscopy technique revealed a homogenous filler dispersion, and oil sorption capabilities of 3.70–10.01 g/g for PVDF were observed (Pascariu Dorneanu *et al.*, 2018). PVDF was modified with ZrO_2-multiwalled carbon nanotubes (MWCNTs) where zirconium dioxide was attached to MWCNT using a hydrothermal method, followed by the preparation of composite by phase inversion method. Scanning electron microscopy (SEM) and atomic force microscopy (AFM) revealed that ZrO_2 was uniformly dispersed on the surface of nanotubes. These membranes had a lower contact angle and better flux than the pristine PVDF membrane. Their high antifouling performance along with the oil–water separating makes them a convenient excipient for this application (Yang *et al.*, 2016). Binding nanoparticles into the complex structure for attaining superhydrophilic property is yet investigated. A mussel-inspired method (Wang *et al.*, 2019) was employed to develop a PVDF membrane by anchoring TiO_2 directly on the surface of the membrane. This makes the membrane of PVDF hydrophilic. The anchoring of the metal nanoparticles is achieved using dopamine. Dopamine is a class of bioglue, which can self-polymerize into any substrate irrespective of its surface morphology or chemical composition. Dopamine along with silane coupling agents was copolymerized as the TiO_2 was directly anchored on the surface of the PVDF membrane. These membranes were evaluated for the separation of oil in water and showed an efficiency of 99% along with antifouling and durable resistance (Shi *et al.*, 2016).

Many inorganic nanoparticles are employed in water purification applications. Many factors, including biofouling, bacterial growth, and colonization, harm membrane performance. These inimical factors are addressed using biocidal silver nanoparticles. PVDF membranes that have been alkaline-treated are reacted with thioglycolic acid (TGA) and pentaerythritol tetrakis (3-mercapto propionate) (PETMP). TGA-PVDF had a cluster morphology, whereas PETMP-PVDF had dispersed Ag^+ nanoparticles. According to with findings, a cluster-based morphology provides superior antibacterial and antifouling properties. This is owing to regulated leaching in the case of the TGA-PVDF membrane, where the major mechanism of antibacterial characteristics is Ag^+ leaching. The trans-membrane flux was measured, and findings revealed that the constructed membranes had a greater flux rate. These findings have crucial implications for the design of water purification membranes and highlight the role of surface assembly of biocidal nanoparticles in antibacterial activities (Sharma *et al.*, 2016). TiO_2/PVDF-based membranes made utilizing a wet chemical process demonstrated antifouling properties and a high flux of 95% (Pang *et al.*, 2016). Under SEM examination, hydrothermally produced nanoparticles embedded in PVDF demonstrated uniform dispersion. This flexible membrane appears to have the potential to function as an MF membrane. Many other inorganic fillers such as graphene oxide, nanotubes, and their application in polymers for water treatment will be discussed in the following sections.

10.5.2 Polyether Sulfone

PES is another membrane polymer. Polymers having sulfonyl groups, such as PES and PSF, are thermoplastic polymers with high oxidative thermal, chemical, and mechanical durability. PES has high mechanical properties, high Tg, environmental endurance, and good heat aging resistance. They are extensively employed for the application of nanofiltration, ultrafiltration, and microfiltration. Their excellent hydrophobicity imparts their application, thus leading to low antifouling properties which are improved by various organic–inorganic reinforcing systems.

The poor hydrophilicity of the PES membranes can be improved by the incorporation of inorganic fillers. SiO_2 can be employed due to its simple synthesis procedure, low cost, and less toxic to aqueous solutions. Monodisperse silica nanoparticles are dispersed in a specific concentration onto the PES membrane which escalates their performance. The hydrophilicity of the membranes is improved resulting in higher permeability. Moreover, the antifouling properties are enhanced with an increased fouling resistance of ca. 70% which can be attributed to improvement in hydrophilicity (Lin *et al.*, 2016). ZnO and SiO_2 nanoparticles were evaluated in the PES membrane and showed promising results. SiO_2 performed well in the hybrid composite compared to ZnO, showing a better flux, due to the enhancement of hydrophilicity in the SiO_2 membrane (Kusworo *et al.*, 2017).

Lin et al. studied the effect of Tungsten disulfide WS_2 explaining the permeability, antifouling porosity, etc. WS2-PES membranes were fabricated using NIPS with nanoparticle concentration in the range of 0.025%–0.25%. Their permeability, solute rejection, and antifouling properties were evaluated and showed an enhancement. An increase in permeability was observed for the modified membrane. This can be attributed to increasing hydrophilicity upon the addition of WS_2 and an increase in pore size. Up to a specific concentration, the pore size is improved along with hydrophilicity which escalates their permeate flux. Upon addition of higher loading of filler, the pores are plugged and a thin layer of skin is formed on the surface of the membranes which redistricts the flow. The increase in pore size also increases the contact angle measurement, as more water is absorbed due to higher pore sizes. The fouling activity is subjected to the hydrophilic property of the membranes. As their hydrophilicity increases, they inhibit hydrophobic interactions with membranes, thus reducing their interactions with organic wastes, oils, and fouling agents. Higher concentrated nanoparticle membranes have better antifouling properties as they have more WS_2 particles on the surface which imparts hydrophilicity. Their rejection is also high but compromises the flux (Lin *et al.*, 2013).

Another inorganic filler mostly employed in membranes is TiO_2. TiO_2-embedded PS membrane via NIPS process shows higher hydrophilicity, and thus antifouling and good mechanical properties. As previously discussed, the contact angle is the key parameter to indicate the flux, hydrophilicity, and antifouling. Embedding nanoparticles (up to 7 wt %) in polymer matrix increased the exchange rate between solvent and non-solvent, resulting in membranes with more finger-like macro voids, greater porosity, bigger mean pore size, and therefore faster water flow. The enhanced water flow for nanocomposite membranes can be attributed to the interaction of improved greater porosity surface hydrophilicity and the creation of macro voids in the membrane structure (Hosseini *et al.*, 2018).

10.6 GRAPHENE-BASED POLYMER COMPOSITES FOR WATER TREATMENT

Graphene is one of the most promising materials for the next generation, because of its diverse tailoring ability, 2D assembly, and associated band structure. These materials were awarded the Noble Prize in 2010 and are thought to be stronger than steel. High electrical conductivity, surface area, fracture strength, mechanical property, mobility as a charge carrier, thermal conductivity, optical transmittance, chemical stability, and other properties distinguish them. Graphene is hydrophobic, and it cannot be used in water application without modification. Chemically modified graphene, such as graphene oxide (GO) and reduced graphene oxide (rGO), is used in water treatment applications as it has adequate surface groups such as hydroxyl, carboxyl, ketone, and epoxy groups at its edges and basal planes. GO is indeed very hydrophilic and has a negative charge, which aids in the removal of cationic dyes and heavy metal atoms. RGO, on the other hand, has a large surface area and a low negative charge, allowing it to effectively remove anionic contaminants. They are prepared using the hummers method and modified hummers method. These three materials can be used in water treatment applications; however, they must first be modified by anchoring to polymer materials (Gandhi *et al.*, 2016).

GO reinforced in polyamide (PA) was evaluated for their susceptibility for water treatment application. They were synthesized using an in situ interfacial polymerization process at various concentrations. Their morphological studies show a uniform dispersion, and their hydrophilic properties improve as concentration increases. The film demonstrated significant NaCl and Na_2SO_4 rejection as well as good flex, which may be attributed to the interlayer gap between GO nanosheets, which may operate as water channels for permeate transport (Yin *et al.*, 2016). GO can be employed to impart hydrophilicity to hydrophobic membranes. PVDF being a hydrophobic material has yielded hydrophily by incorporation of GO. The membrane is prepared by incorporating previously made GO into PVDF followed by electrospinning. Although the pores are higher in PVDF, the flux is less when compared to GO modified. The incorporation of GO into PVDF enhances the hydrophilicity which improves the flux property of the membrane. The membrane also exhibits improved antifouling properties and is a good excipient for water treatment. GO can be used as an agent to enhance hydrophilicity which directly enhances their flux as well as antifouling properties (Jang *et al.*, 2015). Another hydrophobic polymer such as PES is also analyzed by incorporation of GO and showed similar results (Ouyang *et al.*, 2015). Cellulose, a biopolymer, is exploited in many applications such as water purification, electronics (Nizam *et al.*, 2021), drug delivery, as well as reinforcing agents (Rose Joseph *et al.*, 2021) in many applications. Amine-functionalized GO was embedded in a cellulose nanofiber membrane, which was fabricated using simple vacuum filtration. The amine group enhances their adsorption toward anionic dye, and at the same time exhibits improved antibacterial properties as well (Nizam *et al.*, 2020).

Go and rGO incorporated in polyaniline were evaluated for a comparative study of dye adsorption. Both the composites exhibit improved properties for water treatment, and the characterizations revealed that rGO outperforms GO in the case of dye adsorption. rGO has fewer oxygen groups when compared to GO. This result suggests that the RGO is mostly functionalized by other forms, most likely vacancies or -conjugated structures; such active/defect sites can enable a variety of interactions with dye, assuming identical adsorption of dyes regardless of charge state (El-Sharkaway *et al.*, 2020).

Carbon-based nanomaterials, on the other hand, are frequently hampered by low stability, loss of substantial concentration, and the deposition of non-selective pores during interfacial polymerization. These disadvantages limit their application in the absorption of heavy metals and desalination. Cyclodextrin due to its physicochemical properties and porosity can be employed to address these disadvantages. They are recyclable and at the same time improve the chemical stability as well as the surface properties. The efficacy of cyclodextrin to influence stability and porosity is investigated to improve GO stability during interfacial polymerization and to increase overall membrane porosity. Cyclodextrin-incorporated GO-PSF membrane was evaluated for permeability and rejection of contaminants. The membrane removes efficiently whole 100% of heavy metals, salts, and organic contaminants continuously for five hours without loss of flux. The developed membrane possesses good surface characteristics, surface charges, and continually developing hydrophilic qualities, which might be attributed to the included CD and GO's expressed conformational features. The treatment method entails the rejection of hydrophobic functional groups in organic pollutants by the hydrophobic inner region of CD. The b-outer CD's hydrophilic portion attracts water molecules, resulting in increased water flow into the embedded membrane. Similarly, the GO-based nanocomposite's extensive positive and negative charges enhance the rejection of both anions and cations (Badmus *et al.*, 2021).

10.7 CARBON NANOTUBES–POLYMER COMPOSITES FOR WATER TREATMENT

CNT in simple words can be visualized as a graphene sheet rolled into a cylinder with diameter in nanorange and capped with a spherical fullerene. Graphene sheets are made up of an x-y plane monolayer of sp2-bonded carbon atoms. CNTs exhibit distinct electrical characteristics due to the existence of delocalized electrons in the z-axis. Two types of CNTs are single carbon nanotube (SWCNT) and multiwall carbon nanotube (MWCNT) wherein MWCNTs' multilayer of graphene sheets is employed in many applications. The strong forces such as van der walls forces resulting from the high-polarized π-electron clouds enhance their attraction between CNT and graphene sheets. CNT is naturally hydrophobic, which offers advantages in water treatment. For starters, hydrophobicity and capillarity influence sorbate adsorption behavior and orientation in microporous carbons. For unfunctionalized nanomaterials, physio-sorption is the major sorption process (Shawky *et al.*, 2011).

MWCNT/PA nanocomposite membranes synthesized by the grafting process were evaluated for the water treatment process. The basic characterizations revealed a uniform dispersion of CNT in PA matrices. Albeit they have good rejection for NaCl and humic acid and mechanical properties, their flux is reduced due to the good dispersion of the CNT. The CNT dispersion yields a compact structure, which reduces the flux as well as the contact angle. A low-resistance channel for solvent and water transport exists between an individual MWCNT and the surrounding polymer chains (Shawky *et al.*, 2011). PVDF-CNT membranes exhibited some interesting results. The membrane fabricate showed hydrophilic characteristics, albeit PVDF and CNT are hydrophobic. When the membrane is dried at 55°C, all of the solvent moieties are removed and the CNT open ends become reactive, forming strong interactions with the fluoride ions of PVDF and resulting in hydrophilic membranes. Furthermore, this significant transformation can be related to the development of enhanced surface negative charge density and the surface smoothness (skin type shape as measured in SEM) of the membrane, which results in good hydrophilicity (Dhand *et al.*, 2019). The mechanically robust membranes exhibited good salt rejection as well as flux which could be attributed to their enhanced hydrophilicity.

Functionalized CNTs, like nitrogen-doped CNT prepared by conventional vapor deposition method followed by refluxing in HNO_3, improve the flux of the membrane at a low concentration of filler. These materials reinforced in the PES membrane showed an improved mechanical property, rejecting capacity, and improved flux. AFM analysis reveals a reduction in roughness after the incorporation suggesting a good dispersion of the filler. The addition of N–CNT (0.04%) enhances the water flux, which can be attributed to the increase in the hydrophilicity of the membrane; above (0.04%) concentration, a decrease was observed due to the accumulation of the N–CNT in pores (Phao *et al.*, 2013). Oxidized MWCNTs in PSF show a similar result. The water flux for the membranes increases first and with the increase of CNT, they decrease. The water contact angle of the membrane reduced as filler content increased, indicating an improvement in membrane surface hydrophilicity that may have contributed to enhanced membrane performance (Yin *et al.*, 2013). A mixed blending of GO and MWCNTs was studied in the PVDF membrane and exhibited promising results. The membrane broadcasted a high efficiency of removal of all sorts of contaminants from the water throughout 6 times of testing. The size and the number of pores were controlled by varying the ratio of GO and MWCNTs (Chae *et al.*, 2021). It has been reported that carbon materials such as CNTs and graphene have intrinsic hydrophilic nature. Their hydrophobicity is due to the hydrocarbon contaminants present along the surface formed during their preparation. From the standpoint of basic research, the reported findings provide a new perspective to the study of the tailorable nanocarbon surface (Stando *et al.*, 2019).

10.8 METAL-ORGANIC FRAMEWORKS

MOFs are promising materials of inorganic–organic hybrid material which can be used as a novel filler in polymer matrices for water treatment. MOFs can be synthesized targeting the end application, by varying the structure, pre-size, and functionalities. MOFs are organic ligands with inorganic metal-containing nodes formed via coordination bonds. Their combination produces unique properties which include internal surface area, porosity, adjustable pore size versatile structure, and flexibility. Thus, they are being researched to employ them in applications such as adsorbents, heavy metal removal, purification, drug delivery, etc. Mixed matrix membranes are fabricated using MOFs and polymer matrixes using techniques like layer-by-layer, blending, gelatin-based seed growth, etc. It is recommended that the ideal loading of MOF is 10% above which chances of agglomerations are high. Their difficult synthesis condition, stability, and sustainability are some of the main challenges of these materials.

A thin film composite (TFC) was fabricated by coating zeolitic imidazolate framework (ZIF-8)/chitosan on the surface of the PVDF membrane. This aim was to improve the membrane distillation performance for seawater desalination. Albeit there was no improvement in hydrophilicity, as there was no change in contact angle, the water permeate flux was increased by 350% due to the formation of porous ZIF-8 on the surface. An enhanced antifouling property was shown by this membrane which is due to the chitosan layer on PVDF. The presence of the chitosan layer decreased the fouling tendency of the TFC membrane compared to the neat PVDF membrane during the fouling test. The presence of hydrophilic nitrogen- and oxygen-containing functional groups on the structure of chitosan can lead to provide a thin water layer on the membrane surface that would hinder the foulants' adhesion and therefore lead to an improved antifouling property for the modified membrane. Although the incorporation of ZIF-8 with hydrophobic nature into the chitosan layer would compensate the hydrophilic nature of chitosan considerably, these results demonstrate that the unique structure of the chitosan thin layer is the dominant factor to improve the antifouling property of the PVDF membrane (Kebria *et al.*, 2019). Leaf-shaped zeolitic imidazolate frameworks were incorporated in PES ultrafiltration membranes, and the properties were analyzed. The study revealed an increase in water flux which confirms that the incorporation of the MOFs significantly increases the pores and thus the water permeate rate. The antifouling was twice that of the pristine membrane which was attributed to improved hydrophilicity. The combined impacts of surface smoothness and surface pore size distribution should primarily contribute to the increase in hydrophilicity. The extremely small amount of hydrophobic ZIF-L nanoflakes on the membrane surface has a negligible effect on membrane surface wettability (Low *et al.*, 2014). ZIF-8 nanoparticles incorporated into the PA matrix show similar results. The organic ligand in the MOF improves its compatibility with PA. The increased flux was attributed to the hydrophobic passage of the filler incorporated in the polymer matrices. Furthermore, the water contact angle showed a decrease, emphasizing an increase in hydrophilicity. This is due to the reduced cross-linking extent of the PA surface structure; more carboxylic acid groups exposed on the surface result in a lower water contact angle of TFN membranes. The more hydrophilic surfaces might potentially minimize the organic fouling proclivity of these membranes (Duan *et al.*, 2015).

Materials of Institute Lavoisier (MIL) are a class of porous carboxylates along with trivalent metals such as vanadium (III), chromium (III), and iron (III). They have substantially larger channels/cages and topologies comparable to zeolites, except MIL-n having differing surface chemistry, pore sizes, and density. MIL-101 (Cr) was doped into the PA layer on PSF ultrafiltration for water desalination. MIL-101 (Cr) porous structures can provide direct water channels in the thick selective PA layer allowing water molecules to go through fast, enhancing membrane water permeance along with more than 99% NaCl rejection. The contact angle demonstrates that the membranes have hydrophilic surfaces due to the hydrophilic carboxylic acid groups of PA and the hydrophilic hydroxyl groups of MIL-101 (Cr) (Xu *et al.*, 2016). University of Oslo (UiO-66) UiO-66 is a member of the zirconium–carboxylate-based MOFs family, which has significant chemical stability in organic solvents; its stability in water is superior to other classes of MOFs, and it has great thermal stability up to 550C. Their incorporation into PA thin film nanocomposite broadcasted a superior permeability and rejection. The hydrophilic nature of UiO-66 binds well with PA and escalates the water permeability as well as rejection efficiency.

Albeit these materials are excellent for water treatment, their drawbacks such as high cost, dangerous, and harsh synthesis processes may cause potential threats to the environment and human health. New synthesis processes using moderate conditions and non-hazardous chemicals, as well as the selection of a suitable MOF/polymer system and the specific features that must be promoted for a certain application, should be thoroughly researched. Because of the extraordinary qualities supplied by MOFs, MOFs-based membranes have significant potential in many separation applications; nonetheless, their success, competitiveness, and upscaling require further continuous efforts to overcome difficulties identified with their manufacturing and use (Elrasheedy *et al.*, 2019).

10.9 CONCLUSION

The water is polluted in this generation needs an efficient treatment process of which membrane technologies are prime. Various types of membranes along with their uses and mode of fabrications were discussed. Polymer being the traditional material for membranes was analyzed with the incorporation of inorganic fillers. These nanocomposites improve the properties such as antifouling, antibacterials, rejection efficiency of dyes, heavy metal, etc.

REFERENCES

Apel, P. (2001) 'Track etching technique in membrane technology', *Radiation Measurements*, 34(1–6), pp. 559–566. https://doi.org/10.1016/S1350-4487(01)00228-1.

Badmus, S. O., Oyehan, T. A. and Saleh, T. A. (2021) 'Enhanced efficiency of polyamide membranes by incorporating cyclodextrin-graphene oxide for water purification', *Journal of Molecular Liquids*, 340, p. 116991. https://doi.org/10.1016/j.molliq.2021.116991.

Chae, J. *et al.* (2021) 'Graphene oxide and carbon nanotubes-based polyvinylidene fluoride membrane for highly increased water treatment', *Nanomaterials*, 11(10), 2498. https://doi.org/10.3390/nano11102498.

Dhand, V. *et al.* (2019) 'Fabrication of robust, ultrathin and light weight, hydrophilic, PVDF-CNT membrane composite for salt rejection', *Composites Part B: Engineering*, 160(September 2018), pp. 632–643. https://doi.org/10.1016/j.compositesb.2018.12.106.

Duan, J. *et al.* (2015) 'High-performance polyamide thin-film-nanocomposite reverse osmosis membranes containing hydrophobic zeolitic imidazolate framework-8', *Journal of Membrane Science*, 476, pp. 303–310. https://doi.org/10.1016/j.memsci.2014.11.038.

El-Sharkaway, E. A. *et al.* (2020) 'Removal of methylene blue from aqueous solutions using polyaniline/graphene oxide or polyaniline/reduced graphene oxide composites', *Environmental Technology*, 41(22), pp. 2854–2862. https://doi.org/10.1080/09593330.2019.1585481.

Elrasheedy, A., Nady, N. and Bassyouni, M. (2019) 'Matrix membranes: Review on applications in water purification', *Membranes*, 9(7), pp. 1–31.

Gandhi, M. R. *et al.* (2016) 'Graphene and graphene-based composites: A rising star in water purification', *ChemistrySelect*, 1(15), pp. 4358–4385. https://doi.org/10.1002/slct.201600693.

Gao, X. *et al.* (2017) 'Highly permeable and antifouling reverse osmosis membranes with acidified graphitic carbon nitride nanosheets as nanofillers', *Journal of Materials Chemistry A*, 5(37), pp. 19875–19883. https://doi.org/10.1039/c7ta06348b.

Gohil, J. M. and Choudhury, R. R. (2018) *Introduction to Nanostructured and Nano-enhanced Polymeric Membranes: Preparation, Function, and Application for Water Purification, Nanoscale Materials in Water Purification.* Elsevier Inc.

Homem, N. C. *et al.* (2019) 'Surface modification of a polyethersulfone microfiltration membrane with graphene oxide for reactive dyes removal', *Applied Surface Science*, 486(May), pp. 499–507. https://doi.org/10.1016/j.apsusc.2019.04.276.

Hosseini, S. S. *et al.* (2018) 'Fabrication, characterization, and performance evaluation of polyethersulfone/TiO_2 nanocomposite ultrafiltration membranes for produced water treatment', *Polymers for Advanced Technologies*, 29(10), pp. 2619–2631. https://doi.org/10.1002/pat.4376.

Jang, W. *et al.* (2015) 'PVdF/graphene oxide hybrid membranes via electrospinning for water treatment applications', *RSC Advances*, 5(58), pp. 46711–46717. https://doi.org/10.1039/c5ra04439a.

Jose Varghese, R. *et al.* (2019) *Introduction to Nanomaterials: Synthesis and Applications, Nanomaterials for Solar Cell Applications*. Elsevier Inc.

Kebria, M. R. S. *et al.* (2019) 'Experimental and theoretical investigation of thin ZIF-8/chitosan coated layer on air gap membrane distillation performance of PVDF membrane', *Desalination*, 450(September 2018), pp. 21–32. https://doi.org/10.1016/j.desal.2018.10.023.

Kusworo, T. D., Qudratun and Utomo, D. P. (2017) 'Performance evaluation of double stage process using nano hybrid PES/SiO_2-PES membrane and PES/ZnO-PES membranes for oily waste water treatment to clean water', *Journal of Environmental Chemical Engineering*, 5(6), pp. 6077–6086. https://doi.org/10.1016/j.jece.2017.11.044.

Lakhotia, S. R., Mukhopadhyay, M. and Kumari, P. (2019) 'Iron oxide (FeO) nanoparticles embedded thin-film nanocomposite nanofiltration (NF) membrane for water treatment', *Separation and Purification Technology*, 211, pp. 98–107. https://doi.org/10.1016/j.seppur.2018.09.034.

Lin, J. *et al.* (2013) 'Nano-WS2 embedded PES membrane with improved fouling and perm-selectivity', *Journal of Colloid and Interface Science*, 396, pp. 120–128. https://doi.org/10.1016/j.jcis.2013.01.028.

Lin, J. *et al.* (2016) 'Enhancement of polyethersulfone (PES) membrane doped by monodisperse Stöber silica for water treatment', *Chemical Engineering and Processing: Process Intensification*, 107, pp. 194–205. https://doi.org/10.1016/j.cep.2015.03.011.

Liu, F. *et al.* (2011) 'Progress in the production and modification of PVDF membranes', *Journal of Membrane Science*, 375(1–2), pp. 1–27. https://doi.org/10.1016/j.memsci.2011.03.014.

Liu, H. *et al.* (2020) 'Preparation of a hydrophilic and antibacterial dual function ultrafiltration membrane with quaternized graphene oxide as a modifier', *Journal of Colloid and Interface Science*, 562, pp. 182–192. https://doi.org/10.1016/j.jcis.2019.12.017.

Low, Z. X. *et al.* (2014) 'Effect of addition of two-dimensional ZIF-L nanoflakes on the properties of polyethersulfone ultrafiltration membrane', *Journal of Membrane Science*, 460, pp. 9–17. https://doi.org/10.1016/j.memsci.2014.02.026.

Nizam, P. A. *et al.* (2020) 'Mechanically robust antibacterial nanopapers through mixed dimensional assembly for anionic dye removal', *Journal of Polymers and the Environment*, 28(4), pp. 1279–1291. https://doi.org/10.1007/s10924-020-01681-3.

Nizam, P. A. *et al.* (2021) *Nanocellulose-Based Composites, Nanocellulose Based Composites for Electronics*. Elsevier Inc.

Ouyang, G. *et al.* (2015) 'Remarkable permeability enhancement of polyethersulfone (PES) ultrafiltration membrane by blending cobalt oxide/graphene oxide nanocomposites', *RSC Advances*, 5(86), pp. 70448–70460. https://doi.org/10.1039/c5ra11349k.

Pang, Z. *et al.* (2016) 'A room temperature ammonia gas sensor based on cellulose/TiO_2/PANI composite nanofibers', *Colloids and Surfaces A: Physicochemical and Engineering Aspects*, 494, pp. 248–255. https://doi.org/10.1016/j.colsurfa.2016.01.024.

Pascariu Dorneanu, P. *et al.* (2018) 'Novel fibrous composites based on electrospun PSF and PVDF ultrathin fibers reinforced with inorganic nanoparticles: Evaluation as oil spill sorbents', *Polymers for Advanced Technologies*, 29(5), pp. 1435–1446. https://doi.org/10.1002/pat.4255.

Phao, N. *et al.* (2013) 'A nitrogen-doped carbon nanotube enhanced polyethersulfone membrane system for water treatment', *Physics and Chemistry of the Earth*, 66, pp. 148–156. https://doi.org/10.1016/j.pce.2013.09.009.

Rose Joseph, M. *et al.* (2021) 'Development and characterization of cellulose nanofibre reinforced acacia nilotica gum nanocomposite', *Industrial Crops and Products*, 161(December 2020), p. 113180. https://doi.org/10.1016/j.indcrop.2020.113180.

Sharma, M. *et al.* (2016) 'Facile one-pot scalable strategy to engineer biocidal silver nanocluster assembly on thiolated PVDF membranes for water purification', *RSC Advances*, 6(45), pp. 38972–38983. https://doi.org/10.1039/c6ra03143a.

Shawky, H. A. *et al.* (2011) 'Synthesis and characterization of a carbon nanotube/polymer nanocomposite membrane for water treatment', *Desalination*, 272(1–3), pp. 46–50. https://doi.org/10.1016/j.desal.2010.12.051.

Shi, H. *et al.* (2016) 'A modified mussel-inspired method to fabricate TiO_2 decorated superhydrophilic PVDF membrane for oil/water separation', *Journal of Membrane Science*, 506, pp. 60–70. https://doi.org/10.1016/j.memsci.2016.01.053.

Shon, H. K. *et al.* (2013) 'Nanofiltration for water and wastewater treatment - A mini review', *Drinking Water Engineering and Science*, 6(1), pp. 47–53. https://doi.org/10.5194/dwes-6-47-2013.

Stando, G. *et al.* (2019) 'Intrinsic hydrophilic character of carbon nanotube networks', *Applied Surface Science*, 463(August), pp. 227–233. https://doi.org/10.1016/j.apsusc.2018.08.206.

Wang, Z. *et al.* (2019) 'Mussel-inspired surface engineering for water-remediation materials', *Matter*, 1(1), pp. 115–155. https://doi.org/10.1016/j.matt.2019.05.002.

Xu, Y. *et al.* (2016) 'Highly and stably water permeable thin film nanocomposite membranes doped with MIL-101 (Cr) nanoparticles for reverse osmosis application', *Materials*, 9(11). https://doi.org/10.3390/ma9110870.

Xue, J. *et al.* (2019) 'Electrospinning and electrospun nanofibers: Methods, materials, and applications', *Chemical Reviews*, 119(8), pp. 5298–5415. https://doi.org/10.1021/acs.chemrev.8b00593.

Yang, X. *et al.* (2016) 'Novel hydrophilic PVDF ultrafiltration membranes based on a ZrO_2–multiwalled carbon nanotube hybrid for oil/water separation', *Journal of Materials Science*, 51(19), pp. 8965–8976. https://doi.org/10.1007/s10853-016-0147-6.

Yin, J., Zhu, G. and Deng, B. (2013) 'Multi-walled carbon nanotubes (MWNTs)/polysulfone (PSU) mixed matrix hollow fiber membranes for enhanced water treatment', *Journal of Membrane Science*, 437, pp. 237–248. https://doi.org/10.1016/j.memsci.2013.03.021.

Yin, J., Zhu, G. and Deng, B. (2016) 'Graphene oxide (GO) enhanced polyamide (PA) thin-film nanocomposite (TFN) membrane for water purification', *Desalination*, 379, pp. 93–101. https://doi.org/10.1016/j.desal.2015.11.001.

Zahid, M. *et al.* (2018) 'A comprehensive review on polymeric nano-composite membranes for water treatment', *Journal of Membrane Science & Technology*, 08(01). https://doi.org/10.4172/2155-9589.1000179.

11 The Integral Postulation of Inorganic Nanofiller-Derived Polymer Applications in Agriculture

Puspendu Barik
S. N. Bose National Centre for Basic Sciences

Ashis Bhattacharjee
Institute of Science, Visva-Bharati University

CONTENTS

DOI: 10.1201/9781003279389-11

11.1 INTRODUCTION

Polymeric nanocomposites (PNCs) are composed of polymers (act as matrix) and homogeneously reinforced with organic or inorganic nanoparticles (NPs), which act as fillers having at least one of their dimension in the nanoscale (10–100 nm). PNCs are a new class of high-performance materials and have recently been used in packaging, energy, safety, transportation, electromagnetic shielding, defense systems, sensors, catalysis, and the information industry (Darwish et al., 2022; Fu et al., 2019; Hiremath et al., 2021). Some of the typical applications of PNCs in the agricultural industry like alternative packaging materials, controlled-released fertilizer, and pesticide formulation, plant-protecting agents, agriculture delivery agents, soil feature regulation, heavy metal removal in soil and water, and many more (Darwish et al., 2022; Guha et al., 2020; Hashim et al., 2020; Kumar et al., 2021; Puoci et al., 2008; Sikder et al., 2021). The modification of PNCs with proper inorganic fillers may improve the properties applicable in respective fields. Here, the filler dimension plays a vital role in obtaining the desired outcome. Hence, this chapter elucidates the functionalization of PNC's interfacial interaction and its relationship to enhancing the properties so that it can be adapted to the agricultural industry. In this chapter, we will concentrate on the inorganic fillers only, which include layered silicates such as montmorillonite, carbon-based additives (e.g., carbon black, carbon nanotubes, graphene, and fullerenes), semiconductor quantum dots (CdSe, CdS, PbS), nanoscale metal oxides (e.g., SiO_2, TiO_2, and Al_2O_3), metals (e.g., Ag, Au, and Cu) and other inorganic nano-objects (carbides, nitrides). Currently, PNCs with inorganics fillers only play a marginal role in industry sectors due to limitations in good dispersibility, large-scale production, high costs, and uncertainty of the toxicity in humans and the environment. More research needs to be conducted to overcome the considerable gap in knowledge concerning release, exposure, and environmental behavior.

The increasing trend in the global population demands the enhancement of the crop's yield, henceforth fulfilling the needs of people. The recent advancement of research in PNCs reinforced by inorganic nanofillers may open an avenue toward the application in the agriculture industry soon, owing to the rapid advancement of the synthesis of nanomaterials (act as inorganic fillers) with different sizes, shapes, and enhanced biocompatibility. Scientists have revealed that PNCs may play an essential role in promoting sustainable agriculture by providing protection, lowering detrimental effects on the soil, and increasing crop yield. Conventional agriculture solely relies on chemical compounds that have toxic effects on every living being and the entire ecosystem. Thus, PNCs may be superior for the intelligent delivery of desired chemicals in a sustainable manner to crop and maintain soil health in the upcoming years. This chapter elucidates the state-of-the-art technologies based on PNCs in agriculture and will discuss important aspects of applying PNCs as intelligent nanofertilizers, superabsorbents, and pollutant removal agents, which are socially, environmentally, and technically sustainable. First, the chapter presents the design and synthesis techniques in short. The following section and subsections describe the applicability of PNCs with inorganic nanofillers in agriculture. The chapter also includes a short overview of the environmental risk assessment and toxicology study for agricultural purposes.

11.2 POLYMER NANOCOMPOSITES: DESIGN AND SYNTHESIS TECHNIQUES

PNCs are a 3D structure of a combination of polymer matrix and inorganic reinforcement materials (inorganic fillers). Inorganic nanofillers for PNCs applications are mainly classified into three categories according to the number of nanometer dimensions in inorganic nanofillers: one-dimensional (1D; nanodisks, nanoprism, nanosheets, branched structures, nanoplates), two-dimensional 2D; nanotube, graphene) and three-dimensional (3D; NPs, quantum dots). PNCs with 1D nanofillers, including ZnO nanoplates/nanosheets/nanodisks, carbon nanowall, and graphene platelets, display unique shape-dependent characteristics that make them useful in the formation of crucial components in applications, e.g., microelectronics, biosensors, sensors, biomedical, and coatings (Akpan et al., 2019), because of excellent electrical, magnetic, and optic properties. 2D nanofiller-reinforced PNCs are suitable for catalysis, sensors, photocatalysts, nanocontainers, and nanoreactors, and some most commonly used nanofillers are Au/Ag nanotube, graphene, MoS_2, h-BN, graphene oxide, TiO_2 (Akpan et al., 2019). 3D nanofillers, like various NPs (semiconductor nanoclusters, carbon black, SiC, Si, TiO_2, Ag, SiO_2, Fe_3O_4, and ZnO), enable the PNCs suitable for the application areas like coatings, separation, purification, and biomedicine. However, PNCs can be made of different nanofillers other than the nanofillers mentioned above, e.g., metallic NPs, various metal oxide NPs, and nanoclays. To obtain high-performance nanocomposites, researchers must consider the design, material selection (polymers and fillers), synthesis procedure, and fabrication. Designers should maintain the basic functionalities of PNCs – targeted applications with desired properties, functionality, and cost analysis, limiting the design ideas (Akpan et al., 2019).

During the PNC formation, besides the above, designers must consider synthesis parameters (like route, temperature, pressure, and time) and nanofillers parameters (like size, shape, concentration, and orientation) to obtain specific properties oriented toward the applications. In this perspective, the aspect ratio and surface area are two fundamental properties to characterize the 2D nanofillers and their reflection on the overall properties of PNCs. The larger surface area of the nanofillers means more interaction between the polymer matrix and the nanofillers. Aspect ratio (AR), i.e., the ratio between the longest dimension and the shortest dimension of the nanofillers, determines the interfacial area per volume of filler, which has a crucial role in determining the mechanical and physical properties of PNCs. The variation of properties happens owing to the AR, including Young's modulus (Ashrafi et al., 2006; Weon & Sue, 2005), thermal properties (Gojny et al., 2006; Weon & Sue, 2005), electrical properties (Gojny et al., 2006; Kim et al., 2010; Kim & Macosko, 2008), percolation threshold (Ayatollahi et al., 2011), flammability, viscosity (Cipiriano et al., 2007), permeability (Giannelis et al., 1999) and dielectric properties (Gerratt & Bergbreiter, 2013). The mechanical, thermal, and electrical properties of PNCs depend on the interface/interface adhesion between the polymer matrix and the nanofillers. The interaction at the interface will control the effects like altered behavior of the polymer, morphology, space charge distribution, and phonon scattering. Therefore, nanofillers may produce more prominent variation in the properties of PNCs owing to the large surface area of the fillers. The interaction between the surface of the nanofiller and polymer directly influences the wettability (ability to form a polymer layer on the surface of filler), mechanical locking (roughness of the surface), and electrostatic interaction (interaction among the spacial charges at the molecular level), and adhesion strength. The roughness of the surface contributes to the shear strength of the PNCs (Buggy et al., 2005; Levita et al., 1989; Roulin-Moloney et al., 1987; Thio et al., 2004; Wang et al., 2003). The directional properties like thermal and electrical conductivity and diffusion depend on the orientation of the nanofillers. The proper orientation may develop/decrease van der Waals attraction and photoelastic response in PNCs. Therefore, researchers develop various techniques, including shear, flow, magnetic field, and electric field-induced alignment, to make the proper orientation of nanofillers to obtain the preferred properties of PNCs (Akpan et al., 2019).

To date, wide varieties of synthesis methods are available for making PNCs efficiently. This section will briefly discuss the standard methods, and Table 11.1 shows an overview of these methods with some random examples. For a given weight percentage, the number of nanofillers in a PNC is much more significant than the microfillers, owing to their size variations. Due to the massive number of nanofillers, they are indistinguishable from each other, and their interparticle distances are minimal. Therefore, the van der Waals or electrostatic interactions between nanofillers are large. However, there is a tendency to agglomerate for anisotropic nanofillers, e.g., nanotubes, nanowires, and nanoplatelets, with AR ~10^4 in some cases. So, re-entanglement after proper dispersion must be done to obtain a homogeneous mixture. The proper dispersion of nanofillers is the prerequisite for the synthesis of PNCs. The mixing technique like *in situ* polymerization, solution, and melt mixing is the most common straightforward method for PNCs. The in situ polymerization technique involves making a monomer dispersion with nanofillers and carrying out the polymerization by external triggers like temperature, initiator, or light diffusion. In the solution method, the matrix polymer and nanofiller are dispersed in a solvent consecutively, and later, the intercalated PNCs are obtained by evaporating the solvent.

11.3 APPLICATION OF POLYMER NANOCOMPOSITES IN AGRICULTURE

During the last decade, the advancement of PNCs in agriculture has expanded to various fields, including crop protection, food packaging, toxin, and pathogen delivery, biosensors, water purification, antimicrobial activity, enzyme immobilization, and wastewater treatment (Idumah et al., 2020; Kumar et al., 2021; Mistretta et al., 2021; Momina & Ahmad, 2021; Pandey et al., 2017). Inorganic reinforcing nanofillers cause inferior interactions at the NPs and polymer interface in PNCs, improving their thermal, mechanical, and barrier properties. Agriculture uses a significant share (~70%) of global freshwater resources. The world would face a global water deficit under a business-as-usual scenario in the future. According to the United Nations (UN) report, reliable access to clean and affordable water would be one of the most basic humanitarian goals and a major global challenge for the 21st century. Modern agriculture uses nanotechnology in sustainable agriculture to attain

TABLE 11.1
Synthesis Methods for PNCs and Some Random Examples to Understand the Method

Techniques/Methods	Advantages	Drawbacks	Examples
Ultrasonication-assisted solution mixing	• Straightforward fabrication technique • fast screening of new nanofillers in PNCs • homogeneous dispersion depends on sonication power and time	• Low viscous polymer and filler solution needed • a certain level of damage to the nanofillers due to high shear energy • removal of solvent with a high boiling point may be problematic for PNCs • not scalable at the industrial level	(Hatami & Yazdan Panah, 2017; Madhukar et al., 2019; Mahmoudabadi et al., 2021; Preetha et al., 2022)
Shear Mixing (Melt mixing, melt blending, melt extrusion)	• less intensive • scalable to industry level • shear rate depends on the surface energy of the solvent and nanofillers • able to disperse agglomerates of loosely bound nanofillers	• different shear rate is needed for different solvents and nanofillers • not suitable for dispersing high-AR nanofillers	(A. Ávila-Orta et al., 2019; Backes et al., 2017; J. Banerjee & Dutta, 2019; Paton et al., 2014)
Three Roll Milling or calendering	• suitable for high viscosities polymer mixture • generates adjustable, controllable shear forces • best suitable for anisotropic geometries with at least one dimension ~ 1 μm or more • minimal or no use of solvents	• not suitable for isotropic nanofillers, e.g., spherical nanofillers not suitable for thermoplastics	(Ahmadi-Moghadam & Taheri, 2014; Chandrasekaran et al., 2014; Dalir et al., 2012; Kothmann et al., 2015; Park & Lee, 2013; Souri et al., 2015)

(Continued)

TABLE 11.1 (*Continued*)
Synthesis Methods for PNCs and Some Random Examples to Understand the Method

Techniques/Methods	Advantages	Drawbacks	Examples
Ball Milling	• suitable for thermoplastic and thermosetting matrices • useful for in situ exfoliation, dispersion, and functionalization of 2D nanosheets • can be done either in the wet or dry state • Solvent is not required	• Contamination of product from ball and casing • difficult to clean after processing • long milling time	(Almotairy et al., 2020; Ji et al., 2019; F. Yu et al., 2021)
Twin-Screw Extrusion	• effective for thermoplastic-based PNCs • no solvent is required • high-shear forces can be achieved • all types of nanofillers can be dispersed • effective for high loading of nanofillers	• monodispersing of nanofillers is almost impossible • not suitable for nano-sized fillers	(Isayev et al., 2009; Sanes et al., 2020; Vergnes, 2019; L. Wang et al., 2016)
Direct compounding (ex-situ processes, spray drying)	• nanofillers are synthesized separately with high quality and well dispersed	• suitable solvent is needed • solvent evaporation is required	(Silvério et al., 2013; C. Yang et al., 2016; W. Zhao et al., 2016)
In Situ Polymerization Method	• in either a solvent-free system (i.e., a bulk phase) or in a solvent-based system (aqueous phase) • control over the physicochemical properties of PNCs • synthesized with high organic loadings	• the increase of viscosity with the progress of the polymerization process • allows polymer macromolecules to attach to the surface of the nanofillers	(Adnan et al., 2018; Y.-P. Wang et al., 2018; L. Zhang et al., 2018)

(Continued)

TABLE 11.1 (*Continued*)
Synthesis Methods for PNCs and Some Random Examples to Understand the Method

Techniques/Methods	Advantages	Drawbacks	Examples
Sol-Gel Processes	• affordable • easy synthesis technique • high control over homogeneity and dispersibility • precise control over the composition and purity	• precursors are relatively limited or expensive	(El Nahrawy et al., 2020; Khirade, 2019; X. Ma et al., 2018; Z. Ma et al., 2018)
Electrochemical Method	• best method to make PNCs film • formed directly on the electrode surface • controlled by the applied potential or current density • short reaction time and operational simplicity	• the working electrode has a limited surface area • restricted for a few selective PNCs	(Gao et al., 2013)
Template synthesis	• synthesis of inorganic nanofillers inside the polymer matrix • limited control over the nanofiller microstructure or size	• the high temperature is required • agglomeration problem	(Karak, 2019; Rane et al., 2018; Tian et al., 2019)
Intercalation techniques (melt intercalation, polymer solution intercalation, polymerization intercalation, in situ intercalation polymerization)	• producing polymeric chains and alternating inorganic nanofillers layers • effective for 2D nanofillers • a large amount of surface area available for filler and polymer interaction • good electrical conductivity, electromagnetic absorption, mechanical properties, and thermal conductivity	• need to understand the structural evolution of polymers during polymerizing into PNCs	(Guo et al., 2018; Madhumitha et al., 2018; Reddy et al., 2019)

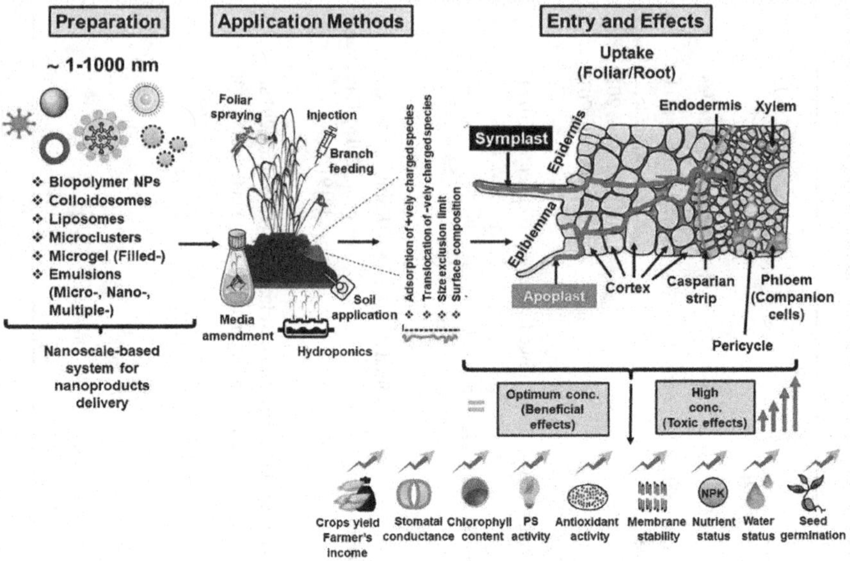

FIGURE 11.1 Illustration shows the NPs transport and their interactions in crop plants. Reproduced from Ref. (Kumar et al., 2021) under Creative Commons Attribution 4.0 International License.

maximum output from limited resources by applying NPs and their nanocomposites in various application aspects, as described in the following sections. Figure 11.1 illustrates the nanoscale-based system for agrochemical delivery, various routes of penetration for nanopesticide, nanoherbicide, nanoherbicides, and nanogrowth-promoting compounds, and the effect on the plants.

This section discusses the application of PNCs in five different fields – controlled agrochemical delivery, control of plant diseases, pollutants removal, food packaging, biosensing, and gene transfection. Engineered PNCs can potentially control the delivery of agrochemicals like macro-and micronutrients, pesticides, and other agrochemicals (An et al., 2022; Dasgupta et al., 2017; Guha et al., 2020). PNCs are biodegradable and eco-friendly, which is the primary concern in plant protection, i.e., controlling weeds, fungi, and bacteria (Hashim et al., 2020; Kalia et al., 2020). Pollutant removal by PNCs includes removing heavy metal ions and dyes from the aqueous solution, typically for wastewater treatment (Momina & Ahmad, 2021; Pandey et al., 2017; Qasem et al., 2021; Zhao et al., 2018). PNCs are potential alternative packaging materials in industrial, food, and agricultural products due to their functionalization, flexibility, biodegradability, and minimal cost (Idumah et al., 2020; Qasim et al., 2021; Rhim et al., 2013; Sarfraz et al., 2020; Vasile, 2018). The biosensor can easily detect plant diseases/pathogens and promising alternatives in practical management steps to minimize production loss (Ali et al., 2021, 2022; Idumah et al., 2020; Kulabhusan et al., 2022). Polymeric materials act as delivery vectors for gene transfection in biomedical fields, whereas their potential for plant systems remains unclear and requires more research (Christiaens et al., 2020; Sangeetha et al., 2019;

Sikder et al., 2021). During the last decades, PNCs have been demonstrated in various fields related to agriculture; however, the discussion of all the applications is out of the scope of the book chapter.

The controlled release (CR) technique is the unique property that are the most desirable in agrochemical delivery. CR has various advantages, including decreased phytotoxicity, reduced agrochemical loss, and retaining soil quality. The CR mechanism can be accomplished via different modes: (i) *Diffusion via relaxation/swelling* – nanocarriers absorb water under irrigation or rainfall events depending on the porosity of the polymer matrix or the polymer encapsulation (El-Hamshary et al., 2015; Ianchis et al., 2017), (ii) *Burst release* – the most common technique and release in an undesirable manner, that is not favorable for targeted applications (Stloukal et al., 2012), (iii) *Degradation* - triggered and achieved by hydrolysis with water, light exposure, temperature, pH, specific stimulus, and enzymatic activities (Chawla & Amiji, 2002; Hou et al., 2012; Ianchis et al., 2017; Lin et al., 2019; Orellana-Tavra et al., 2015). CR technique somehow resolves the problems associated with traditional agrochemicals, allowing synchronized and optimized release of agrochemicals (nutrients, fertilizers, pesticides) at the target site (Morgan et al., 2009; Ni et al., 2011).

11.3.1 Agrochemical Delivery

The use of conventional agrochemicals in fertilizers or pesticides has many disadvantages, including (i) working as per specification initially, but it rapidly declines to a below adequate level with time. Repeated application eventually results in a decline in soil quality and eutrophication of nearby water bodies (Guha et al., 2020), (ii) carcinogenic and mutagenic properties and hazardous effects on human health and the environment, consuming agricultural products (Sarıgül & İnam, 2009; Singh et al., 2009). Many fertilizers as soil nutrient supplements, e.g., ammonium salts, urea, nitrate, or phosphate compounds, are utilized during the plant growth process, negatively impacting the crops' growth. For example, the accumulation of NH^{4+} causes an increase in soil pH; urea is also lost via volatilization and leaching; Fe- and Al-based oxides weaken the soil phosphate levels up to 0.01–1.00 ppm; the excess ammonia and sodium restrict the uptake of potassium from potassium fertilizers. Hence, the development of nanofertilizer is essential to the modern agricultural industry, paving a new era of CR systems, i.e., fertilizers are target-specific and sustained release at a predetermined level (Kalia et al., 2020). Nanofertilizers have several benefits over conventional chemical fertilizers due to – (i) consistently long-term delivery of nutrients to plants and control of the availability of nutrients in crops through slow/control release mechanisms, (ii) requirement of a small amount which reduces the cost of transportation and field application, (iii) minimization of the accumulation of salt in the soil, (iv) the manipulation of synthesis according to the nutrient requirements of planned crops (here, biosensors can control the delivery of the nutrients by assessing the soil nutrient status, growth period or environmental conditions), (v) high reactivity which increases the bioavailability of nutrients due to high surface area of nano-constituents, (vi) the ability to fight various biotic and abiotic stresses (Raliya et al., 2018; Zulfiqar et al., 2019).

Various PNCs are designed as carriers for nanofertilizers and administered through direct delivery, encapsulated delivery (for CR), and nutrient delivery via complex nanocapsule using organic polymers. The encapsulated delivery techniques are attractive, and several control factors like particle size, distribution, solubility, shape, surface area, encapsulation mechanism, and release mechanism must be considered. The release of nanofertilizers can be triggered by external factors like water, the chemical composition of the environment, light, and temperature, as indicated in Figure 11.2. Standard techniques for forming 10–1000 nm nanocapsules are emulsification, coacervation, inclusion complexes, nanoprecipitation or solvent displacement technique, and nanoliposomes (Gouin, 2004; Zarrabi et al., 2020). Micronutrients like nitrogen, phosphorus (for the structural development of plants and synthesis of nucleic acids and protein), and potassium (for maintaining osmotic potential and ion balance) are vital for plant growth and maintenance. Aminopropyl trimethoxysilane (APTMS) surface-modified zeolite is a carrier for urea delivery in plants (Rahmat et al., 2015). Superabsorbent hydrogel-coated slow-release fertilizers have a dual advantage, i.e., they increase soil fertility and reduce the need for water supply by irrigation. Some examples are polyacrylamide (PAAm)/Methyl Cellulose/Montmorillonite clay (MMt) Hydrogel (for urea) (Bortolin et al., 2013), PAAm/carboxymethylcellulose (CMC)/MMt (for urea and boron) (Bortolin et al., 2016), sodium alginate-g-poly(acrylic acid-co-acrylamide) (g-poly(AA-co-AAm))/MMt (for Nitrogen, Phosphorus, and Potassium, i.e., NPK fertilizer) (Rashidzadeh & Olad, 2014), maize bran (MB)-g-poly(AA-co-AAm)/MMT (Hyd/MMT) (Olad, Gharekhani, et al., 2018), sulfonated-carboxymethyl cellulose (SCMC)-g-poly(AA)/polyvinylpyrrolidone (PVP)/Silica (for NPK fertilizers) (Olad, Zebhi, et al., 2018), Kaolinite/gum arabic (for urea) (Sempeho et al., 2015) and many more (Guha et al., 2020). Besides single-nutrient loaded design, multi-nutrient loaded PNCs are also available as controlled-released fertilizers, e.g., chitosan (CS)/graphene oxide (GO) (Li et al., 2019). PNCs play an important role in the proper growth and development of the plants, supplying micronutrients like iron (Chi et al., 2018; Wang et al., 2016), zinc (Rahul Kumar et al., 2018; V. Kumar et al., 2018; Wang & Nguyen, 2018), magnesium, molybdenum, copper (Ashfaq et al., 2017; Rahul Kumar et al., 2018), and boron (Iqbal et al., 2019).

11.3.2 Control of Plant Diseases

The productivity of crops also depends on other biotic factors like weeds, animals, pests, and pathogens, which can cause up to ~40% crop loss globally (Nassar & Ortiz, 2007). Therefore, plant protection is also equally crucial as fertilizer to improve overall productivity and food security. Agrochemicals like pesticides and herbicides are harmful to ecosystems and may contaminate water resources. However, the recent development of intelligent delivery systems may reduce the potential risk by minimizing agrochemical usage and increasing the use of nontoxic PNCs as a carrier of agrochemicals. Nanoherbicide molecules can easily bind with carbon nanotube (CNT), and hence, the CNT or its nanocomposites are helpful for target-specific weed control (Dhillon & Mukhopadhyay, 2015). Biodegradable nanocomposites and

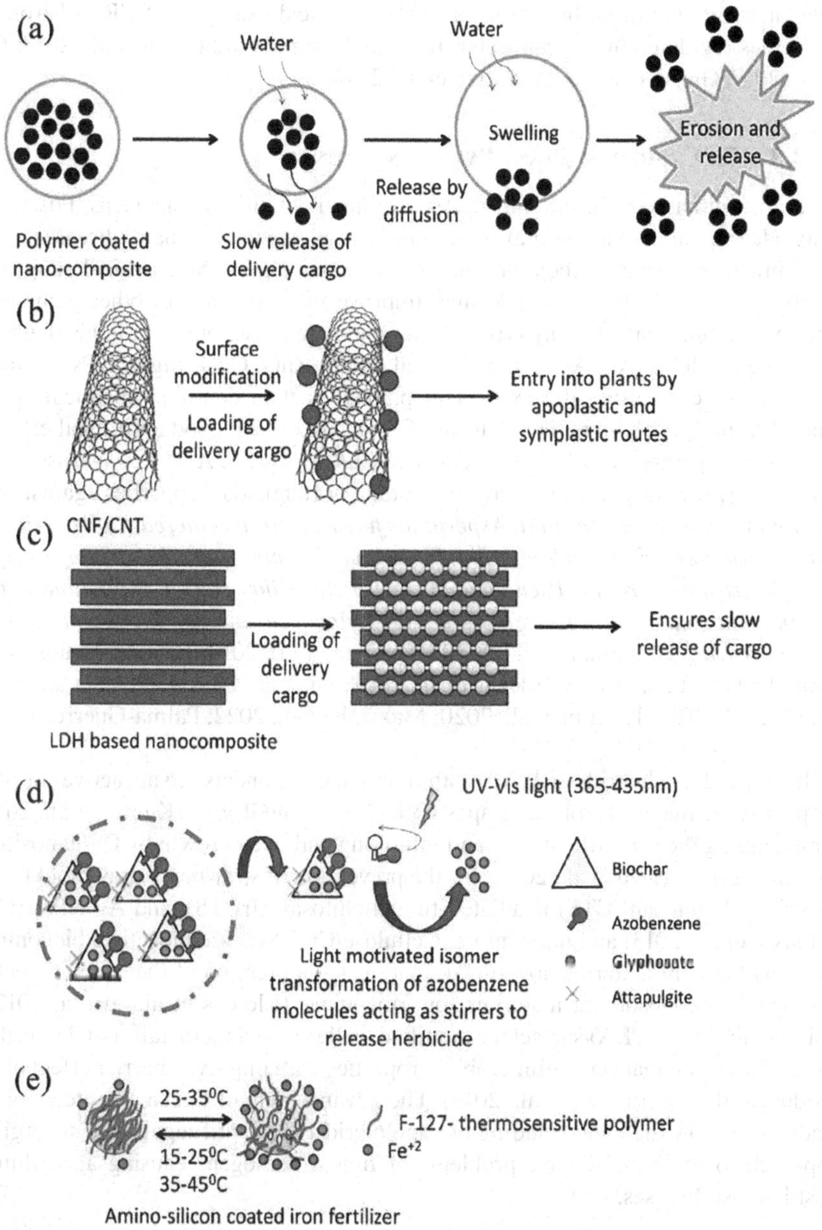

FIGURE 11.2 Different nanocomposites used to deliver agrochemicals (a) PNCs promote the slow release of agrochemicals by absorbing water. (b) Surface-modified carbon nanotube (CNT)-based nanocomposites can cross plant cell barriers and be translocated via symplastic and apoplastic pathways. (c) Hydroxide-based nanocomposites strongly interact with agrochemicals and ensure the sustained release of agrochemicals. (d) Light-induced release. (e) Temperature-induced release of agrochemicals from nanocomposites. Reprinted with permission from Ref. (Guha et al., 2020). Copyright © 2020, American Chemical Society.

nanoencapsulation of herbicides/pesticides are typical examples of CR of herbicide molecules regulated by the rainwater, light, and temperature (Chen et al., 2018; Chi et al., 2017; Giroto et al., 2014; Sarlak et al., 2014).

11.3.3 PNCs Against Plant Pathogenic Fungi

Several plant pathogenic fungi (e.g., Aspergillus, Penicillium, Claviceps, Fusarium, Trichoderma, and many more), dominant causal agents of plant diseases, produce infection in plant seeds, producing poisonous toxins. Mycorrhizal fungi live symbiotically with the plants and may improve the resistance to other pathogens. Endophytic fungi infect many types of grass and are poisonous to animals that consume them. Metal (Au, Ag, Cu) and metal oxide (ZnO, CuO, MgO) NPs are more effective for controlling different plant pathogens than common chemical fungicides. Similarly, polymers like chitosan (CS), agarose have good antifungal efficacy against plant pathogens. PNCs, e.g., clay/CS, Cu/CS, Ag/CS, Ag@AgI/agarose, ZnO/CS, TiO_2/Ag/CS, and silica/CS have fungistatic or fungicidal properties against various fungi (*Alternaria alternata, Aspergillus flavus, Botrytis cinerea, Colletotrichum gloeosporioides, Escherichia coli, Fusarium graminarum, Fusarium oxysporum, Neurospora crassa, Pseudomonas sps, Penicillium steckii, Penicillium spp., Rhizoctonia solani, Rhizopus stolonifer, Staphylococcus aureus, Xanthomonas oryzae pv. oryzae*) (Abdallah et al., 2020; Banerjee et al., 2010; Bautista-Baños et al., 2006; Brunel et al., 2013; Cárdenas et al., 2009; Di et al., 2012; Ebisike et al., 2020; Ghosh et al., 2012; Hashim et al., 2020; Maxwalt et al., 2022; Palma-Guerrero et al., 2009; Xing et al., 2022).

Biodegradable hybrids or PNCs with multiple components are attractive for plastic packaging materials for red grapes with CS/gelatin(GL)/Ag (Kumar et al., 2018), for extending the shelf life of fruit and inhibiting pathogen growth by CS/nanosilica/sodium alginate (Kou et al., 2019), for the prevention of strawberry gray mold by Ag NPs/fungal chitosan (CTS) irradiated fungal chitosan (IrCTS), and Ag NPs-IrCTS (Moussa et al., 2013) and many more. Cellulose/Cu PNCs are bioactive, biocompatible, and brilliant materials for fungal treatment, suppression of the development of pathogens, and absorbent materials for fruit juices (Llorens et al., 2012a, 2012b). Poly(lactic acid) (PLA)/surfactant-modified cellulose nanocrystals (s-CNC)/silver (Ag) NPs show enhanced antimicrobial properties with improved barrier effect of the produced films (Fortunati et al., 2013). The advancement of Green Nanotechnology leads us to grow the understanding of nanohybrid/PNCs antifungals as a synergistic approach to resolving diverse problems of fungal pathogens causing agricultural/post-harvest diseases.

11.3.4 PNCs Against Plant Pathogenic Bacteria

Similar to the antifungal properties, metal, and metal oxide NPs show antibacterial properties. Among them, Ag/CS PNCs offer more significant antifungal properties. CS and metallic Cu nanocomposite also show higher antibacterial action than CS itself or doxycycline (El Zowalaty et al., 2013; Qi et al., 2004). CS-based nanocomposites are low-cost and multipurpose PNCs with numerous desired physicochemical and biological properties and support nutrient uptake by plants, control the release of micro/macronutrients, fungicides, and herbicides, help to fight against different threats (pathogen attack), act as an antimicrobial agent against bacteria, fungi, and viruses (Yu et al., 2021). Some metal NPs show good antibacterial properties; however, inside PNCs, the same metal NPs may not show antibacterial properties with the same efficiency, e.g., Cu NPs in polyvinylmethyl ketone (PVMK), poly(vinyl chloride) (PVC), and polyvinylidenefluoride (PVDF) (Cioffi et al., 2005; Pradeep et al., 2022) and Polypropylene/silver PNCs (Radheshkumar & Münstedt, 2006). PNCs using conducting polymers like polyacetylene (PA), polythiophene (PT), poly[3,4-(ethylenedioxy)thiophene] (PEDOT), polypyrrole (PPy), polyindole (PIN), polyphenylene and polyaniline (PANi) have been demonstrated as promising antimicrobial and antifungal agents due to their biocompatibility, low toxicity, and good environmental stability (Pradeep et al., 2022). In a recent review, authors comprehensively describe the utilities of Cu PNCs as an excellent and cost-effective biocide (Tamayo et al., 2016). Moreover, PNCs with a composition of metal NPs and natural polymers (e.g., cellulose, starch, chitin, chitosan, gelatin, dextran, alginate, and pectin) are advantageous in agriculture for unique properties like high purity and crystallinity, tensile solidity, improved elasticity, and extensive surface area (Zahran & Marei, 2019). With the present understanding of PNCs as antimicrobial and antifungal agents, researchers must focus on the comprehensive *in vivo* study to make PNCs available to farmers.

11.3.5 Pollutants Removal

Water is essential for sustaining all kinds of corps and plants. Agricultural activities are heavily dependent on a clean supply of water from various sources, including underground water, river, and lakes, which is less than ~3% of the entire water volume on Earth. However, the wastewater coming out from numerous industries contaminates the clean water on Earth with pollutants. Therefore, there is a need to consider reusing wastewater and recycling it efficiently. Commonly, diverse effluents (including heavy metal ions, distillate microcystins, and dyes) have been added into the environment or natural bodies due to the improper discharge of wastewater from various industries and agricultural activities. The recent development of PNCs may prove themselves as good adsorber of heavy metal ions and a good alternative for photodegradation of dye molecules in wastewater.

11.3.5.1 Removal of Metal Ions

Toxic inorganic metals including Cr, Hg, Cd, Pb, and As are persistent environmental pollutants, non-transformable or non-biodegradable, contaminating water resources, which are crucial in our living world. Metal ions from the water can be removed via adsorption, electrolysis, electrodialysis, reverse osmosis, ion exchange, conventional coagulation, and chemical precipitation (Amari et al., 2021). However, electrodialysis, electrolysis, ion exchange, and reverse osmosis are expensive methods and cannot regularly be used in the agricultural industry. The adsorption method is the most attractive and cost-effective process used for agriculture. Common adsorbents are carbon-based adsorbents (e.g., graphene, activated carbon, carbon nanotubes), chitosan-based adsorbents, mineral adsorbents (e.g., zeolite, silica, clay), magnetic adsorbents (e.g., Fe_3O_4), bio-sorbents, metal-organic frameworks (MOF) adsorbents, and many more. Various conducting polymers (like polyaniline, PANI, and its derivative) are effective for heavy metal adsorption due to multiple amine groups at the ortho, meta, and para positions for wastewater treatment applications.

Moreover, conducting polymer nanocomposites with inorganic fillers are recently used as heavy metal adsorbents owing to the ease of synthesis, low cost, mechanical flexibility, high electrical conductivity, and environmental stability. Table 11.2 presents the various types of metal ions and their respective effects in agriculture. Metal ions are divided according to their properties and use in agriculture.

Rivers, lakes, and ocean water are directly or indirectly contaminated with heavy metal ions owing to the discharges from various industries like metal plating, mining, batteries, fertilizers, and paper industries. Adsorption is a low-cost industrial separation technique, and good selectivity can be achieved for wastewater purification. Composites/nanocomposites materials have excellent adsorption properties, chemical and thermal stability, reproducibility, and selectivity in separating heavy metals. PNCs have been most effective in removing heavy metal ions from wastewater, using several inorganic fillers, including nanoclay, carbon-based nanomaterials, metal oxide NPs, and magnetic NPs. Table 11.3 summarizes some PNCs for removing heavy metal ions and their efficiency in recent years. Despite the recent development of various PNCs, only a few metallic pollutants can be removed by the current technologies. From the agricultural perspective and wastewater applications, new PNCs must be cost-effective and feasible and can remove many pollutants/ions or a mixture of pollutants. Fulfilling these requirements makes PNCs as alternative adsorbents to accomplish massive technological potential in wastewater treatment.

11.3.5.2 Removal of Dye Molecules

Photodegradation is a common technique for degrading dyes and is a more frequently and effectively used method for treating wastewater in industries. During photodegradation, pollutants are decomposed using a photocatalyst in sunlight or UV as a function of catalyst loading, pH, temperature, dissolved oxygen, pollutant concentration, light wavelength, and light intensity (Chong et al., 2010). PNCs serve the function of a photocatalyst and adsorber for the pollutant in wastewater owing to their high surface area, low cost, high adsorption capacity, compatibility, binding affinity, and chemical and thermal stability. Dyes containing nitrogen (N) groups such as Rhodamine B (RhB) and Methylene Blue (MB) are resistant to photolysis, resulting in carcinogenic products. Hence, removing these dyes from wastewater and converting them into nontoxic components is the primary concern in resolving the environmental and agricultural problems. The significant effluents (more than 50% of the dyestuffs) are Azo dyes from commercial industries; thus, azo dyes are a potential threat to the environment and agriculture. Table 11.4 shows some examples of dye degradation of various dyes. Future development of photocatalysts must consider the fact that a suitable catalyst (here polymer nanocomposites) should have strong photoabsorption efficiency, less electron-hole recombination, low cost, and be less hazardous, which can increase photocatalytic efficiency and avoid further contamination of water.

TABLE 11.2
Different Metal Ions and Their Effects on Agriculture

Types	Examples	Descriptions	Adverse Effects in Agriculture	References
Metal ions, by their properties depending on the atomic number and atomic weight, metal ions have various properties, making them toxic, precious, or radioactive.				
Toxic metals	Hg, Cr, Pb, Zn, Cu, Ni, Cd, As, Co, and Sn	Generally, heavy metals produce a significant toxic impact. However, trace amounts of metals like Cu, Mn, Co, Zn, and Cr are required for plants' metabolic activities.	Directly affect plant growth, food safety, and soil microflora. Metal induced reactive oxygen species (ROS) generation affect various biochemical molecules, membrane lipids, amino acid chain, carbohydrates, proteins, pigments, and nucleic acids	(Mishra et al., 2020; Srivastava et al., 2017)
Radioactive metals	U, Th, Ra, and Am	The contamination happens from agricultural lands and groundwater. The primary sources of radioactive metals in soil and water are unregulated disposal of industrial nuclear waste, wastes from military applications, dumping of medically used radioisotopes, the wastes from globally situated nuclear power plants, and use of phosphate fertilizers in farming.	Radionuclides are well-known potent carcinogens.	(Kumar et al., 2019; Singh et al., 2022)
Metal ions, by their use various metal ions are naturally present in the soil, functioning as essential micronutrients				
Essential micronutrients	Fe, Zn, Mn, Ca, Ni, Na, K, Cu, and Mo	Essential micronutrients are essential in small but critical amounts for their average growth and development. Most metal micronutrients are essential cofactors for enzymes.	It becomes toxic only when a concentration limit is exceeded. The limit has a narrow optimal concentration range.	(Alloway, 2013; Arif et al., 2016; Kolbert et al., 2022; Rengel, 2004)
Beneficial micronutrients	Co, Ni, Cr, Ti, and V	Beneficial elements for plant growth directly participate in plant metabolism.	When present at high enough concentrations, the elements are toxic or at a critical value.	
Nonessential micronutrients	Ag, Cd, Cr, Hg, and Pb	These are not essential and are contaminated from various sources. They can modify the metabolism and hence the growth of plants.	It has adverse effects on plant growth and metabolism. Also, contaminate the soil and water.	

TABLE 11.3
Examples of Heavy Metal Ions Pollutants Removal by Polymer Nanocomposites with Inorganic Fillers

Metal Ions	Polymer Composites	Efficiency	References
Cd (II)	Mn_2O_3/Polyaniline (PANI)	480 mg/g	(Rajakumar et al., 2014)
	Polypyrrole (PPy)/Al_2O_3	9.709 mg/g	(Hasani & Eisazadeh, 2013)
	Silica-based zinc oxide	30.98 mg/g	(Garg et al., 2022)
Cr (V)	PPy/Fe_3O_4	169.4–243.9 mg/g	(Bhaumik et al., 2011)
Cr (VI)	Fe_3O_4/poly(m -phenylenediamine)	240.09 mg/g	(Wang et al., 2015)
	PPy/graphene oxide (GO)	497.1 mg/g	(Li et al., 2012)
	PANI/GO	1149.4 mg/g	(Zhang et al., 2013)
	PANI/ multi-walled carbon nanotubes (MWCNTs)	55.5 mg/g	(Rajeev Kumar et al., 2013)
Cs (I)	PANI/GO	1.39 mmol/g	(Sun et al., 2013)
Cu (II)	Silica-based zinc oxide	32.53 mg/g	(Garg et al., 2022)
	Ethylenediaminetetraacetic acid (EDTA)-GO-chitosan	130 mg/g	(Verma et al., 2022)
	Polyacrylonitrile (PAN)/goethite	45%	(Hossaini-Zahed et al., 2022)
Eu (III)	PANI/GO	1.65 mmol/g	(Sun et al., 2013)
Hg (II)	Ppy/GO	980 mg/g	(Chandra & Kim, 2011)
	Polythiophene (Pth)/GO	113.6 mg/g	(Muliwa et al., 2017)
	Ethylenediaminetetraacetic acid (EDTA)-GO-chitosan	324 mg/g	(Verma et al., 2022)
Ni (II)	Mn_2O_3/PANI	494 mg/g	(Rajakumar et al., 2014)
	Silica-based zinc oxide	32.10 mg/g	(Garg et al., 2022)
Pb (II)	Clay/polymethoxyethylacrylamide (PMEA)	80 mg/g	(Şölener et al., 2008)
	Poly(acrylic acid)/bentonite	93 mg/g	(Rafiei et al., 2016)
	PANI/carbon nanotube	22.2 mg/g	(Shao et al., 2012)
	MnO_2/poly(m-phenylenediamine)	446 mg/g	(Xiong et al., 2020)
	Mn_2O_3/PANI	437 mg/g	(Rajakumar et al., 2014)
	PANI/Sb_2O_3	92%	(Khalili, 2014)
	Magnetite/GO-Activated Carbon	153.2 mg/g	(Yan & Li, 2022)
	Polyacrylonitrile (PAN)/goethite	43%	(Hossaini-Zahed et al., 2022)
Sr (II)	PANI/GO	1.68 mmol/g	(Sun et al., 2013)
U (VI)	Chitosan/GO	30 mg/g	(Liu et al., 2022)
	PANI/GO	1.03 mmol/g	(Sun et al., 2013)

TABLE 11.4
Examples of Dye Removal by Polymer Nanocomposites with Inorganic Fillers

Dye	Polymer Composites	Efficiency	References
Acid yellow (AY)	GO/chitosan	98.18% in 6.48 minutes	(Banerjee et al., 2017)
Acid blue (AB)	GO/chitosan	98.80% in 6.48 minutes	(Banerjee et al., 2017)
	Polyaniline (PANI)/ZnO/ diethylene glycol (DEG)	90% in 60 minutes	(Gilja et al., 2018)
Acid black	Perovskites/polymer	95% in 30 minutes	(Brahmi et al., 2021)
Brilliant green	Zinc oxide/polypyrrole (ZnO/PPy)	95.5% in 20 minutes	(Zhang et al., 2019)
Congo red (CR)	Chitosan/carbon black	88.4% in 180 minutes	(Alshabanat & AL-Anazy, 2018)
	Polyvinyl alcohol/carbon black	84.2% in 180 min	(Alshabanat & AL-Anazy, 2018)
Crystal violet (CV)	Polyaniline/$NiWO_4$	94.5% in 150 minutes	(Kumaresan et al., 2022)
Eosin yellow	Polysulfone (PSf)/N, Pd co-doped TiO_2	97% in 240 minutes	(Kuvarega et al., 2018)
Methylene blue (MB)	Sulfonated polypropylene ether nitrile (SPEN)/TiO_2/ Polyethylene glycol (PEG)	99.9% in 240 minutes	(Liu et al., 2021)
	TiO_2 and triazine-containing conjugated microporous polymers (CMPs)	96% in 60 minutes	(Li et al., 2018)
	Polyaniline/$NiWO_4$	91.5% in 150 minutes	(Kumaresan et al., 2022)
Methyl orange (MO)	SnO_2/ZnO/chitosan	99% in 210 minutes	(Zhu et al., 2011)
Rhodamine B (RhB)	TiO_2/rGO/Poly (styrene-co-glycidyl methacrylate)	96% in 30 minutes	(Fang et al., 2015)
	TiO_2/polydimethylsiloxane (PDMS)	80% in 60 minutes	(Hickman et al., 2018)
Rhodamine 6G (R6G)	Poly(styrene–divinyl benzene)/Ag	100% in 90 minutes	(Yang et al., 2015)

11.3.6 Food Packaging

Some common food packaging materials are soda-lime glass, paper, kraft paper, grease-proof paper, glassine, parchment paper, paperboard, aluminum/aluminum-foil, and tinplate (metals are restricted due to the corrosion process) (Idumah et al., 2020). Nowadays, intelligent/innovative packaging is gaining much attention due to the incorporation of intelligent biosensors (indicating the temperature and quality of the product), resistance toward carbon dioxide (CO_2), oxygen (O_2), moisture; loss reduction, and environmental friendliness. In comparison, active packaging (AP) refers to incorporating additives into packaging systems to extend the shelf life and quality of fresh vegetable or livestock products. Antimicrobial packaging is an AP technique and uses antimicrobial agents (PNCs with carbon nanotubes, metal, and metal oxide NPs) in food packaging materials. The physicomechanical parameters and antibacterial activity of CMC/okra mucilage (OM) blend films containing ZnO NPs may work as an AP material (Mohammadi et al., 2018). Hydroxypropyl methylcellulose (HPMC)/Ag NPs have good tensile strength and act as active antimicrobial internal coatings for AP (de Moura et al., 2012). Bioplastics and PNCs have been essential parts of food packaging and allied industrial products in recent years. Packaging films/containers for food materials must have desired mechanical and barrier properties for light and mass transfer (O_2, CO_2, and moisture). Chitosan (CS) and GL are the most common polymers in food, pharmaceutical, and other product applications (Kumar et al., 2018). Only CS polymer does not have good mechanical and antimicrobial properties for packaging needs; however, it exhibits suitable barrier properties against water vapor (Aljawish et al., 2016; Dutta et al., 2009; Sadeghi & Shahedi, 2016). GL can absorb UV light and have an excellent film-forming ability, but the poor mechanical, thermal, and barrier properties are the significant constraints in using it as a food packaging material (Vieira et al., 2011). Combining CS and GL or incorporating inorganic nanofillers in CS and GL may improve the above properties and have enough strength to make packaging materials (Kumar et al., 2018). Similarly, PNCs with various inorganic nanofillers have been demonstrated for the food packaging industry in recent years. Polylactic acid (PLA)/ZnO PNCs are potential candidates for food-packaging applications (Murariu et al., 2011).

Nano-additives/nano-reinforcements in polymeric nanocomposites modify the PNCs, making them applicable to the food packaging industry. Metal NPs (Ag, Au, Cu, Pt, and their alloys) and metallic oxide NPs (ZnO, Fe_2O_3, and l_2O_3) are considered food safety materials and utilized as food preservatives (Mangaraj et al., 2009). Incorporation of montmorillonite (MMT) in polymeric matrices such as polyolefins, polyurethanes (PU), polyethylene terephthalate (PET), polyamide (PA), polystyrene (PS), and epoxy resins have enhanced barrier properties (Mangaraj et al., 2009; Patiño et al., 2018; Romero-Bastida et al., 2018; H. Wang et al., 2018). PNCs with cellulose fillers are cheap, lightweight, and high-strength and minimize the polymer permeability (Faradilla et al., 2018; Kale & Gorade, 2018; Kim et al., 2021; Pradipasena et al., 2018).

11.3.7 Biosensing

A biosensor is an analytical device detecting a chemical substance indicating plant diseases/pathogens, including selecting bioreceptor material, target analyte identification, signal processing unit, and final sensing unit (Kundu et al., 2019). In the agricultural industry, biosensors are helpful in various fields, including pre-and post-harvest agriculture, artificially ripened fruits and vegetables, intelligent food packaging, detection of heavy metals in soil and water, animal husbandry, and aquaculture. According to bio-recognition systems/elements, the agricultural biosensors can be of various types (e.g., antibody-antigen, enzyme-coenzyme, nucleic acids-complementary sequences, optical, piezoelectric, or magnetic). Harmful residues of agricultural chemicals, such as fertilizer, pesticide, and herbicide have entered the food chain and water cycle, leading to severely challenging pollution sources. The fast and cost-effective detection techniques at the appropriate levels in the food or water bodies remain a challenge. At present, high-performance liquid chromatography (HPLC), gas chromatography (GC), ultraviolet-visible spectroscopy (UV–Vis), Fourier-transform infrared spectroscopy (FTIR), and nuclear magnetic resonance spectroscopy (NMR) are standard detection techniques and methods; however, they are time-consuming, expensive, complex, and bulky requiring skilled technicians to operate (Fang & Ramasamy, 2015). Electrochemical biosensors are essential in the agricultural industry to detect pathogens in food and water safety, environmental monitoring, and bio-threat applications. PNCs have been demonstrated to be an excellent transducer for electrochemical biosensors due to the variation of morphology, facile synthesis techniques and responsibility. Poly(3,4-ethylenedioxythiophene) (PEDOT), polyaniline (PANI), polypyrrole (PPy), and polyindole are common conducting polymers due to their high conductivity in the doped state for electrochemical biosensors. These polymers are utilized to synthesize PNCs with various inorganic fillers, like carbon nanomaterials, metal, and metal oxide NPs. The samples from the agricultural industry are in a very rough and open environment. Therefore, biosensors based on PNCs provide better transduction with the improved current response, enhancing the sensitivity. In addition, PNCs generally have high mechanical strength, resistance to corrosion, high electrical and thermal conduction, and high durability. Some recent examples of biosensors based on PNCs are tabulated in Table 11.5.

11.3.8 Gene Transfection

Gene transfer or transfection is a procedure/analytical tool to introduce foreign nucleic acids into cells, producing genetically modified cells to study gene function, regulation, and protein function (Jat et al., 2020). Depending on the nature of the genetic materials (DNAs and RNAs), they exist in cells either stably or transiently. Plant genetic engineering is a powerful technique to meet the increasing demand for quality food. It helps create crops with higher production yield with resistance to insects and pathogens, increased tolerance to herbicides, and enhanced nutrition profile (Bonny, 2016; Dong & Ronald, 2019; Hilder & Boulter, 1999; Kissoudis et al., 2014). However, the genetic transformation of plant species remains challenging due

TABLE 11.5
Examples of Biosensors for Detecting Plant Pathogens, Heavy Metals, and Pesticide Residues. We Pick Random Examples of Biosensors From a Large Number of Demonstrations

Analyte	Polymer Nanocomposites	Sensitivity/Limit of Detection	References
Organophosphate pesticides in cabbage	Chitosan (CS)-TiO_2- reduced graphene oxide (r-GO)	29 Nm (6.4 ppb)	(Cui et al., 2018)
Indole-3-acetic acid (IAA) of the stem of soybean seedlings	Poly(safranine T) (PST) coated Pt NPs linked r-GO	43 pg mL^{-1}	(Li et al., 2019)
Parathion-methyl (PM), a pesticide residues	palladium-magnetic iron/ poly(3, 4-ethylenedioxythio phene):poly(styrenesulfonic acid)/nitrogen and sulfur co-doped titanium carbide (Pd-Fe_3O_4/PEDOT:PSS/ NS-Ti_3C_2Tx)	3.3 nM	(Deng et al., 2022)
Aflatoxin B1 (AFB1) in peanut	GO/Au NCs	0.03 pg/mL	(Li et al., 2018)
H_2O_2	polypyrrole (Ppy)–r-GO–Au NPs	2.7 μM	(Wu et al., 2016)
Hg^{2+}	Au- dimethyl amino ethanethiol (DMAET) -Single-walled carbon nanotube-poly(m-amino benzenesulfonic acid) (SWCNT-PABSA)	0.06 μM (250 ppb)	(Matlou et al., 2016)
Zn^{2+}, Cd^{2+}, and Pb^{2+}	graphene–polyaniline (G/ PANI)	1, 0.1, and 0.1 μg L^{-1}, respectively	(Ruecha et al., 2015)
Arowana fish DNA	kappa-carrageenan-polypyrrole-gold nanoparticles (KC-PPy-AuNPs)	5×10^{-18} M	(Esmaeili et al., 2017)

to a lack of efficient delivery vectors into plant cells through their rigid and multilayered cell walls, which requires significantly more research focus.

11.4 COMMERCIALIZATION OF POLYMER NANOCOMPOSITES FOR AGRICULTURE

Biopolymer nanocomposites are composed of two or more constituents, in which the matrix is a biopolymer (e.g., gluten, pullulan, curdlan, chitosan, alginates, starch, cellulose, lipids), and the fillers are composed of NPs or inorganic materials. The agricultural industry faces multiple challenges like the exploding population, dietary choices, climate change, and sustainability. Enhancing productivity and yield of crops can overcome the challenges by implementing greenhouses, tunnels, mulch, and silage. They all need polymeric films to increase moisture and prevent weed growth, allowing less dependence on agrochemicals. In general, polyethylene (PE) and its composites offer good mechanical and radiometric properties, though they have several environmental disadvantages. Here bio-nano composites are alternative materials to deal with it. Trademarked PNC like Imperm® (Color Matrix Europe) are now used in multilayer PET sheets and bottles for beverage and food packaging to reduce the permeation of O_2 and the loss of CO_2 from beverages. Duretham® KU 2–2601 (LANXESS Deutschland GmbH) is a PNC film fabricated from PAs with superior barrier properties in packaging juices. Aegis® OX (Honeywell Polymers) is PNC with improved barrier properties.

11.5 ENVIRONMENTAL RISK ASSESSMENT AND TOXICOLOGY STUDY FOR AGRICULTURAL PURPOSES

The presence of NPs in PNCs shows some adverse effects on crop plants directly or indirectly due to toxicity. Common assumptions in ecotoxicology are insufficient to assess the toxicity due to NPs and PNCs. A separate sub-discipline, "nano ecotoxicology," is emerged to understand and comprehensively assess the toxicological effects of NPs/PNCs. As of now, there are no adequate studies on behavior and consequences in the environment of nanoagrochemicals and their toxic effect, restricting their application in agriculture. Thus, a comprehensive toxicological assessment is necessary for all types of PNCs before using and developing a safer agro-product or agro-pesticides. The detection and quantification of nanofillers in complex environmental, biological media, and polymer composites are the primary limitations of toxicological assessment. Then, we have to understand the plant response to engineering PNCs exposure, dose-response, and the effect of the residue after biodegradation. Biodegradability is the prerequisite property of all types of PNCs in most cases. The transport of PNCs of different nanofiller loading in more complex porous media, i.e., soils with various mineralogical compositions, must be understood. The uptake and translocation of different NPs may vary with plant species, cultivar's growth conditions, and chemical composition (Rienzie & Adassooriya, 2018). NPs exhibit low mobility at different ionic concentrations due to strong attachment to the soil colloids with enhanced sorption capacity (L. Zhao et al., 2012). Hence, soil microbial communities can be affected, and they can influence the reactivity of the NPs (Fierer & Jackson, 2006; Gajjar et al., 2009).

NPs may affect humans by consuming plant-derived products obtained via delivery systems or processed foods, as shown in Figure 11.3. The figure details the entry and movement of NPs in agricultural and food systems. With the present understanding, researchers study the toxicological effect of a few NPs and PNCs primarily in the laboratory environment, taking a minimum amount of it. However, a rigorous study is needed with a large sample batch to apply PNCs in the agroindustry. Statistical tools and recently developed artificial intelligence techniques will

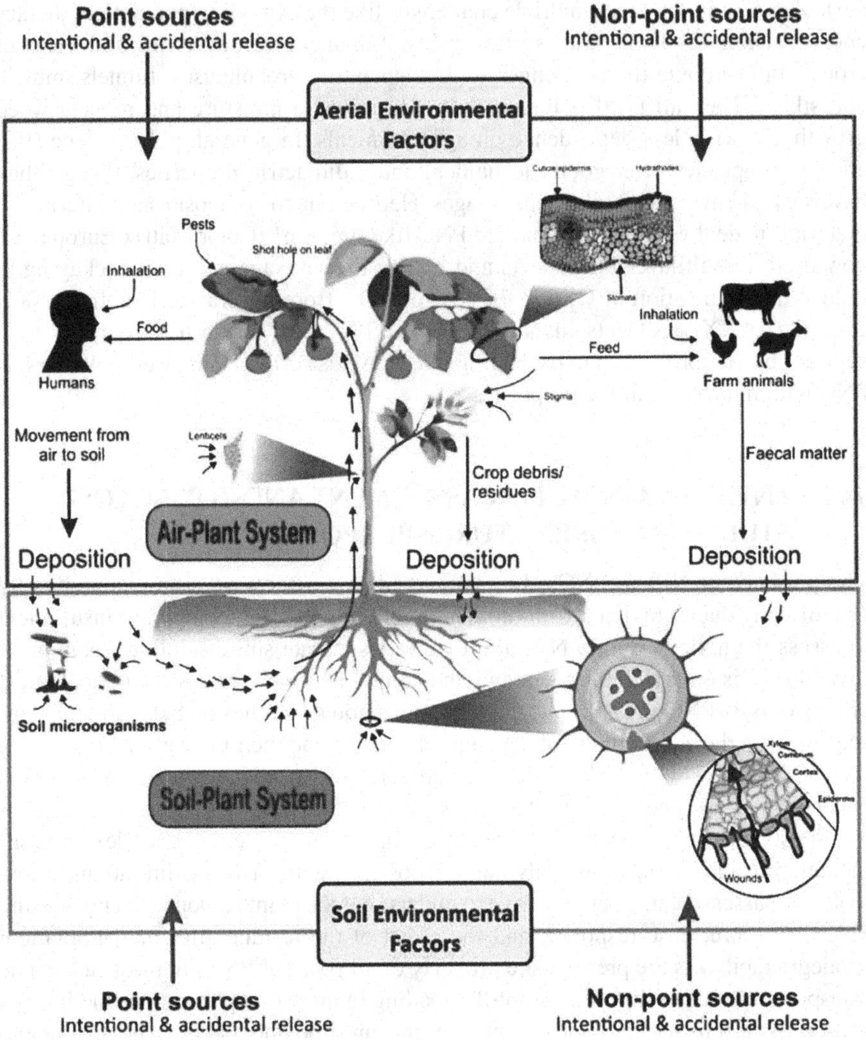

FIGURE 11.3 Schematic diagram showing the entry and movement of NPs in agricultural and food systems. Reproduced with permission from the Ref. (Rienzie & Adassooriya, 2018).

contribute to setting up a predictive ecotoxicological pattern capable of estimating dose concentrations of PNCs and predicting environmental damage and toxicological effects. Various agencies, including the Royal Academy of Engineering, Royal Society, US Environmental Protection Agency, European Commission's Scientific Committee, and other non-governmental organizations, are dynamically studying the potential toxicity and risk assessment of NPs (Chaud et al., 2021; Chowdhury & Bhattacharyya, 2018; Jian et al., 2012). Recent articles comprehensively explain the adverse effect of nanoagrochemicals (Ahmed et al., 2020; Chaud et al., 2021;

Muthukrishnan, 2022; Paramo et al., 2020; L. Yan et al., 2020). Kumar et al. describe in their recent review that smart agrochemicals are used despite their adverse or low toxicity, having the ability to change plant growth, seedling growth, biomass, DNA damage, and many more (Kumar et al., 2021).

Moreover, PNCs and the inorganic nanofillers may adversely affect plant physiology, soil microbiota, and declined enzymatic population (Abdel Latef et al., 2020; Ahmadi et al., 2020; Burklew et al., 2012). Polymers in PNCs with a terminal group -OH and -NH_2 show toxicity to plants and the environment. However, coating with polyethylene glycol (PEG) reduces the potential toxicity. PNCs degrade with time depending on their composition, e.g., biopolymers without treatment degrade faster than synthetic polymers. However, the PNCs degrade faster with external stimuli like sunlight, oxygen, free radicals, enzymes, and temperature. The presence of NPs may alter the degradation of PNCs as NPs behave differently with the external stimuli.

11.6 CONCLUSION AND FUTURE PERSPECTIVE

The green plant's ecological balance and smooth evolvement have been hindered due to contaminants in the environment during industrialization in the last century. A considerable number of hybrid inorganic polymers or PNCs have been demonstrated for different parts of plant science and protection. Now is the right time to develop intelligent nanoagrochemicals to benefit from a simple handling process, low cost, a sharp release system, and a high degradation rate without leaving any residue. However, the scarcity of study in field conditions rather than laboratory conditions is the primary constraint on commercial implementation. Nanostructures or nanoparticles in agrochemicals (fertilizers or pesticides) may expose to the environment through general farm practice, which has potential risk to human health via consuming the agricultural product, depending on the size, shape, solubility, crystal phase, type of material, exposure, and dosage concentrations of NPs. Therefore, a comprehensive toxicological study is needed to address the safety concern and the impact of nanoparticles and nanofertilizers on ecosystem biotic and abiotic components.

REFERENCES

A. Ávila-Orta, C., González-Morones, P., Agüero- Valdez, D., González-Sánchez, A., G. Martínez-Colunga, J., M. Mata-Padilla, J., & J. Cruz-Delgado, V. (2019). Ultrasound-assisted melt extrusion of polymer nanocomposites. In *Nanocomposites: Recent Evolutions*. IntechOpen. https://doi.org/10.5772/intechopen.80216

Abdallah, Y., Liu, M., Ogunyemi, S. O., Ahmed, T., Fouad, H., Abdelazez, A., Yan, C., Yang, Y., Chen, J., & Li, B. (2020). Bioinspired green synthesis of chitosan and zinc oxide nanoparticles with strong antibacterial activity against rice pathogen xanthomonas oryzae pv. oryzae. *Molecules*, *25*(20), 4795. https://doi.org/10.3390/molecules25204795

Abdel Latef, A. A. H., Zaid, A., Abu Alhmad, M. F., & Abdelfattah, K. E. (2020). The impact of priming with Al_2O_3 nanoparticles on growth, pigments, osmolytes, and antioxidant enzymes of egyptian roselle (Hibiscus sabdariffa L.) cultivar. *Agronomy*, *10*(5), 681. https://doi.org/10.3390/agronomy10050681

Adnan, M., Dalod, A., Balci, M., Glaum, J., & Einarsrud, M.-A. (2018). In situ synthesis of hybrid inorganic–polymer nanocomposites. *Polymers*, *10*(10), 1129. https://doi.org/10.3390/polym10101129

Ahmadi-Moghadam, B., & Taheri, F. (2014). Effect of processing parameters on the structure and multi-functional performance of epoxy/GNP-nanocomposites. *Journal of Materials Science*, *49*(18), 6180–6190. https://doi.org/10.1007/s10853-014-8332-y

Ahmadi, S. Z., Ghorbanpour, M., Aghaee, A., & Hadian, J. (2020). Deciphering morpho-physiological and phytochemical attributes of Tanacetum parthenium L. plants exposed to C60 fullerene and salicylic acid. *Chemosphere*, *259*, 127406. https://doi.org/10.1016/j.chemosphere.2020.127406

Ahmed, B., Ameen, F., Rizvi, A., Ali, K., Sonbol, H., Zaidi, A., Khan, M. S., & Musarrat, J. (2020). Destruction of cell topography, morphology, membrane, inhibition of respiration, biofilm formation, and bioactive molecule production by nanoparticles of Ag, ZnO, CuO, TiO_2, and Al_2O_3 toward beneficial soil bacteria. *ACS Omega*, *5*(14), 7861–7876. https://doi.org/10.1021/acsomega.9b04084

Akpan, E. I., Shen, X., Wetzel, B., & Friedrich, K. (2019). Design and synthesis of polymer nanocomposites. In *Polymer Composites with Functionalized Nanoparticles* (pp. 47–83). Elsevier. https://doi.org/10.1016/B978-0-12-814064-2.00002-0

Ali, Q., Ahmar, S., Sohail, M. A., Kamran, M., Ali, M., Saleem, M. H., Rizwan, M., Ahmed, A. M., Mora-Poblete, F., do Amaral Júnior, A. T., Mubeen, M., & Ali, S. (2021). Research advances and applications of biosensing technology for the diagnosis of pathogens in sustainable agriculture. *Environmental Science and Pollution Research*, *28*(8), 9002–9019. https://doi.org/10.1007/s11356-021-12419-6

Ali, Q., Zheng, H., Rao, M. J., Ali, M., Hussain, A., Saleem, M. H., Nehela, Y., Sohail, M. A., Ahmed, A. M., Kubar, K. A., Ali, S., Usman, K., Manghwar, H., & Zhou, L. (2022). Advances, limitations, and prospects of biosensing technology for detecting phytopathogenic bacteria. *Chemosphere*, *296*, 133773. https://doi.org/10.1016/j.chemosphere.2022.133773

Aljawish, A., Muniglia, L., Klouj, A., Jasniewski, J., Scher, J., & Desobry, S. (2016). Characterization of films based on enzymatically modified chitosan derivatives with phenol compounds. *Food Hydrocolloids*, *60*, 551–558. https://doi.org/10.1016/j.foodhyd.2016.04.032

Almotairy, S. M., Boostani, A. F., Hassani, M., Wei, D., & Jiang, Z. Y. (2020). Effect of hot isostatic pressing on the mechanical properties of aluminium metal matrix nanocomposites produced by dual speed ball milling. *Journal of Materials Research and Technology*, *9*(2), 1151–1161. https://doi.org/10.1016/j.jmrt.2019.11.043

Alshabanat, M. N., &<au> AL-Anazy, M. M. (2018). An experimental study of photocatalytic degradation of congo red using polymer nanocomposite films. *Journal of Chemistry*, *2018*, 1–8. https://doi.org/10.1155/2018/9651850

An, C., Sun, C., Li, N., Huang, B., Jiang, J., Shen, Y., Wang, C., Zhao, X., Cui, B., Wang, C., Li, X., Zhan, S., Gao, F., Zeng, Z., Cui, H., & Wang, Y. (2022). Nanomaterials and nanotechnology for the delivery of agrochemicals: strategies towards sustainable agriculture. *Journal of Nanobiotechnology*, *20*(1), 11. https://doi.org/10.1186/s12951-021-01214-7

Ashfaq, M., Verma, N., & Khan, S. (2017). Carbon nanofibers as a micronutrient carrier in plants: efficient translocation and controlled release of Cu nanoparticles. *Environmental Science: Nano*, *4*(1), 138–148. https://doi.org/10.1039/C6EN00385K

Ashrafi, B., Hubert, P., & Vengallatore, S. (2006). Carbon nanotube-reinforced composites as structural materials for microactuators in microelectromechanical systems. *Nanotechnology*, *17*(19), 4895–4903. https://doi.org/10.1088/0957–4484/17/19/019

Ayatollahi, M. R., Shadlou, S., Shokrieh, M. M., & Chitsazzadeh, M. (2011). Effect of multi-walled carbon nanotube aspect ratio on mechanical and electrical properties of epoxy-based nanocomposites. *Polymer Testing*, *30*(5), 548–556. https://doi.org/10.1016/j.polymertesting.2011.04.008

Backes, C., Higgins, T. M., Kelly, A., Boland, C., Harvey, A., Hanlon, D., & Coleman, J. N. (2017). Guidelines for exfoliation, characterization and processing of layered materials produced by liquid exfoliation. *Chemistry of Materials*, *29*(1), 243–255. https://doi.org/10.1021/acs.chemmater.6b03335

Banerjee, J., & Dutta, K. (2019). Melt-mixed carbon nanotubes/polymer nanocomposites. *Polymer Composites*, *40*(12), 4473–4488. https://doi.org/10.1002/pc.25334

Banerjee, M., Mallick, S., Paul, A., Chattopadhyay, A., & Ghosh, S. S. (2010). Heightened reactive oxygen species generation in the antimicrobial activity of a three component iodinated chitosan–silver nanoparticle composite. *Langmuir*, *26*(8), 5901–5908. https://doi.org/10.1021/la9038528

Banerjee, P., Barman, S. R., Mukhopadhayay, A., & Das, P. (2017). Ultrasound assisted mixed azo dye adsorption by chitosan–graphene oxide nanocomposite. *Chemical Engineering Research and Design*, *117*, 43–56. https://doi.org/10.1016/j.cherd.2016.10.009

Bautista-Baños, S., Hernández-Lauzardo, A. N., Velázquez-del Valle, M. G., Hernández-López, M., Ait Barka, E., Bosquez-Molina, E., & Wilson, C. L. (2006). Chitosan as a potential natural compound to control pre and postharvest diseases of horticultural commodities. *Crop Protection*, *25*(2), 108–118. https://doi.org/10.1016/j.cropro.2005.03.010

Bonny, S. (2016). Genetically modified herbicide-tolerant crops, weeds, and herbicides: overview and impact. *Environmental Management*, *57*(1), 31–48. https://doi.org/10.1007/s00267-015-0589-7

Bortolin, A., Aouada, F. A., Mattoso, L. H. C., & Ribeiro, C. (2013). Nanocomposite PAAm/methyl cellulose/montmorillonite hydrogel: Evidence of synergistic effects for the slow release of fertilizers. *Journal of Agricultural and Food Chemistry*, *61*(31), 7431–7439. https://doi.org/10.1021/jf401273n

Bortolin, A., Serafim, A. R., Aouada, F. A., Mattoso, L. H. C., & Ribeiro, C. (2016). Macro- and micronutrient simultaneous slow release from highly swellable nanocomposite hydrogels. *Journal of Agricultural and Food Chemistry*, *64*(16), 3133–3140. https://doi.org/10.1021/acs.jafc.6b00190

Brahmi, C., Benltifa, M., Vaulot, C., Michelin, L., Dumur, F., Airoudj, A., Morlet-Savary, F., Raveau, B., Bousselmi, L., & Lalevée, J. (2021). New hybrid perovskites/polymer composites for the photodegradation of organic dyes. *European Polymer Journal*, *157*, 110641. https://doi.org/10.1016/j.eurpolymj.2021.110641

Brunel, F., El Gueddari, N. E., & Moerschbacher, B. M. (2013). Complexation of copper(II) with chitosan nanogels: Toward control of microbial growth. *Carbohydrate Polymers*, *92*(2), 1348–1356. https://doi.org/10.1016/j.carbpol.2012.10.025

Buggy, M., Bradley, G., & Sullivan, A. (2005). Polymer–filler interactions in kaolin/nylon 6,6 composites containing a silane coupling agent. *Composites Part A: Applied Science and Manufacturing*, *36*(4), 437–442. https://doi.org/10.1016/j.compositesa.2004.10.002

Burklew, C. E., Ashlock, J., Winfrey, W. B., & Zhang, B. (2012). Effects of aluminum oxide nanoparticles on the growth, development, and microRNA expression of tobacco (nicotiana tabacum). *PLoS ONE*, *7*(5), e34783. https://doi.org/10.1371/journal.pone.0034783

Cárdenas, G., Díaz V., J., Meléndrez, M. F., Cruzat C., C., & García Cancino, A. (2009). Colloidal Cu nanoparticles/chitosan composite film obtained by microwave heating for food package applications. *Polymer Bulletin*, *62*(4), 511–524. https://doi.org/10.1007/s00289-008-0031-x

Chandra, V., & Kim, K. S. (2011). Highly selective adsorption of Hg^{2+} by a polypyrrole–reduced graphene oxide composite. *Chemical Communications*, *47*(13), 3942. https://doi.org/10.1039/c1cc00005e

Chandrasekaran, S., Sato, N., Tölle, F., Mülhaupt, R., Fiedler, B., & Schulte, K. (2014). Fracture toughness and failure mechanism of graphene based epoxy composites. *Composites Science and Technology*, *97*, 90–99. https://doi.org/10.1016/j.compscitech.2014.03.014

Chaud, M., Souto, E. B., Zielinska, A., Severino, P., Batain, F., Oliveira-Junior, J., & Alves, T. (2021). Nanopesticides in agriculture: benefits and challenge in agricultural productivity, toxicological risks to human health and environment. *Toxics*, *9*(6), 131. https://doi.org/10.3390/toxics9060131

Chawla, J. S., & Amiji, M. M. (2002). Biodegradable poly(ε-caprolactone) nanoparticles for tumor-targeted delivery of tamoxifen. *International Journal of Pharmaceutics*, *249*(1–2), 127–138. https://doi.org/10.1016/S0378–5173(02)00483–0

Chen, C., Zhang, G., Dai, Z., Xiang, Y., Liu, B., Bian, P., Zheng, K., Wu, Z., & Cai, D. (2018). Fabrication of light-responsively controlled-release herbicide using a nanocomposite. *Chemical Engineering Journal*, *349*, 101–110. https://doi.org/10.1016/j.cej.2018.05.079

Chi, Y., Zhang, G., Xiang, Y., Cai, D., & Wu, Z. (2017). Fabrication of a temperature-controlled-release herbicide using a nanocomposite. *ACS Sustainable Chemistry & Engineering*, *5*(6), 4969–4975. https://doi.org/10.1021/acssuschemeng.7b00348

Chi, Y., Zhang, G., Xiang, Y., Cai, D., & Wu, Z. (2018). Fabrication of reusable temperature-controlled-released fertilizer using a palygorskite-based magnetic nanocomposite. *Applied Clay Science*, *161*, 194–202. https://doi.org/10.1016/j.clay.2018.04.024

Chong, M. N., Jin, B., Chow, C. W. K., & Saint, C. (2010). Recent developments in photocatalytic water treatment technology: A review. *Water Research*, *44*(10), 2997–3027. https://doi.org/10.1016/j.watres.2010.02.039

Chowdhury, P. R., & Bhattacharyya, K. G. (2018). Toxicology and environmental fate of polymer nanocomposites. In *New Polymer Nanocomposites for Environmental Remediation* (pp. 649–677). Elsevier. https://doi.org/10.1016/B978-0-12-811033-1.00039-1

Christiaens, O., Petek, M., Smagghe, G., & Taning, C. N. T. (2020). The use of nanocarriers to improve the efficiency of RNAi-based pesticides in agriculture. In *Nanopesticides* (pp. 49–68). Springer International Publishing. https://doi.org/10.1007/978-3-030-44873-8-3

Cioffi, N., Torsi, L., Ditaranto, N., Tantillo, G., Ghibelli, L., Sabbatini, L., Bleve-Zacheo, T., D'Alessio, M., Zambonin, P. G., & Traversa, E. (2005). Copper nanoparticle/polymer composites with antifungal and bacteriostatic properties. *Chemistry of Materials*, *17*(21), 5255–5262. https://doi.org/10.1021/cm0505244

Cipiriano, B. H., Kashiwagi, T., Raghavan, S. R., Yang, Y., Grulke, E. A., Yamamoto, K., Shields, J. R., & Douglas, J. F. (2007). Effects of aspect ratio of MWNT on the flammability properties of polymer nanocomposites. *Polymer*, *48*(20), 6086–6096. https://doi.org/10.1016/j.polymer.2007.07.070

Cui, H.-F., Wu, W.-W., Li, M.-M., Song, X., Lv, Y., & Zhang, T.-T. (2018). A highly stable acetylcholinesterase biosensor based on chitosan-TiO_2-graphene nanocomposites for detection of organophosphate pesticides. *Biosensors and Bioelectronics*, *99*, 223–229. https://doi.org/10.1016/j.bios.2017.07.068

Dalir, H., Farahani, R. D., Nhim, V., Samson, B., Lévesque, M., & Therriault, D. (2012). Preparation of highly exfoliated polyester–clay nanocomposites: Process–property correlations. *Langmuir*, *28*(1), 791–803. https://doi.org/10.1021/la203331h

Dasgupta, N., Ranjan, S., & Ramalingam, C. (2017). Applications of nanotechnology in agriculture and water quality management. *Environmental Chemistry Letters*, *15*(4), 591–605. https://doi.org/10.1007/s10311-017-0648-9

de Moura, M. R., Mattoso, L. H. C., & Zucolotto, V. (2012). Development of cellulose-based bactericidal nanocomposites containing silver nanoparticles and their use as active food packaging. *Journal of Food Engineering*, *109*(3), 520–524. https://doi.org/10.1016/j.jfoodeng.2011.10.030

Deng, L., Yuan, J., Xie, S., Huang, H., Yue, R., & Xu, J. (2022). A novel Pd-Fe_3O_4/PEDOT:PSS/nitrogen and sulfur doped-Ti_3C_2Tx frameworks as highly sensitive sensing platform toward parathion-methyl residue in nature. *Electrochimica Acta*, *407*, 139897. https://doi.org/10.1016/j.electacta.2022.139897

Dhillon, N. K., & Mukhopadhyay, S. S. (2015). Nanotechnology and allelopathy: Synergism in action. *Journal Crop and Weed*, *11*(2), 187–191.

Di, Y., Li, Q., & Zhuang, X. (2012). Antibacterial finishing of tencel/cotton nonwoven fabric using Ag nanoparticles-chitosan composite. *Journal of Engineered Fibers and Fabrics*, 7(2), 155892501200700. https://doi.org/10.1177/155892501200700205

Dong, O. X., & Ronald, P. C. (2019). Genetic engineering for disease resistance in plants: recent progress and future perspectives. *Plant Physiology*, *180*(1), 26–38. https://doi.org/10.1104/pp.18.01224

Dutta, P. K., Tripathi, S., Mehrotra, G. K., & Dutta, J. (2009). Perspectives for chitosan based antimicrobial films in food applications. *Food Chemistry*, *114*(4), 1173–1182. https://doi.org/10.1016/j.foodchem.2008.11.047

Ebisike, K., Okoronkwo, A. E., & Alaneme, K. K. (2020). Synthesis and characterization of chitosan–silica hybrid aerogel using sol-gel method. *Journal of King Saud University - Science*, *32*(1), 550–554. https://doi.org/10.1016/j.jksus.2018.08.005

El-Hamshary, H., Fouda, M. M. G., Moydeen, M., El-Newehy, M. H., Al-Deyab, S. S., & Abdel-Megeed, A. (2015). Synthesis and antibacterial of carboxymethyl starch-grafted poly(vinyl imidazole) against some plant pathogens. *International Journal of Biological Macromolecules*, *72*, 1466–1472. https://doi.org/10.1016/j.ijbiomac.2014.10.051

El Nahrawy, A. M., Abou Hammad, A. B., Bakr, A. M., Shaheen, T. I., & Mansour, A. M. (2020). Sol–gel synthesis and physical characterization of high impact polystyrene nanocomposites based on Fe_2O_3 doped with ZnO. *Applied Physics A*, *126*(8), 654. https://doi.org/10.1007/s00339-020-03822-w

El Zowalaty, M., Ibrahim, N. A., Salama, M., Shameli, K., Usman, M., & Zainuddin, N. (2013). Synthesis, characterization, and antimicrobial properties of copper nanoparticles. *International Journal of Nanomedicine*, 4467. https://doi.org/10.2147/IJN.S50837

Esmaeili, C., Heng, L. Y., Chiang, C. P., Rashid, Z. A., Safitri, E., & Malon Marugan, R. S. P. (2017). A DNA biosensor based on kappa-carrageenan-polypyrrole-gold nanoparticles composite for gender determination of arowana fish (Scleropages formosus). *Sensors and Actuators B: Chemical*, *242*, 616–624. https://doi.org/10.1016/j.snb.2016.11.061

Fang, R., Liang, Y., Ge, X., Du, M., Li, S., Li, T., & Li, Z. (2015). Preparation and photocatalytic degradation activity of TiO_2/rGO/polymer composites. *Colloid and Polymer Science*, *293*(4), 1151–1157. https://doi.org/10.1007/s00396-015-3507-x

Fang, Y., & Ramasamy, R. (2015). Current and prospective methods for plant disease detection. *Biosensors*, *5*(3), 537–561. https://doi.org/10.3390/bios5030537

Faradilla, R. H. F., Lee, G., Roberts, J., Martens, P., Stenzel, M., & Arcot, J. (2018). Effect of glycerol, nanoclay and graphene oxide on physicochemical properties of biodegradable nanocellulose plastic sourced from banana pseudo-stem. *Cellulose*, *25*(1), 399–416. https://doi.org/10.1007/s10570-017-1537-x

Fierer, N., & Jackson, R. B. (2006). The diversity and biogeography of soil bacterial communities. *Proceedings of the National Academy of Sciences*, *103*(3), 626–631. https://doi.org/10.1073/pnas.0507535103

Fortunati, E., Peltzer, M., Armentano, I., Jiménez, A., & Kenny, J. M. (2013). Combined effects of cellulose nanocrystals and silver nanoparticles on the barrier and migration properties of PLA nano-biocomposites. *Journal of Food Engineering*, *118*(1), 117–124. https://doi.org/10.1016/j.jfoodeng.2013.03.025

Gajjar, P., Pettee, B., Britt, D. W., Huang, W., Johnson, W. P., & Anderson, A. J. (2009). Antimicrobial activities of commercial nanoparticles against an environmental soil microbe, pseudomonas putida KT2440. *Journal of Biological Engineering*, *3*(1), 9. https://doi.org/10.1186/1754-1611-3-9

Gao, Z., Yang, W., Wang, J., Yan, H., Yao, Y., Ma, J., Wang, B., Zhang, M., & Liu, L. (2013). Electrochemical synthesis of layer-by-layer reduced graphene oxide sheets/polyaniline nanofibers composite and its electrochemical performance. *Electrochimica Acta*, *91*, 185–194. https://doi.org/10.1016/j.electacta.2012.12.119

Garg, R., Garg, R., Okon Eddy, N., Ibrahim Almohana, A., Fahad Almojil, S., Amir Khan, M., & Ho Hong, S. (2022). Biosynthesized silica-based zinc oxide nanocomposites for the sequestration of heavy metal ions from aqueous solutions. *Journal of King Saud University - Science*, *34*(4), 101996. https://doi.org/10.1016/j.jksus.2022.101996

Gerratt, A. P., & Bergbreiter, S. (2013). Dielectric breakdown of PDMS thin films. *Journal of Micromechanics and Microengineering*, *23*(6), 067001. https://doi.org/10.1088/0960-1317/23/6/067001

Ghosh, S., Saraswathi, A., Indi, S. S., Hoti, S. L., & Vasan, H. N. (2012). Ag@AgI, core@shell structure in agarose matrix as hybrid: Synthesis, characterization, and antimicrobial activity. *Langmuir*, *28*(22), 8550–8561. https://doi.org/10.1021/la301322j

Giannelis, E. P., Krishnamoorti, R., & Manias, E. (1999). *Polymer-Silicate Nanocomposites: Model Systems for Confined Polymers and Polymer Brushes*. Elsevier. https://doi.org/10.1007/3-540-69711-X_3

Gilja, V., Vrban, I., Mandić, V., Žic, M., & Hrnjak-Murgić, Z. (2018). Preparation of a PANI/ZnO composite for efficient photocatalytic degradation of acid blue. *Polymers*, *10*(9), 940. https://doi.org/10.3390/polym10090940

Giroto, A. S., de Campos, A., Pereira, E. I., Cruz, C. C. T., Marconcini, J. M., & Ribeiro, C. (2014). Study of a nanocomposite starch-clay for slow-release of herbicides: Evidence of synergistic effects between the biodegradable matrix and exfoliated clay on herbicide release control. *Journal of Applied Polymer Science*, *131*(23), 36–45. https://doi.org/10.1002/app.41188

Gojny, F. H., Wichmann, M. H. G., Fiedler, B., Kinloch, I. A., Bauhofer, W., Windle, A. H., & Schulte, K. (2006). Evaluation and identification of electrical and thermal conduction mechanisms in carbon nanotube/epoxy composites. *Polymer*, *47*(6), 2036–2045. https://doi.org/10.1016/j.polymer.2006.01.029

Gouin, S. (2004). Microencapsulation. *Trends in Food Science & Technology*, *15*(7–8), 330–347. https://doi.org/10.1016/j.tifs.2003.10.005

Guha, T., Gopal, G., Kundu, R., & Mukherjee, A. (2020). Nanocomposites for delivering agrochemicals: A comprehensive review. *Journal of Agricultural and Food Chemistry*, *68*(12), 3691–3702. https://doi.org/10.1021/acs.jafc.9b06982

Guo, Y., Peng, F., Wang, H., Huang, F., Meng, F., Hui, D., & Zhou, Z. (2018). Intercalation polymerization approach for preparing graphene/polymer composites. *Polymers*, *10*(1), 61. https://doi.org/10.3390/polym10010061

Hashim, A. F., Youssef, K., Roberto, S. R., & Abd-Elsalam, K. A. (2020). Hybrid inorganic-polymer nanocomposites: Synthesis, characterization, and plant-protection applications. In *Multifunctional Hybrid Nanomaterials for Sustainable Agri-Food and Ecosystems* (pp. 33–49). Elsevier. https://doi.org/10.1016/B978-0-12-821354-4.00003-0

Hatami, M., & Yazdan Panah, M. (2017). Ultrasonic assisted synthesis of nanocomposite materials based on resole resin and surface modified nano CeO_2: Chemical and morphological aspects. *Ultrasonics Sonochemistry*, *39*, 160–173. https://doi.org/10.1016/j.ultsonch.2017.04.028

Hickman, R., Walker, E., & Chowdhury, S. (2018). TiO_2-PDMS composite sponge for adsorption and solar mediated photodegradation of dye pollutants. *Journal of Water Process Engineering*, *24*, 74–82. https://doi.org/10.1016/j.jwpe.2018.05.015

Hilder, V. A., & Boulter, D. (1999). Genetic engineering of crop plants for insect resistance: A critical review. *Crop Protection*, *18*(3), 177–191. https://doi.org/10.1016/S0261-2194(99)00028-9

Hossaini-Zahed, S.-S., Khanlari, S., Bakhtiari, O., Tofighy, M. A., Hadadpour, S., Rajabzadeh, S., Zhang, P., Matsuyama, H., & Mohammadi, T. (2022). Evaluation of process condition impact on copper and lead ions removal from water using goethite incorporated nanocomposite ultrafiltration adsorptive membranes. *Water Science and Technology*, *85*(4), 1053–1064. https://doi.org/10.2166/wst.2022.024

Hou, Y., Hu, J., Park, H., & Lee, M. (2012). Chitosan-based nanoparticles as a sustained protein release carrier for tissue engineering applications. *Journal of Biomedical Materials Research Part A*, *100A*(4), 939–947. https://doi.org/10.1002/jbm.a.34031

Ianchis, R., Ninciuleanu, C., Gifu, I., Alexandrescu, E., Somoghi, R., Gabor, A., Preda, S., Nistor, C., Nitu, S., Petcu, C., Icriverzi, M., Florian, P., & Roseanu, A. (2017). Novel hydrogel-advanced modified clay nanocomposites as possible vehicles for drug delivery and controlled release. *Nanomaterials*, *7*(12), 443. https://doi.org/10.3390/nano7120443

Idumah, C. I., Zurina, M., Ogbu, J., Ndem, J. U., & Igba, E. C. (2020). A review on innovations in polymeric nanocomposite packaging materials and electrical sensors for food and agriculture. *Composite Interfaces*, *27*(1), 1–72. https://doi.org/10.1080/09276440.2019.1600972

Iqbal, M., Umar, S., & Mahmooduzzafar. (2019). Nano-fertilization to enhance nutrient use efficiency and productivity of crop plants. In *Nanomaterials and Plant Potential* (pp. 473–505). Springer International Publishing. https://doi.org/10.1007/978-3-030-05569-1-19

Isayev, A. I., Kumar, R., & Lewis, T. M. (2009). Ultrasound assisted twin screw extrusion of polymer–nanocomposites containing carbon nanotubes. *Polymer*, *50*(1), 250–260. https://doi.org/10.1016/j.polymer.2008.10.052

Jat, S. K., Bhattacharya, J., & Sharma, M. K. (2020). Nanomaterial based gene delivery: A promising method for plant genome engineering. *Journal of Materials Chemistry B*, *8*(19), 4165–4175. https://doi.org/10.1039/D0TB00217H

Ji, H., Hu, S., Jiang, Z., Shi, S., Hou, W., & Yang, G. (2019). Directly scalable preparation of sandwiched MoS_2/graphene nanocomposites via ball-milling with excellent electrochemical energy storage performance. *Electrochimica Acta*, *299*, 143–151. https://doi.org/10.1016/j.electacta.2018.12.188

Jian, F., Zhang, Y., Wang, J., Ba, K., Mao, R., Lai, W., & Lin, Y. (2012). Toxicity of biodegradable nanoscale preparations. *Current Drug Metabolism*, *13*(4), 440–446. https://doi.org/10.2174/138920012800166517

Kale, R. D., & Gorade, V. G. (2018). Preparation of acylated microcrystalline cellulose using olive oil and its reinforcing effect on poly(lactic acid) films for packaging application. *Journal of Polymer Research*, *25*(3), 81. https://doi.org/10.1007/s10965-018-1470-1

Kalia, A., Sharma, S. P., Kaur, H., & Kaur, H. (2020). Novel nanocomposite-based controlled-release fertilizer and pesticide formulations: Prospects and challenges. In *Multifunctional Hybrid Nanomaterials for Sustainable Agri-Food and Ecosystems* (pp. 99–134). Elsevier. https://doi.org/10.1016/B978-0-12-821354-4.00005-4

Karak, N. (2019). Fundamentals of nanomaterials and polymer nanocomposites. In *Nanomaterials and Polymer Nanocomposites* (pp. 1–45). Elsevier. https://doi.org/10.1016/B978-0-12-814615-6.00001-1

Khirade, P. P. (2019). Structural, microstructural and magnetic properties of sol–gel-synthesized novel $BaZrO_3$–$CoFe_2O_4$ nanocomposite. *Journal of Nanostructure in Chemistry*, *9*(3), 163–173. https://doi.org/10.1007/s40097-019-0307-8

Kim, H. J., Jeong, J. H., Choi, Y. H., & Eom, Y. (2021). Review on cellulose nanocrystal-reinforced polymer nanocomposites: Processing, properties, and rheology. *Korea-Australia Rheology Journal*, *33*(3), 165–185. https://doi.org/10.1007/s13367-021-0015-z

Kim, H., & Macosko, C. W. (2008). Morphology and properties of polyester/exfoliated graphite nanocomposites. *Macromolecules*, *41*(9), 3317–3327. https://doi.org/10.1021/ma702385h

Kim, H., Miura, Y., & Macosko, C. W. (2010). Graphene/polyurethane nanocomposites for improved gas barrier and electrical conductivity. *Chemistry of Materials*, *22*(11), 3441–3450. https://doi.org/10.1021/cm100477v

Kissoudis, C., van de Wiel, C., Visser, R. G. F., & van der Linden, G. (2014). Enhancing crop resilience to combined abiotic and biotic stress through the dissection of physiological and molecular crosstalk. *Frontiers in Plant Science*, *5*. https://doi.org/10.3389/fpls.2014.00207

Kothmann, M. H., Ziadeh, M., Bakis, G., Rios de Anda, A., Breu, J., & Altstädt, V. (2015). Analyzing the influence of particle size and stiffness state of the nanofiller on the mechanical properties of epoxy/clay nanocomposites using a novel shear-stiff nano-mica. *Journal of Materials Science*, *50*(14), 4845–4859. https://doi.org/10.1007/s10853-015-9028-7

Kou, X., He, Y., Li, Y., Chen, X., Feng, Y., & Xue, Z. (2019). Effect of abscisic acid (ABA) and chitosan/nano-silica/sodium alginate composite film on the color development and quality of postharvest Chinese winter jujube (Zizyphus jujuba Mill. cv. Dongzao). *Food Chemistry*, *270*, 385–394. https://doi.org/10.1016/j.foodchem.2018.06.151

Kulabhusan, P. K., Tripathi, A., & Kant, K. (2022). Gold nanoparticles and plant pathogens: An overview and prospective for biosensing in forestry. *Sensors*, *22*(3), 1259. https://doi.org/10.3390/s22031259

Kumar, A., Choudhary, A., Kaur, H., Mehta, S., & Husen, A. (2021). Smart nanomaterial and nanocomposite with advanced agrochemical activities. *Nanoscale Research Letters*, *16*(1), 156. https://doi.org/10.1186/s11671-021-03612-0

Kumar, Rahul, Ashfaq, M., & Verma, N. (2018). Synthesis of novel PVA–starch formulation-supported Cu–Zn nanoparticle carrying carbon nanofibers as a nanofertilizer: controlled release of micronutrients. *Journal of Materials Science*, *53*(10), 7150–7164. https://doi.org/10.1007/s10853-018-2107-9

Kumar, Rajeev, Ansari, M. O., & Barakat, M. A. (2013). DBSA doped polyaniline/multi-walled carbon nanotubes composite for high efficiency removal of Cr(VI) from aqueous solution. *Chemical Engineering Journal*, *228*, 748–755. https://doi.org/10.1016/j.cej.2013.05.024

Kumar, S., Shukla, A., Baul, P. P., Mitra, A., & Halder, D. (2018). Biodegradable hybrid nanocomposites of chitosan/gelatin and silver nanoparticles for active food packaging applications. *Food Packaging and Shelf Life*, *16*, 178–184. https://doi.org/10.1016/j.fpsl.2018.03.008

Kumar, V., Sachdev, D., Pasricha, R., Maheshwari, P. H., & Taneja, N. K. (2018). Zinc-supported multiwalled carbon nanotube nanocomposite: A synergism to micronutrient release and a smart distributor to promote the growth of onion seeds in arid conditions. *ACS Applied Materials & Interfaces*, *10*(43), 36733–36745. https://doi.org/10.1021/acsami.8b13464

Kumaresan, A., Arun, A., Kalpana, V., Vinupritha, P., & Sundaravadivel, E. (2022). Polymer-supported $NiWO_4$ nanocomposites for visible light degradation of toxic dyes. *Journal of Materials Science: Materials in Electronics*, *33*(12), 9660–9668. https://doi.org/10.1007/s10854-021-07643-2

Kundu, M., Krishnan, P., Kotnala, R. K., & Sumana, G. (2019). Recent developments in biosensors to combat agricultural challenges and their future prospects. *Trends in Food Science & Technology*, *88*, 157–178. https://doi.org/10.1016/j.tifs.2019.03.024

Kuvarega, A. T., Khumalo, N., Dlamini, D., & Mamba, B. B. (2018). Polysulfone/N,Pd co-doped TiO2 composite membranes for photocatalytic dye degradation. *Separation and Purification Technology*, *191*, 122–133. https://doi.org/10.1016/j.seppur.2017.07.064

Levita, G., Marchetti, A., & Lazzeri, A. (1989). Fracture of ultrafine calcium carbonate/polypropylene composites. *Polymer Composites*, *10*(1), 39–43. https://doi.org/10.1002/pc.750100106

Li, H., Wang, C., Wang, X., Hou, P., Luo, B., Song, P., Pan, D., Li, A., & Chen, L. (2019). Disposable stainless steel-based electrochemical microsensor for in vivo determination of indole-3-acetic acid in soybean seedlings. *Biosensors and Bioelectronics*, 126, 193–199. https://doi.org/10.1016/j.bios.2018.10.041

Li, J., Wen, X., Zhang, Q., & Ren, S. (2018). Adsorption and visible-light photodegradation of organic dyes with TiO 2 /conjugated microporous polymer composites. *RSC Advances*, *8*(60), 34560–34565. https://doi.org/10.1039/C8RA06491A

Li, T., Gao, B., Tong, Z., Yang, Y., & Li, Y. (2019). Chitosan and graphene oxide nanocomposites as coatings for controlled-release fertilizer. *Water, Air, & Soil Pollution, 230*(7), 146. https://doi.org/10.1007/s11270-019-4173-2

Li, Z., Xue, N., Ma, H., Cheng, Z., & Miao, X. (2018). An ultrasensitive and switch-on platform for aflatoxin B1 detection in peanut based on the fluorescence quenching of graphene oxide-gold nanocomposites. *Talanta, 181*, 346–351. https://doi.org/10.1016/j.talanta.2018.01.039

Lin, G., Chen, X., Zhou, H., Zhou, X., Xu, H., & Chen, H. (2019). Elaboration of a feather keratin/carboxymethyl cellulose complex exhibiting pH sensitivity for sustained pesticide release. *Journal of Applied Polymer Science, 136*(10), 47160. https://doi.org/10.1002/app.47160

Liu, X., Wang, L., Zhou, X., He, X., Zhou, M., Jia, K., & Liu, X. (2021). Design of polymer composite-based porous membrane for in-situ photocatalytic degradation of adsorbed organic dyes. *Journal of Physics and Chemistry of Solids, 154*, 110094. https://doi.org/10.1016/j.jpcs.2021.110094

Liu, Y., Tang, X., Zhou, L., Liu, Z., Ouyang, J., Dai, Y., Le, Z., & Adesina, A. A. (2022). Nanofabricated chitosan/graphene oxide electrodes for enhancing electrosorptive removal of U(VI) from aqueous solution. *Separation and Purification Technology, 290*, 120827. https://doi.org/10.1016/j.seppur.2022.120827

Llorens, A., Lloret, E., Picouet, P. A., Trbojevich, R., & Fernandez, A. (2012). Metallic-based micro and nanocomposites in food contact materials and active food packaging. *Trends in Food Science & Technology, 24*(1), 19–29. https://doi.org/10.1016/j.tifs.2011.10.001

Llorens, A., Lloret, E., Picouet, P., & Fernandez, A. (2012). Study of the antifungal potential of novel cellulose/copper composites as absorbent materials for fruit juices. *International Journal of Food Microbiology, 158*(2), 113–119. https://doi.org/10.1016/j.ijfoodmicro.2012.07.004

Ma, X., Peng, C., Zhou, D., Wu, Z., Li, S., Wang, J., & Sun, N. (2018). Synthesis and mechanical properties of the epoxy resin composites filled with sol–gel derived ZrO_2 nanoparticles. *Journal of Sol-Gel Science and Technology, 88*(2), 442–453. https://doi.org/10.1007/s10971-018-4827-3

Ma, Z., Jiang, Y., Xiao, H., Jiang, B., Zhang, H., Peng, M., Dong, G., Yu, X., & Yang, J. (2018). Sol-gel preparation of Ag-silica nanocomposite with high electrical conductivity. *Applied Surface Science, 436*, 732–738. https://doi.org/10.1016/j.apsusc.2017.12.101

Madhukar, P., Selvaraj, N., Gujjala, R., & Rao, C. S. P. (2019). Production of high performance AA7150–1% SiC nanocomposite by novel fabrication process of ultrasonication assisted stir casting. *Ultrasonics Sonochemistry, 58*, 104665. https://doi.org/10.1016/j.ultsonch.2019.104665

Madhumitha, G., Fowsiya, J., Mohana Roopan, S., & Thakur, V. K. (2018). Recent advances in starch–clay nanocomposites. *International Journal of Polymer Analysis and Characterization, 23*(4), 331–345. https://doi.org/10.1080/1023666X.2018.1447260

Mahmoudabadi, Z. S., Rashidi, A., Tavasoli, A., Esrafili, M., Panahi, M., Askarieh, M., & Khodabakhshi, S. (2021). Ultrasonication-assisted synthesis of 2D porous MoS_2/GO nanocomposite catalysts as high-performance hydrodesulfurization catalysts of vacuum gasoil: Experimental and DFT study. *Ultrasonics Sonochemistry, 74*, 105558. https://doi.org/10.1016/j.ultsonch.2021.105558

Mangaraj, S., Goswami, T. K., & Mahajan, P. V. (2009). Applications of plastic films for modified atmosphere packaging of fruits and vegetables: A review. *Food Engineering Reviews, 1*(2), 133–158. https://doi.org/10.1007/s12393-009-9007-3

Matlou, G. G., Nkosi, D., Pillay, K., & Arotiba, O. (2016). Electrochemical detection of Hg(II) in water using self-assembled single walled carbon nanotube-poly(m -amino benzene sulfonic acid) on gold electrode. *Sensing and Bio-Sensing Research, 10*, 27–33. https://doi.org/10.1016/j.sbsr.2016.08.003

Maxwalt, S., Rahupathy, K., Suresh, S. N., Subramani, R., & Pushparaj, C. (2022). Synthesis of eco-friendly nanocomposite with silver nanoparticle to increase the antimicrobial activity. *Materials Today: Proceedings*. https://doi.org/10.1016/j.matpr.2022.02.374

Mistretta, M. C., Botta, L., La Mantia, F. P., Di Fiore, A., & Cascone, M. (2021). Film blowing of biodegradable polymer nanocomposites for agricultural applications. *Macromolecular Materials and Engineering*, *306*(9), 2100177. https://doi.org/10.1002/mame.202100177

Mohammadi, H., Kamkar, A., & Misaghi, A. (2018). Nanocomposite films based on CMC, okra mucilage and ZnO nanoparticles: Physico mechanical and antibacterial properties. *Carbohydrate Polymers*, *181*, 351–357. https://doi.org/10.1016/j.carbpol.2017.10.045

Momina, S., & Ahmad, K. (2021). Study of different polymer nanocomposites and their pollutant removal efficiency: Review. *Polymer*, *217*, 123453. https://doi.org/10.1016/j.polymer.2021.123453

Morgan, K. T., Cushman, K. E., & Sato, S. (2009). Release mechanisms for slow- and controlled-release fertilizers and strategies for their use in vegetable production. *HortTechnology*, *19*(1), 10–12. https://doi.org/10.21273/HORTSCI.19.1.10

Moussa, S. H., Tayel, A. A., Alsohim, A. S., & Abdallah, R. R. (2013). Botryticidal activity of nanosized silver-chitosan composite and its application for the control of gray mold in strawberry. *Journal of Food Science*, *78*(10), M1589–M1594. https://doi.org/10.1111/1750-3841.12247

Muliwa, A. M., Onyango, M. S., Maity, A., & Ochieng, A. (2017). Batch equilibrium and kinetics of mercury removal from aqueous solutions using polythiophene/graphene oxide nanocomposite. *Water Science and Technology*, *75*(12), 2841–2851. https://doi.org/10.2166/wst.2017.165

Murariu, M., Doumbia, A., Bonnaud, L., Dechief, A., Paint, Y., Ferreira, M., Campagne, C., Devaux, E., & Dubois, P. (2011). High-performance polylactide/ZnO nanocomposites designed for films and fibers with special end-use properties. *Biomacromolecules*, *12*(5), 1762–1771. https://doi.org/10.1021/bm2001445

Muthukrishnan, L. (2022). An overview on the nanotechnological expansion, toxicity assessment and remediating approaches in agriculture and food industry. *Environmental Technology & Innovation*, *25*, 102136. https://doi.org/10.1016/j.eti.2021.102136

NASSAR, N. M. A., & ORTIZ, R. (2007). Cassava improvement: Challenges and impacts. *The Journal of Agricultural Science*, *145*(02), 163. https://doi.org/10.1017/S0021859606006575

Ni, B., Liu, M., Lü, S., Xie, L., & Wang, Y. (2011). Environmentally friendly slow-release nitrogen fertilizer. *Journal of Agricultural and Food Chemistry*, *59*(18), 10169–10175. https://doi.org/10.1021/jf202131z

Olad, A., Gharekhani, H., Mirmohseni, A., & Bybordi, A. (2018). Superabsorbent nanocomposite based on maize bran with integration of water-retaining and slow-release NPK fertilizer. *Advances in Polymer Technology*, *37*(6), 1682–1694. https://doi.org/10.1002/adv.21825

Olad, A., Zebhi, H., Salari, D., Mirmohseni, A., & Reyhani Tabar, A. (2018). Slow-release NPK fertilizer encapsulated by carboxymethyl cellulose-based nanocomposite with the function of water retention in soil. *Materials Science and Engineering: C*, *90*, 333–340. https://doi.org/10.1016/j.msec.2018.04.083

Orellana-Tavra, C., Baxter, E. F., Tian, T., Bennett, T. D., Slater, N. K. H., Cheetham, A. K., & Fairen-Jimenez, D. (2015). Amorphous metal–organic frameworks for drug delivery. *Chemical Communications*, *51*(73), 13878–13881. https://doi.org/10.1039/C5CC05237H

Palma-Guerrero, J., Huang, I.-C., Jansson, H.-B., Salinas, J., Lopez-Llorca, L. V., & Read, N. D. (2009). Chitosan permeabilizes the plasma membrane and kills cells of neurospora crassa in an energy dependent manner. *Fungal Genetics and Biology*, *46*(8), 585–594. https://doi.org/10.1016/j.fgb.2009.02.010

Pandey, N., Shukla, S. K., & Singh, N. B. (2017). Water purification by polymer nanocomposites: An overview. *Nanocomposites*, *3*(2), 47–66. https://doi.org/10.1080/20550324.2017.1329983

Paramo, L. A., Feregrino-Pérez, A. A., Guevara, R., Mendoza, S., & Esquivel, K. (2020). Nanoparticles in agroindustry: Applications, toxicity, challenges, and trends. *Nanomaterials*, *10*(9), 1654. https://doi.org/10.3390/nano10091654

Park, J.-J., & Lee, J.-Y. (2013). Effect of nano-sized layered silicate on AC electrical treeing behavior of epoxy/layered silicate nanocomposite in needle-plate electrodes. *Materials Chemistry and Physics*, *141*(2–3), 776–780. https://doi.org/10.1016/j.matchemphys.2013.06.003

Patiño, L. S., Castellanos, D. A., & Herrera, A. O. (2018). Influence of 1-MCP and modified atmosphere packaging in the quality and preservation of fresh basil. *Postharvest Biology and Technology*, *136*, 57–65. https://doi.org/10.1016/j.postharvbio.2017.10.010

Paton, K. R., Varrla, E., Backes, C., Smith, R. J., Khan, U., O'Neill, A., Boland, C., Lotya, M., Istrate, O. M., King, P., Higgins, T., Barwich, S., May, P., Puczkarski, P., Ahmed, I., Moebius, M., Pettersson, H., Long, E., Coelho, J., … Coleman, J. N. (2014). Scalable production of large quantities of defect-free few-layer graphene by shear exfoliation in liquids. *Nature Materials*, *13*(6), 624–630. https://doi.org/10.1038/nmat3944

Pradeep, H., M., B., Suresh, S., Thadathil, A., & Periyat, P. (2022). Recent trends and advances in polyindole-based nanocomposites as potential antimicrobial agents: A mini review. *RSC Advances*, *12*(13), 8211–8227. https://doi.org/10.1039/D1RA09317G

Pradipasena, P., Chollakup, R., & Tantratian, S. (2018). Formation and characterization of BC and BC-paper pulp films for packaging application. *Journal of Thermoplastic Composite Materials*, *31*(4), 500–513. https://doi.org/10.1177/0892705717712633

Preetha, S., Ramamoorthy, S., Pillai, R., Narasimhamurthy, B., & Lekshmi, I. C. (2022). Synthesis of rGO-nanoTiO_2 composite mixture via ultrasonication assisted mechanical mixing method and their photocatalytic studies. *Materials Today: Proceedings*. https://doi.org/10.1016/j.matpr.2022.04.816

Qasem, N. A. A., Mohammed, R. H., & Lawal, D. U. (2021). Removal of heavy metal ions from wastewater: A comprehensive and critical review. *Npj Clean Water*, *4*(1), 36. https://doi.org/10.1038/s41545-021-00127-0

Qasim, U., Osman, A. I., Al-Muhtaseb, A. H., Farrell, C., Al-Abri, M., Ali, M., Vo, D.-V. N., Jamil, F., & Rooney, D. W. (2021). Renewable cellulosic nanocomposites for food packaging to avoid fossil fuel plastic pollution: A review. *Environmental Chemistry Letters*, *19*(1), 613–641. https://doi.org/10.1007/s10311-020-01090-x

Qi, L., Xu, Z., Jiang, X., Hu, C., & Zou, X. (2004). Preparation and antibacterial activity of chitosan nanoparticles. *Carbohydrate Research*, *339*(16), 2693–2700. https://doi.org/10.1016/j.carres.2004.09.007

Radheshkumar, C., & Münstedt, H. (2006). Antimicrobial polymers from polypropylene/silver composites—Ag+ release measured by anode stripping voltammetry. *Reactive and Functional Polymers*, *66*(7), 780–788. https://doi.org/10.1016/j.reactfunctpolym.2005.11.005

Rahmat, H., Ganjar, F., Uswatul, C., Sayekti, W., & Ari, H. R. (2015). Effectiveness of urea nanofertilizer based aminopropyltrimethoxysilane (APTMS)-zeolite as slow release fertilizer system. *African Journal of Agricultural Research*, *10*(14), 1785–1788. https://doi.org/10.5897/AJAR2014.8940

Raliya, R., Saharan, V., Dimkpa, C., & Biswas, P. (2018). Nanofertilizer for precision and sustainable agriculture: Current state and future perspectives. *Journal of Agricultural and Food Chemistry*, *66*(26), 6487–6503. https://doi.org/10.1021/acs.jafc.7b02178

Rane, A. V., Kanny, K., Abitha, V. K., & Thomas, S. (2018). Methods for synthesis of nanoparticles and fabrication of nanocomposites. In *Synthesis of Inorganic Nanomaterials* (pp. 121–139). Elsevier. https://doi.org/10.1016/B978-0-08-101975-7.00005-1

Rashidzadeh, A., & Olad, A. (2014). Slow-released NPK fertilizer encapsulated by NaAlg-g-poly(AA-co-AAm)/MMT superabsorbent nanocomposite. *Carbohydrate Polymers*, *114*, 269–278. https://doi.org/10.1016/j.carbpol.2014.08.010

Reddy, K. R., Venkata Reddy, C., Babu, B., Ravindranadh, K., Naveen, S., & Raghu, A. V. (2019). Recent advances in layered clays–intercalated polymer nanohybrids. In *Modified Clay and Zeolite Nanocomposite Materials* (pp. 197–218). Elsevier. https://doi.org/10.1016/B978-0-12-814617-0.00013-X

Rhim, J.-W., Park, H.-M., & Ha, C.-S. (2013). Bio-nanocomposites for food packaging applications. *Progress in Polymer Science*, *38*(10–11), 1629–1652. https://doi.org/10.1016/j.progpolymsci.2013.05.008

Rienzie, R., & Adassooriya, N. M. (2018). Toxicity of nanomaterials in agriculture and food. In *Nanomaterials: Ecotoxicity, Safety, and Public Perception* (pp. 207–234). Springer International Publishing. https://doi.org/10.1007/978-3-030-05144-0-11

Romero-Bastida, C. A., Chávez Gutiérrez, M., Bello-Pérez, L. A., Abarca-Ramírez, E., Velazquez, G., & Mendez-Montealvo, G. (2018). Rheological properties of nanocomposite-forming solutions and film based on montmorillonite and corn starch with different amylose content. *Carbohydrate Polymers*, *188*, 121–127. https://doi.org/10.1016/j.carbpol.2018.01.089

Roulin-Moloney, A. C., Cantwell, W. J., & Kausch, H. H. (1987). Parameters determining the strength and toughness of particulate-filled epoxy resins. *Polymer Composites*, *8*(5), 314–323. https://doi.org/10.1002/pc.750080506

Ruecha, N., Rodthongkum, N., Cate, D. M., Volckens, J., Chailapakul, O., & Henry, C. S. (2015). Sensitive electrochemical sensor using a graphene–polyaniline nanocomposite for simultaneous detection of Zn(II), Cd(II), and Pb(II). *Analytica Chimica Acta*, *874*, 40–48. https://doi.org/10.1016/j.aca.2015.02.064

Sadeghi, K., & Shahedi, M. (2016). Physical, mechanical, and antimicrobial properties of ethylene vinyl alcohol copolymer/chitosan/nano-ZnO (ECNZn) nanocomposite films incorporating glycerol plasticizer. *Journal of Food Measurement and Characterization*, *10*(1), 137–147. https://doi.org/10.1007/s11694-015-9287-7

Sanes, J., Sánchez, C., Pamies, R., Avilés, M.-D., & Bermúdez, M.-D. (2020). Extrusion of polymer nanocomposites with graphene and graphene derivative nanofillers: An overview of recent developments. *Materials*, *13*(3), 549. https://doi.org/10.3390/ma13030549

Sangeetha, J., Sarim, K. M., Thangadurai, D., Gupta, A., Renu, Mundaragi, A., Sheth, B. P., Wani, S. A., Baqual, M. F., & Habib, H. (2019). Nanoparticle-mediated plant gene transfer for precision farming and sustainable agriculture. In *Nanotechnology for Agriculture* (pp. 263–284). Springer. https://doi.org/10.1007/978-981-32-9370-0-14

Sarfraz, J., Gulin-Sarfraz, T., Nilsen-Nygaard, J., & Pettersen, M. K. (2020). Nanocomposites for food packaging applications: An overview. *Nanomaterials*, *11*(1), 10. https://doi.org/10.3390/nano11010010

Sarıgül, T., & İnam, R. (2009). A direct method for the polarographic determination of herbicide triasulfuron and application to natural samples and agrochemical formulation. *Bioelectrochemistry*, *75*(1), 55–60. https://doi.org/10.1016/j.bioelechem.2008.11.009

Sarlak, N., Taherifar, A., & Salehi, F. (2014). Synthesis of nanopesticides by encapsulating pesticide nanoparticles using functionalized carbon nanotubes and application of new nanocomposite for plant disease treatment. *Journal of Agricultural and Food Chemistry*, *62*(21), 4833–4838. https://doi.org/10.1021/jf404720d

Sempeho, S. I., Kim, H. T., Mubofu, E., Pogrebnoi, A., Shao, G., & Hilonga, A. (2015). Encapsulated urea-kaolinite nanocomposite for controlled release fertilizer formulations. *Journal of Chemistry*, *2015*, 1–17. https://doi.org/10.1155/2015/237397

Sikder, A., Pearce, A. K., Parkinson, S. J., Napier, R., & O'Reilly, R. K. (2021). Recent trends in advanced polymer materials in agriculture related applications. *ACS Applied Polymer Materials*, *3*(3), 1203–1217. https://doi.org/10.1021/acsapm.0c00982

Silvério, H. A., Flauzino Neto, W. P., Dantas, N. O., & Pasquini, D. (2013). Extraction and characterization of cellulose nanocrystals from corncob for application as reinforcing agent in nanocomposites. *Industrial Crops and Products*, *44*, 427–436. https://doi.org/10.1016/j.indcrop.2012.10.014

Singh, B., Sharma, D. K., & Gupta, A. (2009). A study towards release dynamics of thiram fungicide from starch–alginate beads to control environmental and health hazards. *Journal of Hazardous Materials*, *161*(1), 208–216. https://doi.org/10.1016/j.jhazmat.2008.03.074

Souri, H., Nam, I. W., & Lee, H. K. (2015). Electrical properties and piezoresistive evaluation of polyurethane-based composites with carbon nano-materials. *Composites Science and Technology*, 121, 41–48. https://doi.org/10.1016/j.compscitech.2015.11.003

Stloukal, P., Kucharczyk, P., Sedlarik, V., Bazant, P., & Koutny, M. (2012). Low molecular weight poly(lactic acid) microparticles for controlled release of the herbicide metazachlor: preparation, morphology, and release kinetics. *Journal of Agricultural and Food Chemistry*, *60*(16), 4111–4119. https://doi.org/10.1021/jf300521j

Sun, Y., Shao, D., Chen, C., Yang, S., & Wang, X. (2013). Highly efficient enrichment of radionuclides on graphene oxide-supported polyaniline. *Environmental Science & Technology*, *47*(17), 9904–9910. https://doi.org/10.1021/es401174n

Tamayo, L., Azócar, M., Kogan, M., Riveros, A., & Páez, M. (2016). Copper-polymer nanocomposites: An excellent and cost-effective biocide for use on antibacterial surfaces. *Materials Science and Engineering: C*, *69*, 1391–1409. https://doi.org/10.1016/j.msec.2016.08.041

Thio, Y. S., Argon, A. S., & Cohen, R. E. (2004). Role of interfacial adhesion strength on toughening polypropylene with rigid particles. *Polymer*, *45*(10), 3139–3147. https://doi.org/10.1016/j.polymer.2004.02.064

Tian, J., Shao, Q., Zhao, J., Pan, D., Dong, M., Jia, C., Ding, T., Wu, T., & Guo, Z. (2019). Microwave solvothermal carboxymethyl chitosan templated synthesis of TiO_2/ZrO_2 composites toward enhanced photocatalytic degradation of Rhodamine B. *Journal of Colloid and Interface Science*, *541*, 18–29. https://doi.org/10.1016/j.jcis.2019.01.069

Vasile, C. (2018). Polymeric nanocomposites and nanocoatings for food packaging: A review. *Materials*, *11*(10), 1834. https://doi.org/10.3390/ma11101834

Vergnes, B. (2019). Influence of processing conditions on the preparation of clay-based nanocomposites by twin-screw extrusion. *International Polymer Processing*, *34*(5), 482–501. https://doi.org/10.3139/217.3827

Verma, M., Lee, I., Oh, J., Kumar, V., & Kim, H. (2022). Synthesis of EDTA-functionalized graphene oxide-chitosan nanocomposite for simultaneous removal of inorganic and organic pollutants from complex wastewater. *Chemosphere*, *287*, 132385. https://doi.org/10.1016/j.chemosphere.2021.132385

Vieira, M. G. A., Da Silva, M. A., Dos Santos, L. O., & Beppu, M. M. (2011). Natural-based plasticizers and biopolymer films: A review. *European Polymer Journal*, *47*(3), 254–263. https://doi.org/10.1016/j.eurpolymj.2010.12.011

Wang, H., Chen, M., Jin, C., Niu, B., Jiang, S., Li, X., & Jiang, S. (2018). Antibacterial [2-(methacryloyloxy) ethyl] trimethylammonium chloride functionalized reduced graphene oxide/poly(ethylene- co -vinyl alcohol) multilayer barrier film for food packaging. *Journal of Agricultural and Food Chemistry*, *66*(3), 732–739. https://doi.org/10.1021/acs.jafc.7b04784

Wang, K., Wu, J., Ye, L., & Zeng, H. (2003). Mechanical properties and toughening mechanisms of polypropylene/barium sulfate composites. *Composites Part A: Applied Science and Manufacturing*, *34*(12), 1199–1205. https://doi.org/10.1016/j.compositesa.2003.07.004

Wang, L., Qiu, J., Sakai, E., & Wei, X. (2016). The relationship between microstructure and mechanical properties of carbon nanotubes/polylactic acid nanocomposites prepared by twin-screw extrusion. *Composites Part A: Applied Science and Manufacturing*, *89*, 18–25. https://doi.org/10.1016/j.compositesa.2015.12.016

Wang, M., Zhang, G., Zhou, L., Wang, D., Zhong, N., Cai, D., & Wu, Z. (2016). Fabrication of pH-controlled-release ferrous foliar fertilizer with high adhesion capacity based on nanobiomaterial. *ACS Sustainable Chemistry & Engineering*, *4*(12), 6800–6808. https://doi.org/10.1021/acssuschemeng.6b01761

Wang, S.-L., & Nguyen, A. D. (2018). Effects of Zn/B nanofertilizer on biophysical characteristics and growth of coffee seedlings in a greenhouse. *Research on Chemical Intermediates*, *44*(8), 4889–4901. https://doi.org/10.1007/s11164-018-3342-z

Wang, Y.-P., Zhou, P., Luo, S.-Z., Guo, S., Lin, J., Shao, Q., Guo, X., Liu, Z., Shen, J., Wang, B., & Guo, Z. (2018). In situ polymerized poly(acrylic acid)/alumina nanocomposites for Pb^{2+} adsorption. *Advances in Polymer Technology*, *37*(8), 2981–2996. https://doi.org/10.1002/adv.21969

Weon, J.-I., & Sue, H.-J. (2005). Effects of clay orientation and aspect ratio on mechanical behavior of nylon-6 nanocomposite. *Polymer*, *46*(17), 6325–6334. https://doi.org/10.1016/j.polymer.2005.05.094

Wu, B., Zhao, N., Hou, S., & Zhang, C. (2016). Electrochemical synthesis of polypyrrole, reduced graphene oxide, and gold nanoparticles composite and its application to hydrogen peroxide biosensor. *Nanomaterials*, *6*(11), 220. https://doi.org/10.3390/nano6110220

XING, Y., TANG, J., LI, X., HUANG, R., WU, L., XU, Q., LIU, X., & BI, X. (2022). Photo-induced antifungal activity of chitosan composite film solution with nano–titanium dioxide and nano-silver. *Journal of Food Protection*, *85*(4), 597–606. https://doi.org/10.4315/JFP-21-290

Yan, J., & Li, R. (2022). Simple and low-cost production of magnetite/graphene nanocomposites for heavy metal ions adsorption. *Science of The Total Environment*, *813*, 152604. https://doi.org/10.1016/j.scitotenv.2021.152604

Yan, L., Li, P., Zhao, X., Ji, R., & Zhao, L. (2020). Physiological and metabolic responses of maize (Zea mays) plants to Fe_3O_4 nanoparticles. *Science of The Total Environment*, *718*, 137400. https://doi.org/10.1016/j.scitotenv.2020.137400

Yang, C., Yu, D.-G., Pan, D., Liu, X.-K., Wang, X., Bligh, S. W. A., & Williams, G. R. (2016). Electrospun pH-sensitive core–shell polymer nanocomposites fabricated using a tri-axial process. *Acta Biomaterialia*, *35*, 77–86. https://doi.org/10.1016/j.actbio.2016.02.029

Yang, Y., Liao, H., Tong, Z., & Wang, C. (2015). Porous Ag/polymer composite microspheres for adsorption and catalytic degradation of organic dyes in aqueous solutions. *Composites Science and Technology*, *107*, 137–144. https://doi.org/10.1016/j.compscitech.2014.12.015

Yu, F., Tian, F., Zou, H., Ye, Z., Peng, C., Huang, J., Zheng, Y., Zhang, Y., Yang, Y., Wei, X., & Gao, B. (2021). ZnO/biochar nanocomposites via solvent free ball milling for enhanced adsorption and photocatalytic degradation of methylene blue. *Journal of Hazardous Materials*, *415*, 125511. https://doi.org/10.1016/j.jhazmat.2021.125511

Yu, J., Wang, D., Geetha, N., Khawar, K. M., Jogaiah, S., & Mujtaba, M. (2021). Current trends and challenges in the synthesis and applications of chitosan-based nanocomposites for plants: A review. *Carbohydrate Polymers*, *261*, 117904. https://doi.org/10.1016/j.carbpol.2021.117904

Zahran, M., & Marei, A. H. (2019). Innovative natural polymer metal nanocomposites and their antimicrobial activity. *International Journal of Biological Macromolecules*, *136*, 586–596. https://doi.org/10.1016/j.ijbiomac.2019.06.114

Zarrabi, A., Alipoor Amro Abadi, M., Khorasani, S., Mohammadabadi, M.-R., Jamshidi, A., Torkaman, S., Taghavi, E., Mozafari, M. R., & Rasti, B. (2020). Nanoliposomes and tocosomes as multifunctional nanocarriers for the encapsulation of nutraceutical and dietary molecules. *Molecules*, *25*(3), 638. https://doi.org/10.3390/molecules25030638

Zhang, L., Tu, S., Wang, H., & Du, Q. (2018). Preparation of polymer/graphene oxide nanocomposites by a two-step strategy composed of in situ polymerization and melt processing. *Composites Science and Technology*, *154*, 1–7. https://doi.org/10.1016/j.compscitech.2017.10.030

Zhang, M., Chang, L., Zhao, Y., & Yu, Z. (2019). Fabrication of zinc oxide/polypyrrole nanocomposites for brilliant green removal from aqueous phase. *Arabian Journal for Science and Engineering*, *44*(1), 111–121. https://doi.org/10.1007/s13369-018-3258-3

Zhang, S., Zeng, M., Xu, W., Li, J., Li, J., Xu, J., & Wang, X. (2013). Polyaniline nanorods dotted on graphene oxide nanosheets as a novel super adsorbent for Cr(vi). *Dalton Transactions*, *42*(22), 7854. https://doi.org/10.1039/c3dt50149c

Zhao, G., Huang, X., Tang, Z., Huang, Q., Niu, F., & Wang, X. (2018). Polymer-based nanocomposites for heavy metal ions removal from aqueous solution: a review. *Polymer Chemistry*, *9*(26), 3562–3582. https://doi.org/10.1039/C8PY00484F

Zhao, L., Peralta-Videa, J. R., Ren, M., Varela-Ramirez, A., Li, C., Hernandez-Viezcas, J. A., Aguilera, R. J., & Gardea-Torresdey, J. L. (2012). Transport of Zn in a sandy loam soil treated with ZnO NPs and uptake by corn plants: Electron microprobe and confocal microscopy studies. *Chemical Engineering Journal*, *184*, 1–8. https://doi.org/10.1016/j.cej.2012.01.041

Zhao, W., Kong, J., Liu, H., Zhuang, Q., Gu, J., & Guo, Z. (2016). Ultra-high thermally conductive and rapid heat responsive poly(benzobisoxazole) nanocomposites with self-aligned graphene. *Nanoscale*, *8*(48), 19984–19993. https://doi.org/10.1039/C6NR06622D

Zhu, H.-Y., Xiao, L., Jiang, R., Zeng, G.-M., & Liu, L. (2011). Efficient decolorization of azo dye solution by visible light-induced photocatalytic process using SnO_2/ZnO heterojunction immobilized in chitosan matrix. *Chemical Engineering Journal*, *172*(2–3), 746–753. https://doi.org/10.1016/j.cej.2011.06.053

Zulfiqar, F., Navarro, M., Ashraf, M., Akram, N. A., & Munné-Bosch, S. (2019). Nanofertilizer use for sustainable agriculture: Advantages and limitations. *Plant Science*, *289*, 110270. https://doi.org/10.1016/j.plantsci.2019.110270

Index

Printed in the United States
by Baker & Taylor Publisher Services